Organic Phototransformations in Nonhomogeneous Media

ACS SYMPOSIUM SERIES 278

Organic Phototransformations in Nonhomogeneous Media

Marye Anne Fox, EDITOR
University of Texas—Austin

Based on a symposium sponsored by
the Division of Organic Chemistry
at the 188th Meeting
of the American Chemical Society,
Philadelphia, Pennsylvania,
August 26–31, 1984

American Chemical Society, Washington, D.C. 1985

Library of Congress Cataloging in Publication Data

Organic phototransformations in nonhomogeneous media.

(ACS symposium series, ISSN 0097-6156; 278)

Includes bibliographies and indexes.

1. Photochemistry—Congresses. 2. Chemistry, Physical organic—Congresses.

I. Fox, Marye Anne, 1947- . II. American Chemical Society. Division of Organic Chemistry. III. American Chemical Society. Meeting (188th: 1984: Philadelphia, Pa.) IV. Series.

QD701.O74 1985 547.1'35 85-7471
ISBN 0-8412-0913-8

PRINTED IN THE UNITED STATES OF AMERICA

ACS Symposium Series

M. Joan Comstock, *Series Editor*

FOREWORD

The ACS SYMPOSIUM SERIES was founded in 1974 to provide a medium for publishing symposia quickly in book form. The format of the Series parallels that of the continuing ADVANCES IN CHEMISTRY SERIES except that, in order to save time, the papers are not typeset but are reproduced as they are submitted by the authors in camera-ready form. Papers are reviewed under the supervision of the Editors with the assistance of the Series Advisory Board and are selected to maintain the integrity of the symposia; however, verbatim reproductions of previously published papers are not accepted. Both reviews and reports of research are acceptable, because symposia may embrace both types of presentation.

CONTENTS

INDEXES

PREFACE

RECENTLY, THE STUDY OF THE PHOTOCHEMISTRY and photophysics of nonhomogeneous systems has become one of the most active areas in organic photochemistry. The motivation for investigations of these systems has centered on the possibility of discovering new reactions unique to the nonhomogeneous environment, of controlling chemical reactivity as a function of the polarity or molecular associations enhanced or inhibited by the microscopic surface, of learning about the nature and physical properties of the ordered medium, and of describing molecular movement and dispersion in these three-dimensional structures. Aside from their basic chemical interest, these multidimensional systems are relevant to biosynthetic pathways and to mechanistic understanding of molecular catalysis.

The symposium upon which this book is based was organized in order to show in a coherent fashion the wide variety of experimental approaches used to study organic transformations in nonhomogeneous media. I hope that the reader of this book can capture some of the scientific excitement of the symposium and will use it as a starting point for further investigations into this fascinating field.

MARYE ANNE FOX
The University of Texas—Austin
Austin, Texas

October 1984

1

Surface Photochemistry: Temperature Effects on the Emission of Aromatic Hydrocarbons Adsorbed on Silica Gel

P. DE MAYO, L. V. NATARAJAN, and W. R. WARE

Photochemistry Unit, Department of Chemistry, The University of Western Ontario, London, Ontario, Canada N6A 5B7

The emission of aromatic hydrocarbons, in particular that of pyrene, adsorbed on silica gel serves as a surface probe. The changes in silica gel structure imposed by heat treatment in vacuum (700°C) are revealed in poorly resolved spectra, multiple exponential fluorescence decay, ground state association and shortened lifetimes for pyrene emission. Coadsorbed water or alcohols, after heat treatment, render the surface more homogeneous, as indicated by well resolved spectra, an approach to single exponential decay, longer lifetimes, diminished ground state association and even formation of dynamic excimers. For a decanol-covered surface, cooling results in the disappearance of dynamic excimers, and a single exponential decay of pyrene with a life time of 600 nsec. Quenching studies have yielded an activation energy for diffusion on the dry silica gel surface of around 4 Kcal/mol.

The photochemistry and photophysics of molecules adsorbed on solid substrates have hitherto received little attention and, compared with their gas phase and solution counterparts, are very poorly understood. Many interesting questions await experimental and theoretical investigation. For example, what is the effect of an asymmetrical interaction where only a portion of a molecule interacts with a surface group and the remainder is essentially in the vapor phase? What is the effect of the absence of rapidly time averaged interactions such as one has in solution? How fast do molecules move on a surface and what is the activation energy for diffusion? How does the surface influence the course of a photochemical transformation, which may be either inter- or intramolecular[1-2] or occur via free radicals?[3-6] This partial list of questions is given to illustrate the considerable potential in surface photophysics and photochemistry for the exhibition of unique phenomena, for interesting modifications of the behavior of excited states as compared with behavior in other media, and for studies of molecular dynamics on the surface.

The excited molecule is a unique surface probe. The decay time provides a clock which can be used to study such dynamics. Since both singlets and triplets are potential probes, the time scale extends from seconds to picoseconds. In addition, the nature of any photo-

0097-6156/85/0278-0001$06.00/0

chemical products may give information about translational radical motion. The occurrence of photosensitization and quenching may also allow one to examine questions concerning mobility.

This contribution will deal mainly with recent photophysical studies of the behavior of aromatic hydrocarbons on silica gel and modified silica gel surfaces.

The surface of silica gel consists of siloxane bridges between tetracovalent silicon atoms. The presence of geminal, vicinal and isolated silanol groups, sometimes together with tightly bound water, render the surface inhomogeneous and provide binding sites, via hydrogen bonding, for the π-electron systems or non-bonding electron pairs of polar groups of adsorbed molecules (Fig. 1,2). The number of silanol groups on the silica gel surface depends on the temperature pretreatment. When silica gel is heated in vacuum, adsorbed water is removed first and at temperatures above 250°C the concentration of purely physisorbed water is negligible. As the temperature is raised above this point, the number of silanols is reduced from 6/100 $Å^2$ to approximately 2/100 $Å^2$ at 500°C and 1/100 $Å^2$ at 800°C.[7,8] In the temperature range of 500 to 700°C the surface concentration of geminal hydroxyl groups remains roughly constant but the ability to regenerate silanols by exposure to water vapor decreases rather suddenly.

Thus, heating the silica gel to temperatures above 500°C followed by rehydration changes the proportion and distribution of isolated geminal silanol sites. At higher temperatures only isolated silanol groups remain.

Prior adsorption of water and alcohols onto the dehydrated surface can drastically change the interaction of that surface with the adsorbed species of interest. These additives generally render the surface more homogeneous and can be used to influence dynamic behavior and static interactions.

Examples taken from recent work in our laboratories will now be given to illustrate these phenomena. In all cases, the samples were evacuated. Details of sample preparation are to be found in the references cited below.

Evidence for Surface Inhomogeneity

That all binding sites are not equivalent with respect to their influence on the ground and excited state behavior of an adsorbed molecule is indicated by several different types of evidence: (a) spectral band width, (b) complexity of the fluorescence decay, (c) the influence of coadsorbates (d) the influence of surface treatment and (e) ground state association. These will be discussed in turn.

Pyrene absorption and emission spectra exhibit much vibronic structure. If pyrene is adsorbed on silica gel pretreated so as to have only isolated silanol groups, the spectra are considerably broadened with loss of structure as compared to the spectra obtained when the surface is either "wet" or contains physi- and/or chemisorbed methanol.[9] This is illustrated in Fig. 3. Thus, it would seem that the isolated silanols provide a variety of sites with different interactions, an effect which is modified by coadsorbed alcohols or water.

More dramatic evidence for an inhomogeneous surface is provided by the common observation of multiexponential decay. It is, in fact, very unusual to observe single exponential decay when aromatic hydrocarbons are absorbed on silica gel, quite independently of

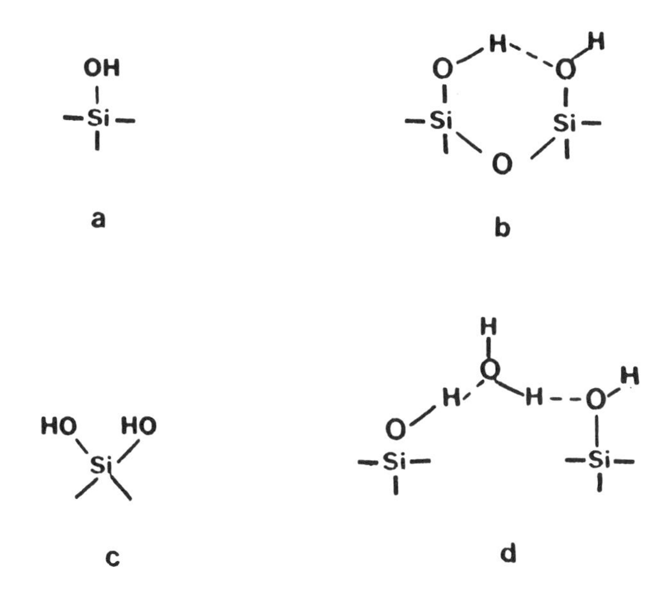

Figure 1. Types of silanol functions on the surface of silica gel. (a) lone; (b) vicinal; (c) geminal and (d) bonded water. (Reproduced by permission from *Pure and Applied Chemistry*, **54**, 1623, 1982.)

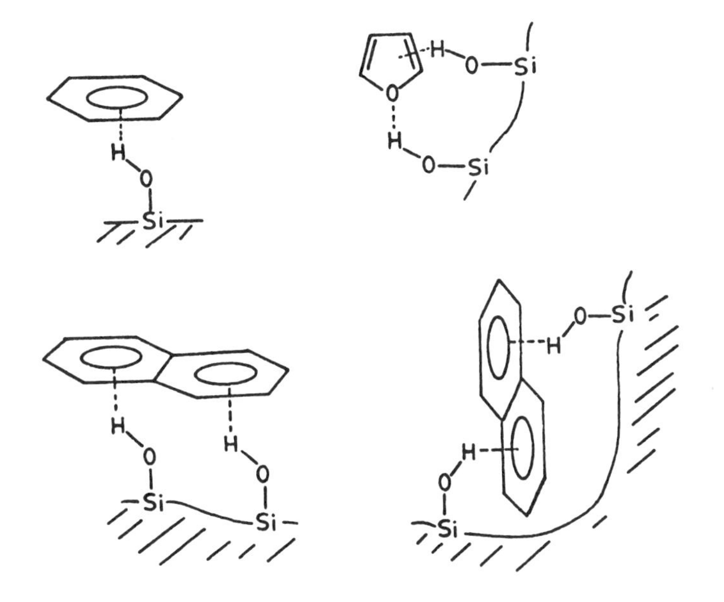

Figure 2. Bonding of silanol functions with aromatic hydrocarbons. (Reproduced by permission from *Pure and Applied Chemistry*, **54**, 1623, 1982.)

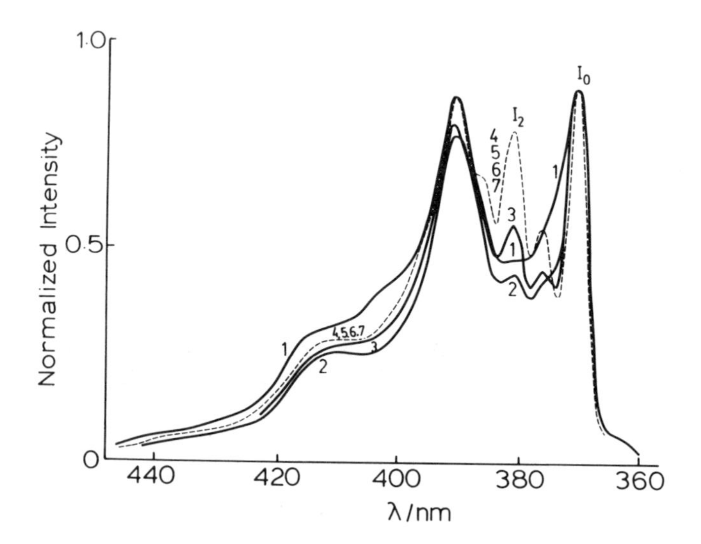

Figure 3. Effect of methanol addition on the emission spectra of pyrene adsorbed on silica gel. The silica gel was heated in vacuum to 700°C for 4 hours.
1) O MeOH; 2) 3.375×10^{-5}; 3) 7.5×10^{-5}; 4) 1.25×10^{-4}; 5) 3.75×10^{-4}; 6) 5×10^{-4}; 7) 7.5×10^{-4}.
Methanol concentrations are in mole/g SiO_2. The excitation wavelength: 332 nm. The pyrene coverage is 0.05 mg/g SiO_2. (Reproduced by permission from *Canadian Journal of Chemistry*, **62**, 1279, 1984.)

whether or not they are excimer-forming species in solution. It is also quite common to achieve reasonable fits (as measured by the reduced χ^2) with a four parameter model:

$$I_F(t) = A_1 e^{-t/\tau_1} + A_2 e^{-t/\tau_2}$$

although this is in many cases only a useful approximation to a time evolution that is more complex, and the number of parameters is not a direct indication of the number of environmentally different photoactive species.

If the surface is modified by coadsorbed alcohols or water, the decay times of pyrene and naphthalene, for example, lengthen dramatically and become essentially single exponentials.[9] We interpret this as a change to a homogeneous surface although, as will be discussed below, increased motion during the excited state lifetime may contribute to the collapse of double exponential decay to give a single exponential decay function. This phenomenon is illustrated in Fig. 4 for naphthalene.

As has been previously reported, when pyrene is adsorbed on silica gel there is evidence for ground state association which is not present in solution or the vapor phase,[9-13] but which has been described as being present when pyrene is dissolved in a plastic medium.[14] This is also a manifestation of surface inhomogeneity - some sites enhance the tendency to form a ground state bimolecular complex, whereas other sites contain isolated pyrene molecules. The interaction differences are sufficient to yield significant spectral shifts in absorption and the ground state complex emits with the characteristic pyrene excimer fluorescence. Fig. 5 shows a typical set of spectra illustrating this association and Fig. 6 presents evidence that this observation is not due to microcrystal formation.

Evidence for Motion on the Surface

Photophysical and photochemical studies have provided conclusive evidence for extensive motion on a microsecond or even submicrosecond time scale for excited and/or ground state molecules adsorbed on silica gel surfaces. Three lines of argument have been presented: (a) A single photosensitizer molecule can sensitize many more molecules than are initialy nearest neighbors. (b) Both dynamic and static quenching of fluorescence are observed. (c) Surface modification strongly influences the dynamics on the surface and under the appropriate conditions dynamic excimer formation, analogous to the behavior in solution, is seen.

The evidence from photosensitization derives from the Rose Bengal (RB) sensitized dimerization of acenaphthylene.[1,15] It is observed that one RB molecule can yield, by repeated excitation, dimeric molecules with a high conversion. This requires that dimers move away from the RB molecules and that monomers move to occupy these vacant places. Evidence has been presented[1] that the RB molecules are firmly fixed to the surface and thus the acenaphthylene monomers and dimers must be undergoing extensive motion on the surface during

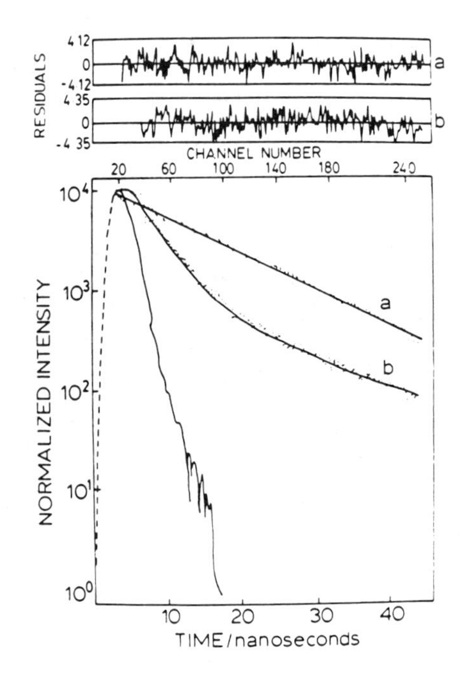

Figure 4. The decay of naphthalene fluorescence. Emission wavelength is 333 nm (a) silica gel heated in open air for 4 hours at 800°C; (b) silica gel was heated in vacuum for 4 hours at 800°C. (Reproduced by permission from *Canadian Journal of Chemistry*, **62**, 1279, 1984.)

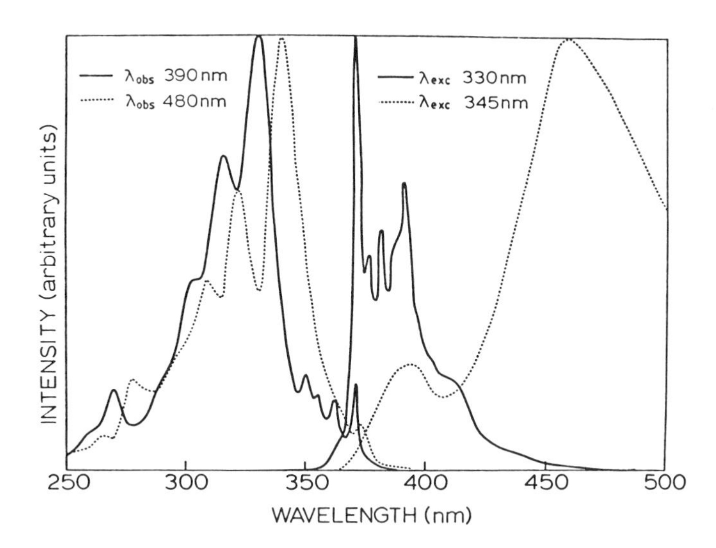

Figure 5. Ground state association of pyrene and pyrene "excimer" emission. Excitation and emission opectra of pyrene on silica gel (———); excitation (observed at 390 nm) and emission (excited with 331 nm for a θ=1 surface coverage sample (...); excitation (θ=1, observed at 480 nm) and emission (θ=3, excited with 345 nm). (Reproduced by permission from the Journal of Physical Chemistry, 86, 3781, 1982.)

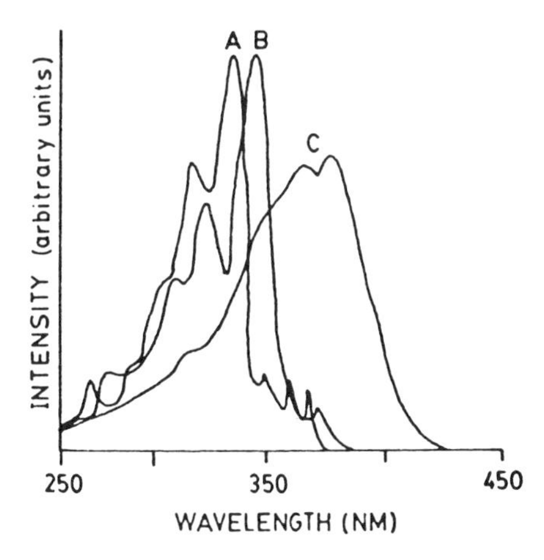

Figure 6. Excitation spectra of pyrene in different states of association. A) Excitation spectrum of pyrene monomer (emission at 393 nm). B) Excitation spectrum of pyrene dimer (emission at 470 nm). C) Excitation spectrum of pyrene microcrystal.

the time of irradiation. The dimerization is also observed upon direct irradiation.[1] It is known that the cis/trans ratio is sensitive to the relative importance of the participation of the singlet and triplet in the dimerization. Since the singlet, which yields the cis dimer is presumably short lived ($\tau \leq 1$ nsec), the cis yield is a measure of nearest neighbor dimerization. The triplet yields both cis and trans dimers and it is observed that surface modification by the coadsorption of decanol strongly decreases the C/T ratio,[11] thus suggesting enhanced triplet dimerization, attributable to enhanced molecular mobility. Here we are dealing with motion during the lifetime of the triplet, presumed to be, as is found in solution, of the order of microseconds.

Pyrene excimer formation provides additional evidence.[11] If pyrene and decanol are coadsorbed, one can observe the excimer-like emission at 470 nm growing in with the same kinetics as is seen in solution, ie.,

$$I_F = c(e^{-\lambda_1 t} - e^{-\lambda_2 t})$$

This is illustrated in Fig. 7. This implies motion during the pyrene singlet lifetime, which is of the order of several hundred nanoseconds. Further evidence is seen in the quenching experiment to be described below.

Dynamics on the Surface

The question of dynamic behavior of adsorbed molecules will now be discussed in more detail. In solution, so called Stern-Volmer behavior is commonplace, but it was surprising to observe this same linear behavior when quenching was studied in the adsorbed state. Both quantum yield ratios and lifetime ratios have been observed to give linear plots against the surface concentration of quenchers. For example, ferrocene quenches the dimerization of acenaphthylene that originates from the triplet state.[1] Fig. 8 illustrates this linear Stern-Volmer plot obtained from a study of the cis and trans dimer yield as a function of ferrocene concentrations. Linear Stern-Volmer plots have also been observed for quenching of pyrene monomer fluorescence by halonaphthalenes. Linear plots of both I_F°/I_F and $\bar{\tau}_0/\bar{\tau}$ have been observed, where $\bar{\tau}$ is the average lifetime calculated from eq. 1:

$$\bar{\tau} = \frac{A_1\tau_1^2 + A_2\tau_2^2}{A_1\tau_1 + A_2\tau_2} \qquad (1)$$

Arguments have been advanced elsewhere that the quenching of pyrene on silica gel by 2-halonaphthalenes is diffusion controlled.[1] This is in contrast to the behavior in solution where the quenching is inefficient. It was postulated[1] that on the surface the rate of separation of encounter pairs is slowed to the point where the rate of quenching is determined by the encounter rate. Central to this argument is the

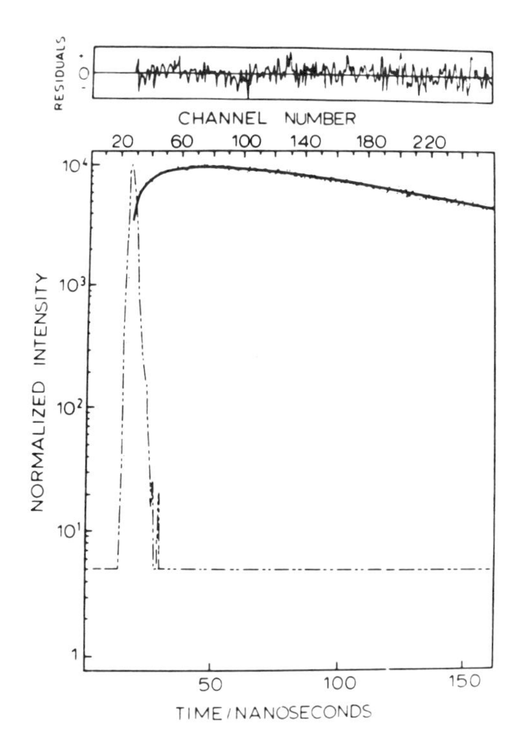

Figure 7. Emission decay profile of pyrene excimer at λ=470 nm coadsorbed with 1-decanol (pyrene 8.1 x 10^{-6} mol/g; 1-decanol, 1.8 x 10^{-4} mol/g of silica gel). (Reproduced by permission from the *Journal of Physical Chemistry*, **87**, 460, 1983.)

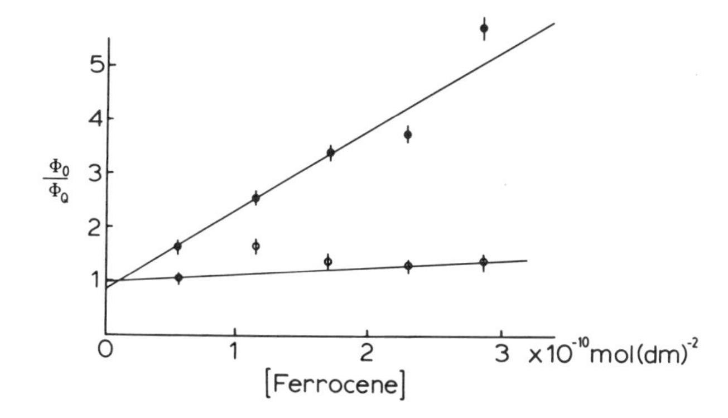

Figure 8. Stern-Volmer plot of the quenching of acenaphthylene in cis (O) and trans (O) dimer formation by ferrocene. (Reproduced by permission from the *Journal of American Chemical Society*, **104**, 4635, 1982).

observation that 2-chloro, bromo and iodonapthalenes quench on the surface with approximately the same rate whereas in solution the rates are: I > Br > Cl, with a range of approximately ten. Thus, this quenching reaction is assumed to be a useful probe of surface diffusion.

In connection with modeling the dynamics on surfaces the related solution model is of interest.

If one has the scheme:

$$A \xrightarrow{h\nu} A^* \underset{k_4}{\overset{k_3}{\rightleftharpoons}} (AQ)^* \xrightarrow{k_6} \text{Prod}$$

$$A^* \xrightarrow{k_1} A + h\nu_F, \quad A^* \xrightarrow{k_2} A, \quad (AQ)^* \xrightarrow{k_5} A + Q + h\nu_E$$

Then it follows that

$$I_F = A_1 e^{-\lambda_1 t} + A_2 e^{-\lambda_2 t} \tag{2}$$

$$I_E = A_3 (e^{-\lambda_1 t} - e^{-\lambda_2 t}) \tag{3}$$

where

$$2\lambda_{1,2} = k_1 + k_2 + k_3[Q] + k_4 + k_5 + k_6 \pm \left[(k_1 + k_2 + k_3[Q] - k_4 - k_5 - k_6)^2 + 4k_3k_4[Q]\right]^{1/2} \tag{4}$$

If $k_4 = 0$, or if $k_4 \ll k_3$, one observes single exponential decay. Also, in general

$$\frac{I_F^\circ}{I_F} = 1 + k_q \tau_0 [Q] \tag{5}$$

i.e., one expects linear Stern-Volmer plots from intensity measurements but not from τ/τ_0 plots except in special cases such as when $k_3 \gg k_4$.

If Q = A in the scheme, then one has the excimer case. It should be noted that when excimer kinetics are followed, the coefficents of the two terms in I_E are equal and opposite in sign giving one just one preexponential A_3, and that the difference of exponentials implies a "growing in" of the excimer emission, provided that at t = 0, $[A_2^*] = 0$. Otherwise, one has

$$I_E = A_4 e^{-\lambda_1 t} + A_5 e^{-\lambda_2 t} \tag{6}$$

and one can observe either sign for A_5. In addition, it is implicit in this simple analysis that I_F and I_E can be separated experimentally. With this background it is possible to discuss the results obtained with pyrene on silica gel when the pyrene monomer and excimer emission is studied. At all but quite low temperatures one observes decay behavior which can be represented by Eq. 2. It is impossible to separate the effects of multiple absorption sites, i.e., surface inhomogeneity, from the effect of dynamic excimer formation and feedback, the latter effect being responsible for the two component decay in solution. However, if the surface is decanol covered, then one observes emission in the excimer region that fits Eq. 3 rather well at room temperature suggesting that the dynamics are dominated by the growth of the excimer population, starting with $[A_2^*] = 0$ at $t = 0$. However, when one compares the values of λ for the decay in the monomer and excimer regions, the λ_1 and λ_2 values do not correspond as required by the simple model.

The monomer-excimer behavior has been studied as a function of temperature on both dry silica gel and silica gel to which various coadsorbates have been added. On dry (700°) silica gel, where we expect only isolated silanol groups to be present, the following behavior is observed:

(a) The lifetime lengthens on cooling but is still a double exponential at 10 K.

(b) The I_{470}/I_{390} ratio of emission intensities increases by lowering of temperature. This may be due to increase in the concentration of ground state dimers which are responsible for the excimer like 470 nm emission. The red shift in the excitation spectrum corresponding to 470 nm emission increases with decrease of temperature.

(c) One never observes A_4/A_5 to be negative.

(d) The data is reproducible as one goes up and down in temperature.

A typical plot of fluorescence intensity for two temperatures is shown in Fig. 9.

Of greater interest is the behavior of pyrene on decanol covered (~ monolayer) silica gel as the temperature is lowered. The following observations are relevant:

(a) $A_4/A_5 \cong -1$, that is the excimer growth is observed and the solution behavior is roughly duplicated.

(b) The observed spectral resolution of the pyrene vibronic structures suggests a quite homogeneous surface.

(c) At low temperature the monomer emission is quite accurately described by a single exponential and the lifetime is ~ 600 nsec. This long monomer lifetime is consistent with that observed in low temperature glasses.[16,17]

(d) The λ values obtained by analysis of the monomer and excimer decay do not correspond, as required by the solution model.

The data relevant to these observations are presented in Tables 1 and 2.

Thus, the decanol coadsorption appears to increase significantly the dynamic excimer formation but the residual surface inhomogeneity still prevents one from observing all the features of solution-like behavior.

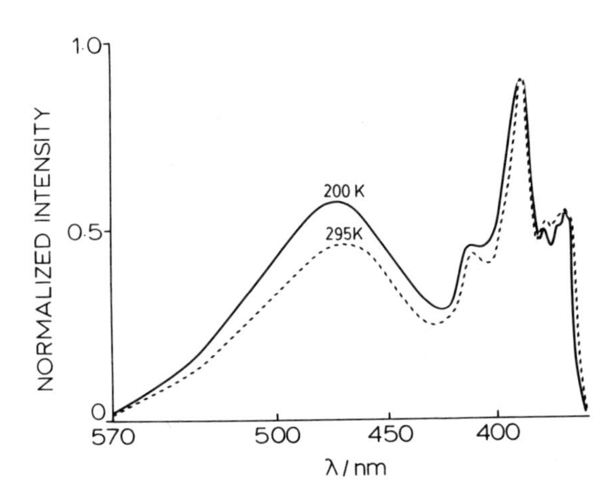

Figure 9. Effect of temperature on the pyrene "excimer" intensity. Excitation wavelength: 345 nm. The pyrene concentration is 1 mg/g dry SiO_2.

Table I. Pyrene Monomer Decay Time on Decanol-Covered SiO_2 as a Function of Temperature

T_K	A_1	τ_1	A_2	τ_2	REMARKS
293	0.05	228	0.135	66	λ_{EX} = 331, λ_{EM} = 393 nm
270	0.14	263	0.09	90	
250	0.13	375	0.06	152	
225	0.10	575	0.04	327	
200	--	540	--	-	Single exponential
150	--	586	--	-	
80	--	623	--	-	
10	--	620	--	-	
80	--	613	--	-	
150	--	588	--	-	
293	0.09	237	0.138	60	

Note: Pyrene coverage is 2.5 mg/g SiO_2.

Decanol coverage is 1 x 10^{-3} mol/g SiO_2.

Table II. Pyrene Excimer-like Emission on Decanol-Covered SiO_2 as a Function of Temperature

T	A_1	τ_1	A_2	τ_2	A_2/A_1	REMARKS
293	0.43	104	-0.44	26	-1.02	λ_{EX} = 370 λ_{EM} = 470 nm
270	0.45	166	-0.39	37	-0.87	
250	0.26	283	-0.19	47	-0.73	
225	0.022	485	0.018	124	--	
150	0.013	533	0.02	61	--	
80	0.016	550	0.023	60	--	
10	0.014	560	0.025	54	--	
250	0.31	260	-0.22	47	-0.71	
293	0.49	109	-0.49	27	-1.0	

Note: Pyrene coverage is 2.5 mg/g SiO_2.

Decanol coverage is 1×10^{-3} mol/g SiO_2.

The rapid decrease in the excimer-like emission with temperature on decanol-covered silica gel, illustrated in Figs. 10, 11, is consistent with a temperature coefficient for pyrene diffusion of 4 to 6 Kcal/mol, but exact calculation is model dependent and not justified at this time. Nevertheless, the qualitative behavior is reasonable.

The quenching of pyrene monomer emission by 2-bromonaphthalene on dry silica gel has also been studied as a function of temperature. Linear Stern-Volmer plots are obtained either with $\bar{\tau}_0/\bar{\tau}$ or with τ_1^0/τ_1 and τ_2^0/τ_2 vs [Q] where [Q] is a surface concentration. Fig. 12 illustrates the $\bar{\tau}/\bar{\tau}_0$ plot. The rate constants derived from these Stern-Volmer plots give a remarkably good Arrhenius plot as shown in Fig. 13, with an activation energy of ~4 Kcal/mol. This activation energy is interpreted as that associated with diffusion of the two molecules on the surface. In this context it is significant that the energy is of the order of hydrogen bond energies.

The Arrhenius plots from τ_1 and τ_2 data give activation energies above and below this value of 4 Kcal/mol which is considered to be an average. The low temperature studies on decanol covered silica gel also allowed of the observation of pyrene phosphorescence (Fig. 14); the phosphorescence disappeared at temperatures above 200° K.

Summary and Conclusions

The most inhomogeneous surface as judged by the pyrene probe is that which results from heating in a vacuum at high temperature (700°C). One sees multiple exponential decay, poorly resolved spectra, bimolecular ground state association and shortened lifetimes. If the surface is prepared by coadsorbing alcohols or water, a much more homogeneous surface results, as indicated by the approach of the decay to one component, longer lifetimes, and diminished bimolecular ground state association. In addition, the surface containing coadsorbed alcohols or water allows more rapid diffusion and one is able to observe dynamic excimer formation on a time scale of several hundred nanoseconds.

Quenching studies have yielded an activation energy for diffusion on the dry silica gel surface of around 4 Kcal/ mol.

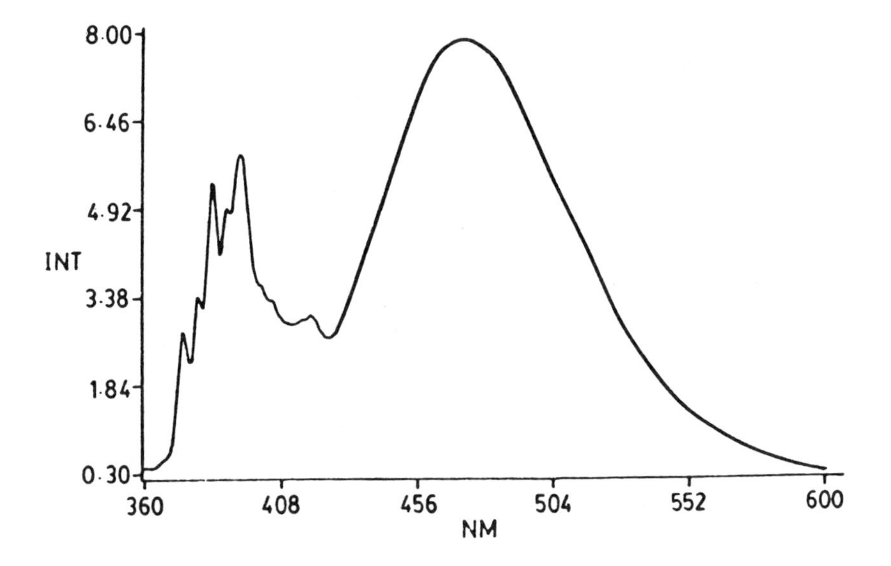

Figure 10. Pyrene excimer formation on a decanol covered silica gel surface. Decanol concentration is 1×10^{-3} mol/g SiO_2. Pyrene concentration: 2.5 mg/g SiO_2. Emission spectrum of pyrene at room temperature (295° K). Excitation wavelength: 345 nm.

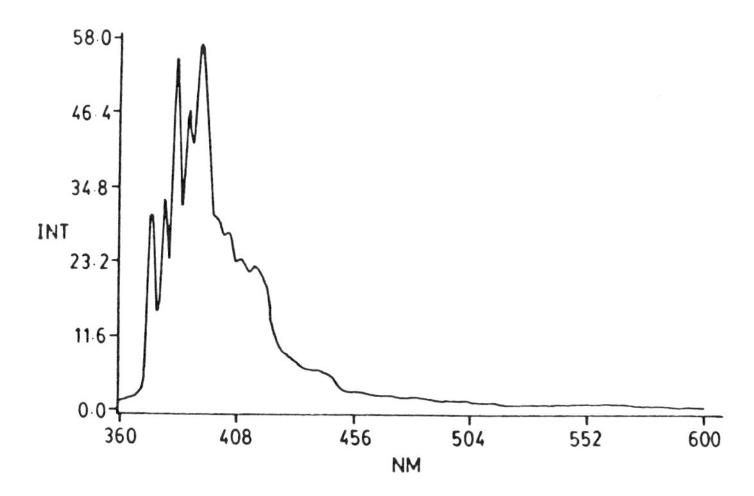

Figure 11. Effect of lowering the temperature on the pyrene excimer formation. The silica gel is covered with decanol (1×10^{-3} mol/g SiO_2). Pyrene is 2.5 mg/g SiO_2. Emission spectrum of pyrene at 200° K. The excitation wavelength: 345 nm.

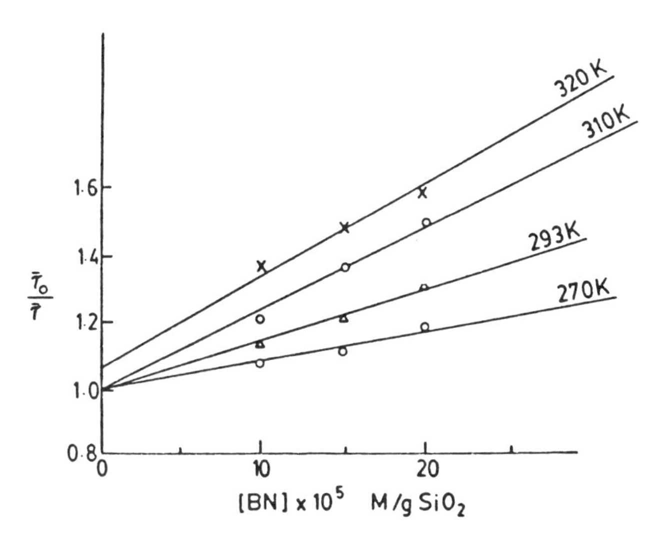

Figure 12. Effect of temperature on the quenching of pyrene monomer emission by 2-bromonaphthalene. Stern-Volmer plots at different temperatures.

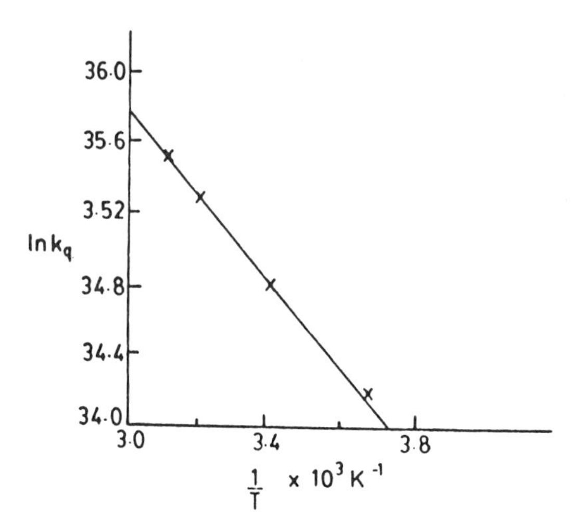

Figure 13. Arrhenius plot of quenching of pyrene monomer emission by 2-bromonaphthalene.

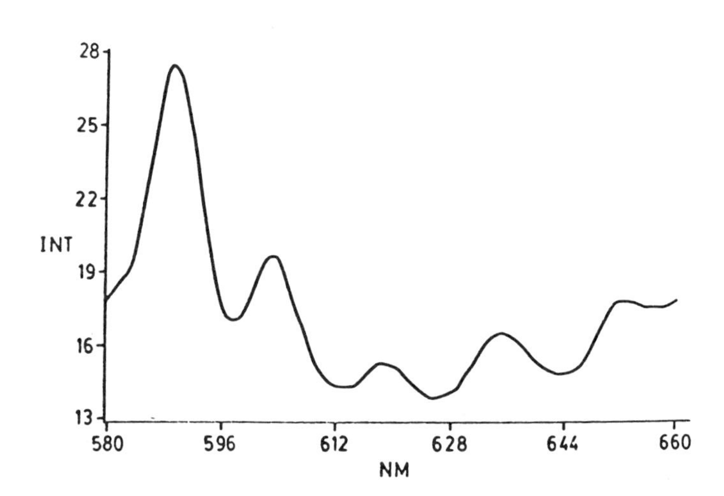

Figure 14. Phosphorescence emission of pyrene on decanol covered silica gel surface at 60°K. The excitation wavelength: 360 nm.

Literature Cited

1. Bauer, R.K.; Borenstein, R.; de Mayo, P.; Okada, K.; Rafalska, M.; Ware, W.R.; Wu, K.C. *J. Am. Chem. Soc.*, **1982**, *104*, 4635.
2. Nicholls, C.H.; Leermakers, P.A. *Adv. Photochem.*, **1971**, *8*, 315.
3. Frederick, B.; Johnston, L.J.; de Mayo, P.; Wong, S.K., *Can. J. Chem.*, **1984**, *62*, 403.
4. Johnston, L.J.; de Mayo, P.; Wong, S.K. *J. Org. Chem.*, **1984**, *49*, 20.
5. Leffler, J.E.; Zupancic, J.J. *J. Am. Chem. Soc.*, **1980**, *102*, 259.
6. Leffler, J.E.; Barbas, J.T., *J. Am. Chem. Soc.*, **1981**, *103*, 7768.
7. Fripiat, J.J.; Uytterhoeven, J., *J. Phys. Chem.*, **1962**, *66*, 800.
8. Kiselev, A.V.; Lygin, V.I., "Infrared Spectra of Surface Compounds", Chapters 4-7, John Wiley and Sons, New York, **1975**.
9. Bauer, R.K.; de Mayo, P.; Natarajan, L.V.; Ware, W.R., *Can. J. Chem.*, **1984**, *62*, 1279.
10. Hara, K.; de Mayo, P.; Ware, W.R.; Weedon, A.C.; Wong, G.S.K.; Wu, K.C. *Chem. Phys. Lett.*, ***1980***, *69*, 105.
11. Bauer, R.K.; de Mayo, P.; Ware, W.R.; Wu, K.C. *J. Phys. Chem.*, **1982**, *86*, 3781.
12. Bauer, R.K.; de Mayo, P.; Okada, K.; Ware, W.R.; Wu, K.C. *J. Phys. Chem.*, **1983**, *87*, 460.
13. de Mayo, P.; Natarajan, L.V.; Ware, W.R. *Chem. Phys. Lett.*, **1984**, *107*, 187.
14. Avis, P.; Porter, G. *Trans. Faraday. Soc.*, **1974**, 1057.
15. Okada, K., unpublished observations.
16. Kawski, A.; Weyna, I.; Kojro, Z.; Kubicki, A. *Z. Naturforsch*, **1983**, *38a*, 1103.
17. Barradas, I.; Ferreira, J.A. and Thomaz, M.F. *Trans. Faraday Soc.*, **1973**, 388.

RECEIVED January 10, 1985

2

Room Temperature Oxidations, Isotopic Exchanges, and Dehydrogenations over Illuminated Neat or Metal-Supporting Semiconductor Catalysts

PIERRE PICHAT

Institut de Recherches sur la Catalyse, CNRS, 69626 Villeurbanne, Cêdex, France

The possibilities of using photonic excitation of large area powder semiconductor catalysts in organic synthesis are surveyed principally by considering the title types of reactions. It is first underlined thatthe effects of parameters, such as texture, defects, impurities, on the creation and separation of charges within ghe semiconductor should be taken into account, in addition to their catalytic influence, when choosing or modifying a solid, which increases the difficulties. In presence of O_2 (air), alkanes, alkenes and alcohols are oxidized to aldehydes or ketones. The selectivity depends on the molecule, the photocatal yst and the conditions, since unspecific attack and cleavages can occur. Aromatic rings withstand oxidation and alkyltoluenes are converted to alkylbenzaldehydes. At least in gas phase, these oxidations over semiconductor oxides involve dissociated adsorbed/surface oxygen species activated by photoproduced holes as inferred from photoconductance and oxygen isotope exchange measurements and from the replacement of O_2 by NO. Group VIII metal deposition on semiconductors allows the extension of heterogeneous photocatalysis to reactions involving H_2, either endergonic (dehydrogenations) or exergonic (acid decarboxylations), as well as hydrogen isotopic exchange. However, charge recombination caused by these deposits determines an optimal metal amount for each metal-semiconductor system. Various liquid alcohols, saturated or unsaturated (allyl, cinnamyl, citronellol, geraniol), are dehydrogenated without over-oxidation and with quantum yields in the 0.025-0.8 range for Pt/TiO_2

Current address : Ecole Centrale de Lyon, B.P. 163, 69131 Ecully Cedex, France.

0097-6156/85/0278-0021$06.50/0

samples. A C = C bond is partially reduced only when adjacent to the hydroxy. In situ hydrogenation with an alcohol as reductant can be performed. In solution high chemical yields are reached. Cyclopentane-deuterium exchange only gives rise to the monodeuterated compound, whereas polydeuteration occurs by thermally activated catalysis. In conclusion, practical applications of heterogeneous photocatalysis for organic syntheses pose a challenge. In view of the background reviewed in this Symposium, concerted efforts of specialists in organic chemistry, chemical engineers and industry, in connection with scientists already involved in this field, are now needed.

Over the last decade, intensive research has appeared on the use of illuminated solid semiconductors and redox reactions to convert light energy (specially solar energy) to electrical (photoregenerative cells) or chemical energy (1-7). Unlike the conversion to electricity, the generation of valuable chemical compounds through photonic excitation of semiconductors does not require electrodes ; accordingly, numerous studies have been concerned with the use of powdered or colloidal semiconductors (3,6, 7). Because of their large surface area, these divided solids enhance the photon capture and the contact with the reactants. Their surface properties can be modified by the methods usually employed in catalysis, and several techniques exist to investigate electron transfer processes and reaction intermediates at their interfaces with gases or liquids. In addition, these divided semiconductors can be manufactured by simpler and therefore much less expensive means than semiconductor electrodes.

Many reports have been written on endergonic reactions ($\Delta G > 0$), involving an abundant reactant, such as water decomposition, and, to a lesser extent, nitrogen or carbon dioxide reductions. These reactions are outside the scope of this Symposium. Moreover, their yields, in the absence of electrical assistance or of a sacrificial compound are, as yet, very low. Dehydrogenations of organic compounds are also endergonic at room temperature but with a smaller change in free energy, and consequently they are easier to perform. For example, the hydrogen production from C_1-C_4 primary or secondary aliphatic alcohols has been studied(8-11). If more complex alcohols or other dehydrogenable organic compounds of interest were considered and selectively transformed, this dehydrogenation method might have an impact in organic synthesis.

On the other hand, oxidations of organic reactants at room temperature might also be interesting (although in that case the radiant energy only serves to overcome the energy of activation), provided one product has a high added value, and provided a reasonable quantum yield and, above all, a high chemical yield can be reached. Indeed studies of photocatalytic oxidations have preceded those on the storage of light energy, but more systematic researches, which will benefit from the improved understanding of photocatalytic processes, should be undertaken. A large variety of organic compounds have redox potentials allowing their oxidation by semiconductor oxides illuminated with photons corresponding to an energy at least equal to their band gap (12). Besides, the use of air and of inexpensive materials, such as titanium dioxide, as well as the absence of polluting spills, constitute advantages of this oxidation method. The exploration of its possibilities should be envisaged in connection with other potential developments for heterogeneous photocatalysis, such as the photocatalytic degradation of pollutants (13) and the recovery of precious and/or toxic metals in diluted solutions (14).

After some general remarks on the relations between semiconductor properties and their use as photocatalysts , this text will first deal with oxidations of organic compounds.The interactions of illuminated semiconductors with gaseous O_2 (and, for comparison, with gaseous NO) will be then presented, whereas the last part will consider metal/semiconductor photocatalysts and the organic reactions they allow. In this presentation, the results of this laboratory will be highlighted.

General remarks on the n-type semiconductors used in photocatalytic reactions

The nature of the semiconductor intervenes in various respects. From the energetic viewpoint, the locations of the valence band and of the conduction band respectively determine the oxidations and reductions which are thermodynamically allowed, and the band gap indicates the light frequencies required for activating the semiconductor, i.e. for generating holes in the valence band. As in thermal catalysis, surface/catalytic properties, which control the structure of the adsorbed species, depend on the chemical nature of the solid and on the presence of various defects. For instance, acid-base surface sites (specially the coverage in OH groups) are critical for the adsorption of organic molecules which contain acidic or basic functional groups. Defects, such as impurities (which can behave as substitutional doping levels) and oxygen vacancies, can have the same role as in thermal catalysis in forming adsorption sites, but furthermore they can act as recombination centers of photoproduced charges and/or they can change the mobility of the charge carriers.Profound changes in photocatalytic activity can thus be observed according to the stoichiometry and purity of the semiconductor oxide.

The texture of the semiconductor powder is also very important. The role of the surface area in determining the extent of contact of the solid with the gaseous or liquid or dissolved reactants is easily understandable. Pores should be of a size comptabile with easy penetration of the reactants and recovery of the products. However, both these factors have supplementary effects in the case of illuminated solid grains. For porous materials, at least part of the internal surface of the pores cannot be directly reached by the light. For nonporous materials, at a given wavelength the depth of penetration of the photons, which is related to the absorption coefficient α, should be compared with the average size of the grains. If α varies in the 10^7-$10^8 m^{-1}$ range, the Beer-Lambert law indicates that 99 % of the incident flux is absorbed within 20 to 200 nm thickness. For anatase (density 3.85 g cm^{-3}) these particles diameters correspond to surface area of about 78 or 7.8 m^2g^{-1}, respectively ; in other words, inner regions of the particles of samples of lower specific area are not illuminated (15). In this respect, the type of photoreactor plays a role (fixed-bed catalyst, suspension, illumination from one side or annular illumination etc...). In addition, the radius of the semiconductor particle can be smaller than the width of the space charge layer which would be formed in an infinite sample of the same semiconductor, and this will affect the charge separation. These latter effects of particle size only begin to be theoretically considered (16). One should notice that they can cause differences in photocatalytic behavior.

As a result, though comparisons between various semiconductors for a given photocatalytic reaction are useful, the classifications thus derived must not be regarded as definitive, since the effects of the texture, of the impurities and of other structural defects, are even more crucial than in thermal catalysis.

Although economical factors are not of primary importance in the early stages of fundamental researches if results obtained with rare materials allow our knowledges to progress, it is nevertheless of interest to mention that industry produces large quantities of some powder semiconductors, either pure of modified, and, among them, the cheapest are Fe_2O_3 and TiO_2. Many studies on photocatalytic organic conversions have been carried out with this latter catalyst because of its higher efficiency. Note that when prepared in a flame reactor it is nonporous.

Photocatalytic oxidations of organic compounds

The review will be limited to relatively simple molecules. Other, and generally more complex examples, will be found in M-A. Fox's and K. Tokumaru's papers in this Symposium.

Alkanes. Detailed studies of the photocatalytic oxidation of C_2-C_8 alkanes, linear or ramified, in gas phase using a differential-flow photoreactor with a fixed bed of TiO_2, have been reported (17-18). Under such conditions, which implies low conversions (0.6 - 2.5 %), a high selectivity (55 -95 %) in aldehyde(s) and/or ketone(s) as compared with total oxidation to CO_2, was found, and a quantum yield (number of molecules oxidized for each quantum of radiation absorbed by the semiconductor) of 0.1 was indicated. This selectivity to partial oxidation products is remarkable with regard to catalytic alkane oxidations at high temperatures. The reactivity of the carbon atoms followed the sequence $C_{tert} > C_{quat} > C_{sec} > C_{prim}$, the carbon atom which was preferentially, but not exclusively, attacked being that with the highest electron density together with the least steric obstruction. It ensued that, unfortunately, a variety of aldehydes and ketones was obtained if the alkane structure allowed it (scheme I). To the author's knowledge, no equivalent studies have been hitherto published for neat-liquid alkanes.

Alkyltoluenes. We have studied the oxidation of gaseous alkyltoluenes $RC_6H_4CH_3$ ($R = C_2H_5$, $(CH_3)_2CH$, $(CH_3)_3C$) in a differential-flow photoreactor with a fixed bed of TiO_2 (19). The ratio O_2/hydrocarbon should not be too great, otherwise the surface coverage in hydrocarbon limited the conversion. In all cases, the selectivity in RC_6H_4CHO was high. For $R = (CH_3)_3C$, no CO_2 was detected. Under the same conditions, toluene yielded only traces of benzaldehyde. This shows the stability of the aromatic ring and of the methyl group directly attached to it.

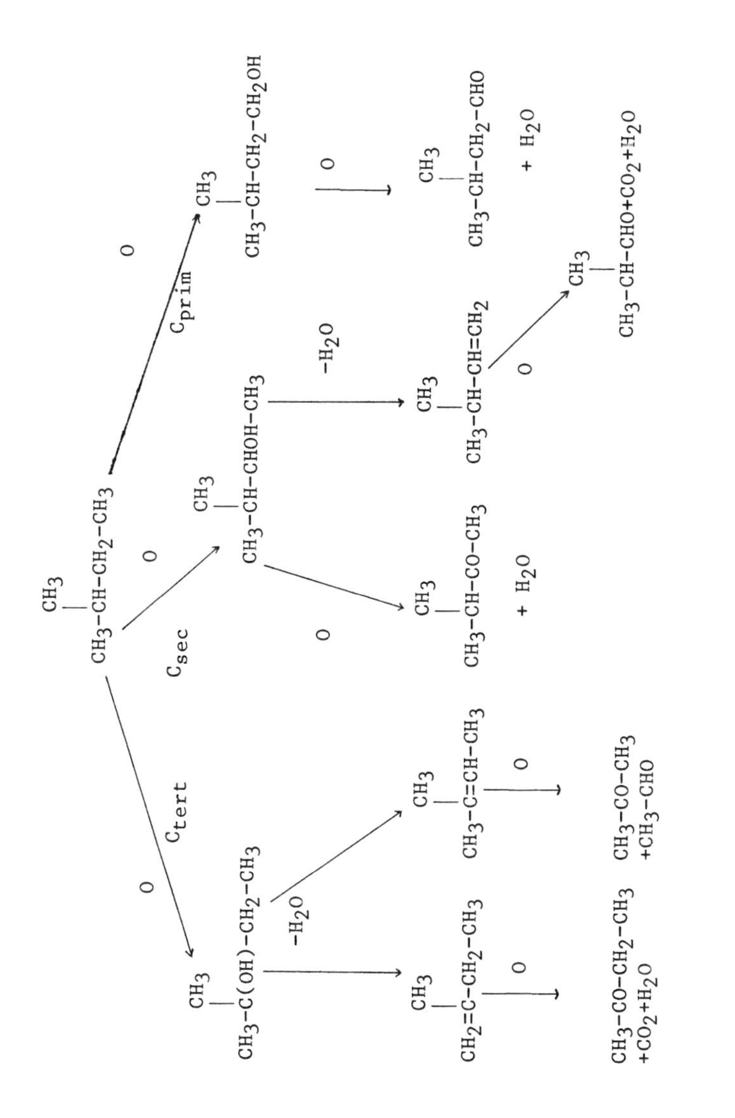
CH3
CH3-CH-CH2-CH3
Cprim
Csec
Ctert
O
CH3
CH3-CH-CH2-CH2OH
CH3
CH3-CH-CHOH-CH3
CH3
CH3-C(OH)-CH2-CH3
-H2O
CH3
CH3-CH-CH2-CHO
+ H2O
CH3
CH3-CH-CH=CH2
CH3
CH3-CH-CO-CH3
+ H2O
CH3
CH3-C=CH-CH3
CH3
CH2=C-CH2-CH3
CH3
CH3-CH-CHO+CO2+H2O
CH3-CO-CH3
+CH3-CHO
CH3-CO-CH2-CH3
+CO2+H2O

Scheme I (17)

These oxidations are not classical, since, usually, the secondary or tertiary hydrogen atomsare preferentially eliminated to give rise to hydroperoxides, which could mean that these latter intermediates are not formed in the photocatalytic process. These oxidations can also be performed with the neat-liquid in a static photoreactor. The methyl group is oxidized, whereas the other alkyl group withstands oxidation. However, part of the photons required to excite TiO_2 is also absorbed by the alkylbenzaldehyde, producing free radicals in the presence of O_2. This accelerates the aldehyde transformation into the corresponding acid. If a high selectivity in aldehyde is sought for, this imposes to operate either with dilute solutions or with a reactor avoiding that the photons pass through a thick layer of the liquid. The acid formation is also enhanced by raising the temperature ; room temperature appears as an optimum compromise between kinetics and selectivity. For neat-liquids, initial quantum yields of about 0.2 were found.

Alkenes. Gas phase propene oxidation has been studied at 320 K over TiO_2, ZrO_2, V_2O_5, ZnO, SnO_2, Sb_2O_4, CeO_2, WO_3 and a mixed Sn-O-Sb sample (20). These solids were not active in the dark at this temperature. Under band-gap illumination, only the V_2O_5 sample was found inactive. Unfortunately, only one anatase sample had a quantum yield sufficient for a possible practical use. The selectivity greatly depended on the particular specimen used. Among the products of partial oxidation, ethanal was generally formed in higher amounts than acrolein, the percentage of acetone being small. Propene oxide was detected, particularly at low conversions, and, in the case of TiO_2, could be the primary product resulting from the attack of adsorbed propene by the active oxygen species.

Under similar conditions, the photocatalytic oxidation of 2-or 3-methyl-1-butene and of 2-methyl-2-butene over TiO_2 yielded carbonyl compounds as partial oxidation products (18). However, the selectivity to a particular aldehyde or ketone was reduced by cleavages not only of the double bond but also of the C_β-C_γ bond.

Alcohols. The oxidation of 2-propanol principally to acetone has been chosen to test the photostability of pigments. The mechanism has been studied in detail, however mainly by performing this reaction in gas phase, essentially over TiO_2 (21-24,). Samples of other oxides like V_2O_5, Cr_2O_3, MnO_2, Fe_2O_3, Co_3O_4, NiO and CuO do not exhibit a photocatalytic activity (23). This gas (23, 25) or liquid (24) phase oxidation was extended to C_4-C_5 aliphatic alcohols. In gas phase, the ease of oxidation followed the sequence : secondary > tertiary > primary and oxidative C_α-C_β scission was found for 2-propanol, 2-butanol, and 2 and 3-methyl-2-butanol(scheme II) . Whether or not this scission is related to the formation of an olefin intermediate has been disputed (23, 25). Also, though 3-methyl-1-butanol is oxidized chiefly to the corresponding aldehyde, acetone, 2-methylpropanal and ethanal are supposed to result from the subsequent cleavage of this aldehyde (25).

Oxidation of ethanol, 1- and 2-propanol was investigated in liquid phase and in aqueous solutions (24). The same apparent energy of activation was found regardless of the alcohol. However, the primary alcohols were found to react preferentially to 2-propanol when mixed to it. Quantum yields in the range 0.1-0.5 at near room temperature have been reported.

NO, which supplies oxygen atoms to TiO_2 on forming N_2 and N_2O, (26) can replace O_2 in such oxidations as already mentioned. For example, 1-butanol was oxidized to butanal, 2-butanol to butanone and 2-methyl-2-propanol to acetone (2-methyl propene being also detected), with the expected order in the ease of oxidation, and a average quantum yield of about 0.1 (27, 28).

Oxalic acid. Its oxidation to CO_2 in aqueous solution was used as a test reaction to examine the photosensitive properties of TiO_2, Fe_2O_3, ZnO, ZrO_2, Sb_2O_4, CeO_2 and WO_3 powder samples in the visible as well as in the UV region (29). In fact, all of these specimens, except Sb_2O_4 and WO_3, were partially degraded in the dark in an oxygenated 5×10^{-3} M solution (pH : 2.34) and CO_2 was evolved. The slight photosensitivities observed in the visible present no practical interest. Under UV-illumination, a quantum yield of ~ 0.35 was found for the anatase sample which was the most efficient. The presence of dissolved oxygen was needed to allow the C-C bond cleavage.

Conclusion. As pointed out in the introduction, heterogeneous photocatalysis offers economical advantages for oxidations. The preceding paragraphs, by showing that at room temperature, alkanes can be oxidized and alkyltoluenes transformed to alkylbenzaldehydes, give an idea of the capabilities of this method. Generally, aldehydes or ketones are the final degree of oxidation, provided that the conditions avoid photochemical oxidations in the presence of O_2 ; this underlines the necessity of researches on photocatalytic reactors. However, depending on the molecular structure of the oxidizable compound, the oxidation can be unspecific and, furthermore, undesired cleavages can occur as a result of the method efficiency. In this respect, no systematic study of the effect of temperature, namely below the ambient, on the selectivity has been reported. Furthermore, a deeper understanding of oxidation catalysts might help to design more selective photocatalysts, with nevertheless the added difficulty of obtaining a high photosensitivity. Finally, by considering molecules more complex than those cited herein (see M.A. Fox's and K. Tokumaru's presentations in this Symposium) and by changing the conditions, applications might be uncovered.

Interaction of illuminated n-type semiconductor oxides with O_2 and NO

Oxygen adsorption on semiconductor oxides has given rise to a large number of studies, mainly by ESR (30). Various ionosorbed species have been found depending upon the conditions and the sample pretreatment. Continuous illumination with photons of energy at least equal to the semiconductor band gap should change the equilibrium between gaseous oxygen and negatively charged chemisorbed oxygen species, since the concentrations of both electron and holes are modified. This effect is very marked for the minority carriers (1). As a result, photodesorption and photoadsorption phenomena have been reported (31). In the following paragraphs we briefly show how we have tried to gain information on the oxygen active species at the surface of various illuminated semiconductor oxides, by using isotopic exchange experiments and photoconductance measurements.

O_2 and NO isotopic exchange over illuminated semiconductor oxide. Oxygen isotope exchange (OIE), in the dark and at temperatures generally largely above the ambient, has been employed to determine the lability of adsorbed/surface oxygen of various metal oxides in connection with the catalytic activity of these solids in oxidations (32). For illuminated metal oxides, these OIE experiments are particularly relevant, since OIE and oxidation reaction can both occur at room temperature, by contrast with thermally activated exchanges which often require higher temperatures than oxidations. The type of OIE can be derived from the changes in the various isotopic species in the gas phase and the competition between a given oxidation and OIE enables one to know whether the same oxygen species are involved in both reactions.

For instance, illumination of preoxidized (treatment in $^{16}O_2$ at 723 K) samples of TiO_2 (33), SnO_2, ZnO and ZrO_2 (34), previously exposed to 4 Pa $^{18}O_2$ in the dark, caused immediate OIE, provided photons of an energy sufficient to create electron-hole pairs were employed, whereas a V_2O_5 specimen was inactive (34). The evolution of the gas isotopic composition perfectly corresponded to an exchange-type represented by the equation

$$^{18}O_2(g) + {}^{16}O(s) \rightleftarrows {}^{18}O^{16}O(g) + {}^{18}O(s) \qquad (1)$$

where one adsorbed/surface oxygen atom is exchanged at a time, while the type involving two such atoms for each exchange act did not intervene, in contrast with most thermally activated OIE (32). On the other hand, the same order of activities was observed for several TiO_2 specimens (33), as well as for the SnO_2, ZnO and ZrO_2 samples (34) in both OIE and isobutane oxidation, chosen as an example of alkane oxidation. In addition, competition between OIE and this reaction was examined by introducing a mixture of $^{18}O_2$ and isobutane over TiO_2 in the dark (33). On illuminating, a decrease in gaseous isobutane was observed, as expected, whereas OIE occurred only after the disappearance of the alkane in the gas phase (Figure 1). From these results it was concluded that the labile adsorbed/surface atomic oxygen species which took part in OIE, also participate in isobutane oxidation.

Moreover, the isotopic heteroexchange of oxygen is a very sensitive photoassisted gas-solid reaction. For example, it is much more affected by the texture (33) or by the doping (35) of TiO_2 samples than the oxidation of different organic compounds in gas (33, 35), liquid or aqueous solutions (35). It can therefore be used to evaluate more precisely the effect of various treatments.

Analogous results were found with $N^{18}O$ (26). On UV-illuminated TiO_2, heteroexchange occurred at room temperature with a much higher initial rate than in the case of $^{18}O_2$ (Figure 2). Instantaneous exchange with TiO_2 surfaces prereduced in H_2 at 723 K showed that illumination renders labile surface oxygen atoms which have not been removed by such a treatment. However, a higher initial efficiency for preoxidized samples indicated that adsorbed atomic oxygen species also participated. Finally, in the presence of 2-butanol the exchange of $N^{18}O$ was suppressed, while butanone was formed. This phenomenon was similar to the suppression of OIE by isobutane. The role of NO in the oxidation of 2-butanol and other butanols (vide supra) is to replenish the coverage of TiO_2 in removable atomic species as evidenced by the formation of N_2O and N_2 (26). This confirms the importance of atomic oxygen species in the photocatalytic reactions performed with gaseous oxygen.

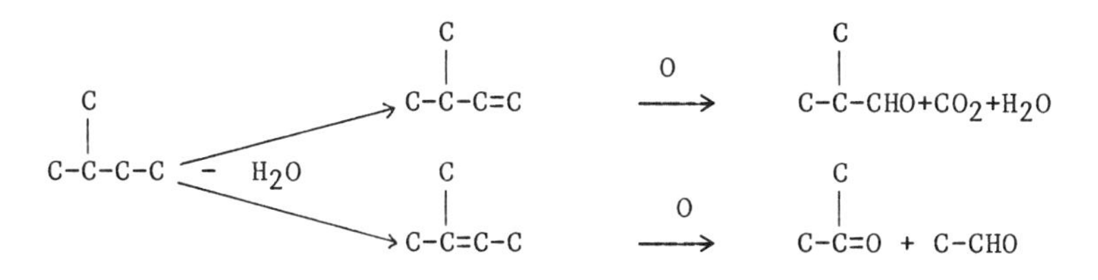

Scheme II (25)

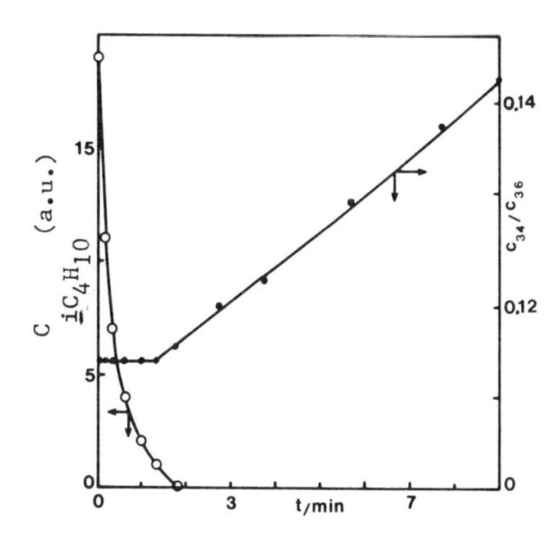

Figure 1. Changes in the $^{18}O^{16}O/^{18}O_2$ ratio and in the isobutane concentration in gas phase over TiO_2 as a function of illumination time

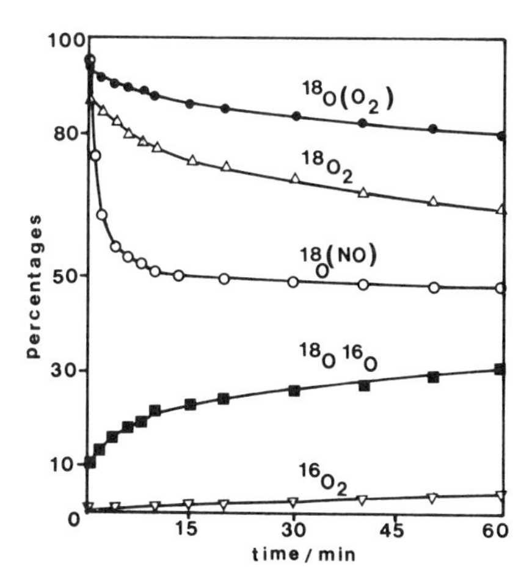

Figure 2. Changes in the isotopic composition of gas phase oxygen and declines of ^{18}O isotope content of gaseous oxygen or nitric oxide as a function of illumination time for preoxidized TiO_2 samples.

$^{18}O_2/^{16}O_2$ equilibrated or non-equilibrated mixtures have been used by other research groups to investigate the effects of illumination principally on ZnO and TiO_2 (36, 37). In particular, illumination can produce homoexchange of gaseous $^{16}O_2$ and $^{18}O_2$ molecules. For instance, over preoxidized ZnO this exchange, which does not involve adsorbed or labile surface oxygen species, was induced by low-intensity illumination with wavelengths $\geqslant$ 460 nm (i.e. $<$ band-gap energy), so that an originally non-equilibrated mixture became equilibrated. This phenomenon was attributed to the location of electrons on coordinatively unsaturated Zn cation and subsequent formation of O_4^- species on these sites. For as long as the electrons remain so located, conditions exist for turnover of several molecules of the gaseous mixture to $^{16}O\ ^{18}O$ on the same site (7, 37). This example further demonstrates the interest of OIE for studying the reactivity of variously pretreated oxide surfaces under different illumination conditions.

Photoconductance of n-type semiconductor oxides exposed to O_2 or NO. Photocatalytic reactions involve the exchange of electrons between an excited semiconductor and at least one adsorbed reactant or surface species. Consequently, photoconductance measurements appear as a discriminating method to investigate these reactions.

Electrophilic gases, such as O_2 (38) and NO (27,28), decrease the photoconductance of preevacuated n-type semiconductor oxides. With O_2 this effect was found for TiO_2, ZnO, ZrO_2, CeO_2, Sb_2O_4, SnO_2 and WO_3 samples (38). By contrast, a V_2O_5 specimen was insensitive to exposure to O_2, in agreement with inactivity for OIE and propene oxidation. Also, an anatase sample, homogeneously doped with Cr^{3+} ions (0.85 at.%), had a much lower photoconductance than an equivalent pure anatase sample, and furthermore was almost unaffected by varying the wavelength at constant photonic flux in an O_2 atmosphere (35). This Cr-doped solid exhibited poor aptitudes as oxidation photocatalyst, and, above all, for OIE as already mentioned. Undoubtedly, the photoconductance variations in O_2 allow one to predict the efficiency of n-type semiconductor oxides for photocatalytic oxidations with O_2.

In addition, from the slope of these variations as a function of O_2 pressure in a log-log plot, the nature of the oxygen species controlling the adsorption equilibrium between the semiconductor free electrons and the gas (for given illumination, temperature and pressure range) can be deduced, provided the ways of formation of these oxygen species from gaseous O_2 are assumed. For example, the predominance of

$$O_2(g) \rightleftarrows O_2\ (ads)$$

$$O_2\ (ads) + e^- \rightleftarrows O_2^-\ (ads)$$

gives rise to a - 1 slope, while a - 1/2 slope can result from

$$1/2\ O_2\ (g) \rightleftarrows O(ads)$$

$$O\ (ads) + e^- \rightleftarrows O^-\ (ads).$$

Other possibilities exist and should be discussed when interpreting the photoconductance data (38). Also, note that the fact that one species governs the adsorption equilibrium do not necessarily imply that other species are not present, but can mean that they saturate the surface. Moreover, if the semiconductor samples previously evacuated under illumination to remove the labile oxygen species under these conditions, are exposed to organic compounds involved in oxidations, as reactants or products, the effect on the photoconductance indicates whether or not these compounds compete with O_2 for electron capture.

Simultaneous measurements of the photocatalytic activity A and of the photoconductance σ in a specially designed cell yield information on the participation of the oxygen species in the oxidation process. For instance, in the case of isobutane (IS) oxidation over TiO_2, the following relations were found (39) :

$$\sigma = k_o \, P_{O_2}^{-\frac{1}{2}} \, P_{IS}^{0} \qquad\qquad A = k_A \, P_{O_2}^{0} \, P_{IS}^{0.35}$$

The independence of σ on IS pressure confirms that IS does not capture nor release electrons, whereas the fractional kinetic order 0.35 shows that IS reacts in an adsorbed phase, since this value is very close to the apparent order of adsorption 0.3 found for the surface coverage in IS according to a Langmuir model in the pressure range investigated (13-60 kPa). The σ-P_{O_2} relationship corroborates that $O_2^{\cdot-}$ species control the adsorption equilibrium for the pressures chosen, while $O^{\cdot-}$ sites are saturated. Since A is unaffected by oxygen pressure, it is deduced that the active oxygen species are associated with $O^{\cdot-}$ ion-radicals.

Conclusion. The photoconductance measurements are thus in agreement with the OIE experiments concerning the role of dissociated oxygen species in photocatalytic oxidations in gas phase. In the absence of other electrophilic substances, this can also be the case for oxidations of organic compounds either as neat-liquids or diluted in an organic solvent. The role of atomic oxygen species has also been stressed in studies by Cunningham et al. (21) dealing with the interactions of vapors of C_3-C_4 aliphatic alcohols with various metal oxides (principally ZnO and rutile). This active species would result from hole trapping at coordinatively unsaturated O^{2-} surface anions.

The formation of $OH^{\cdot}$, HO_2 species and of H_2O_2 for semiconductor oxides in contact with aqueous solutions or exposed to water vapor has often been proposed (22,23, 40-42). As metal oxide surfaces carry OH groups and as the oxidation of organic compounds produces water, the formation of the above species cannot be excluded even in the absence of added water.

Reactions over metal/semiconductor photocatalysts

Preparation and characterization of the metal deposits. Pt deposit was made by impregnation with H_2PtCl_6 and reduction in H_2 at 753 K. The Pt particle size distribution was determined by transmission electron microscopy (TEM) (8, 9) (Figure 3) and H_2, O_2 chemisorptions and titrations (43). The Pt particle size distribution was narrow with a surface weighted mean diameter of ca. 2 mn, almost independent of the Pt content between 0.5 and 10 wt % (9), provided the preparation method, which includes a treatment in O_2 before the reduction, was thoroughly followed.

Ni was deposited by impregnation with nickel hexamine nitrate and reduction in H_2 at 753 K. The Ni particle size was assessed from magnetic measurements. For Ni contents from 0.1 to 14 wt %, the mean diameter varied from ca 6.5 to 15 nm, which explains that the Ni particles were difficult to distinguish from the TiO_2 grains (Degussa P-25, 15-30 nm dia.) by TEM (10). However, on employing another TiO_2 specimen (170 nm dia.) it was possible to confirm by TEM the values deduced from magnetic measurements (44). Nickel was chosen above all for economical reasons, but also for its different characteristics (ease of oxidation, larger crystallites).

Photoconductance measurements were used to determine the effect of the metal deposits on the density of the semiconductor free electrons. At 295 K, after an overnight evacuation in the dark, all M/TiO_2 samples had a high electrical resistance. Ultraviolet illumination under vacuum caused important decreases (10, 45).Figure 4 shows the values of the titania photoconductance σ at equilibrium. Metal deposits decreased σ, which reflects the decrease in the density of free electron n_s of the n-type semiconductor. This effect can be explained by the alignment of the Fermi levels of TiO_2 and of the deposited metal. The works functions of Pt and of Ni have values of about 5.36 and 5.03 eV, respectively. Note that the former value can however be slightly lower for small particles (~ 280 Pt atoms in the present case). Values of 4.6 or 5.5 eV have been found for a (110) rutile surface in two different states : argon bombarded or well annealed, respectively (46). From a difference in conductance under vacuum at 298 K of a least 3 orders of magnitude for a pure TiO_2 sample either preoxidized in O_2 at 723 K or evacuated at 423 K, it can be inferred that the work function of these samples varied over a similar range of values than that of the rutile single crystal. Therefore for non-preoxidized samples, an electron transfer from the semiconductor to the metal particles was possible (47). But the situation changed in the presence of H_2. Whereas H_2 did not affect σ of neat TiO_2, σ of the Pt/TiO_2 samples increased when they were exposed to H_2 at 295 K (45). This is consistent with a decrease in the work function of Pt (48). A similar effect of H_2 for rutile single crystals onto which dots of Pt (or Rh or Ru) films had been evaporated has been recently found. The rectifying air-exposed metal-semiconductor electrical contact became ohmic on flooding with H_2 at 1 atm pressure (5, 49). More generally, the importance of surface preparation of TiO_2 and treatment of the rutile-single crystal/Pt evaporated film for the behavior of the electrical contact has been emphasized (50). In chemistry terms, the half-order dependence of σ on H_2 pressure we observed can be considered to result from

$$Pt_s\text{-}H + O^{2-} \rightarrow Pt_s + OH^- + e^-$$

which symbolizes the migration (spillover) of H atoms from Pt to TiO_2 (45).

These measurements cannot be used to quantify the electron transfer from the semiconductor to the metal deposit, but an estimate has been drawn from studies of oxygen photoadsorption on Pt/TiO_2 samples in a pressure range such that nearly all of the free electrons are captured to form adsorbed O_2^- ion-radicals. Increasing Pt contents corresponded to decreasing amounts of photoadsorbed oxygen, which corroborates the effect of deposited Pt on the TiO_2 free electron density. For TiO_2 samples evacuated at 423 K and

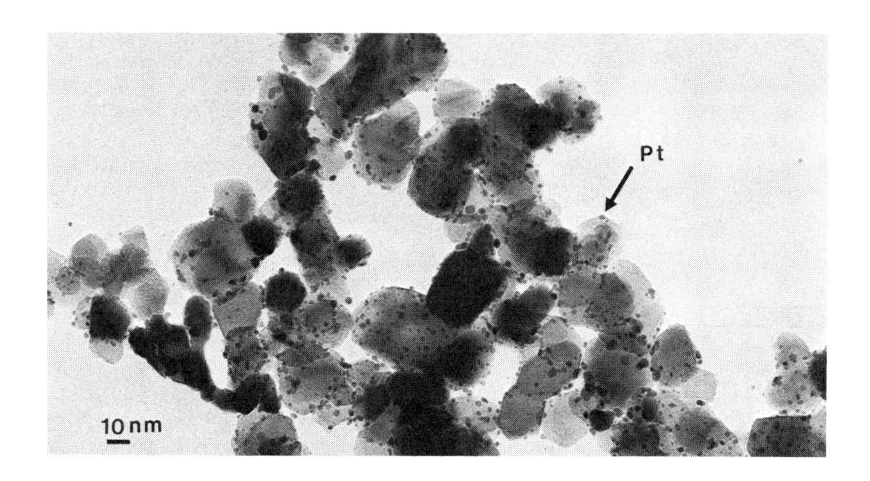

Figure 3. Transmission electron micrograph of a 10 wt % Pt/TiO_2 sample.

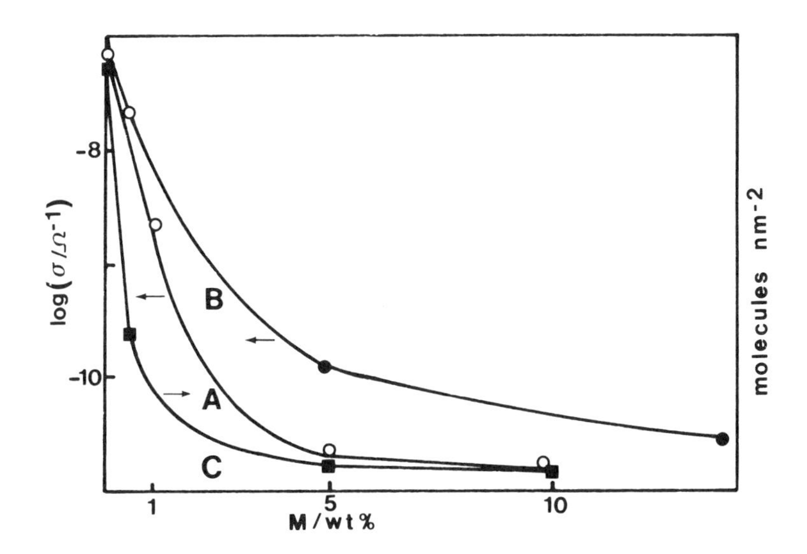

Figure 4. Correlations between the Pt(A, C) or Ni (B) content and (i) the logarithm of the photoconductance σ at equilibrium under vacuum (A, B) or (ii) the amounts of oxygen photo-adsorbed (C).

exposed to O_2 at 295 K up to a final pressure of about 2.66 x 10^2Pa, average electron enrichments of ca. 0.7, 0.1 and 0.05 electron per Pt atom were calculated for the samples containing 0.5, 5 and 10 wt % Pt, respectively (44).

In conclusion, the electron transfer from the semiconductor to the metal depends on the state of both components, i.e., namely, on the medium surrounding the sample and this is certainly an element to consider when using these solids as photocatalysts.

Cyclopentane-deuterium isotopic exchange in gaseous phase (43). Without metal deposit and under UV-illumination, the exchange was limited and occurred only if the sample was previously covered with OD groups. In other words, it was not photocatalytic and corresponded to the consumption of OD groups. With deposited Pt and in the dark, the exchange required temperatures > 273 K and polydeuterated cyclopentane molecules were obtained (Figure 5). It was inferred, namely by replacing TiO_2 by other supports, that the dark reaction took place on the metal particles where the residence time of the chemisorbed cyclopentane molecules enabled a multiple exchange. With deposited Pt and under UV-illumination, the exchange occurred at lower temperatures, yielding C_5H_9D (after an induction period, weak amounts of multiply exchanged molecules progressively appear) at a constant rate (Figure 5) . An experiment carried out with the same catalyst sample showed that, in successive runs, a large amount of C_5H_{10} was exchanged without decrease in activity. It was concluded that the exchange was photocatalytic and that it took place on TiO_2 where the short residence time allowed only simple exchange for each adsorption act. In addition, an optimal Pt content, which will be discussed later on, was found (Figure 6). Although this reaction was chosen for mechanistic reasons, its extension to appropriate molecules might be useful when incomplete deuteration is desired.

Alcohol dehydrogenation in liquid phase (8-11, 51). H_2 evolution from aqueous methanol solutions with Pt/TiO_2 (or Pt and TiO_2) had been reported (52). Moreover, it was claimed that H_2 also resulted from water decomposition (52). From a set of experiments we have established that in that case the dehydrogenation of methanol accounts for the H_2 produced (8).

In the absence of deposited metal, the rate of H_2 generation from an alcohol progessively decreased, while TiO_2 turned to a blue color. By contrast, in the presence of a M/TiO_2 sample (M = Pt, Rh, Ni), the dehydrogenation was photocatalytic (8-10). An optimal Pt content of about 0.5 wt % (ca. 1 Pt particle per TiO_2 grain) was found (8) (Figure 6) and has been confirmed (11). For Ni, the optimal value is about 5 wt % and the maximum initial activity is ca. 7 times lower than that obtained with Pt/TiO_2 (10). This emphasizes the fact that direct comparison of the effects of various metals should be considered with care if the metal-content activity dependence has not been determined. The differences between Ni and Pt can arise from distinct catalytic properties. Most probably, they also related to the metal particle sizes. The reverse reaction exists and as expected, increases with the metal loading (9). The Ni/TiO_2 samples make clear that zerovalent metal atoms are required (10). A study of the temperature influence indicated that the reaction rate can be controlled by H_2 desorption below room temperature (9).

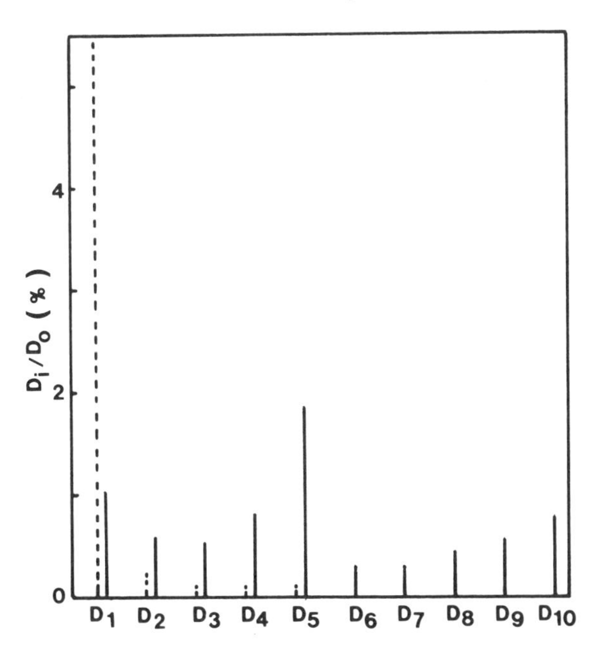

Figure 5. Number of deuterium atoms exchanged per cyclopentane molecule over a 5 wt % Pt/TiO_2 sample : dotted lines, after illuminating for 15 min at 263 K, solid lines after 2.5 min in the dark at 343 K.

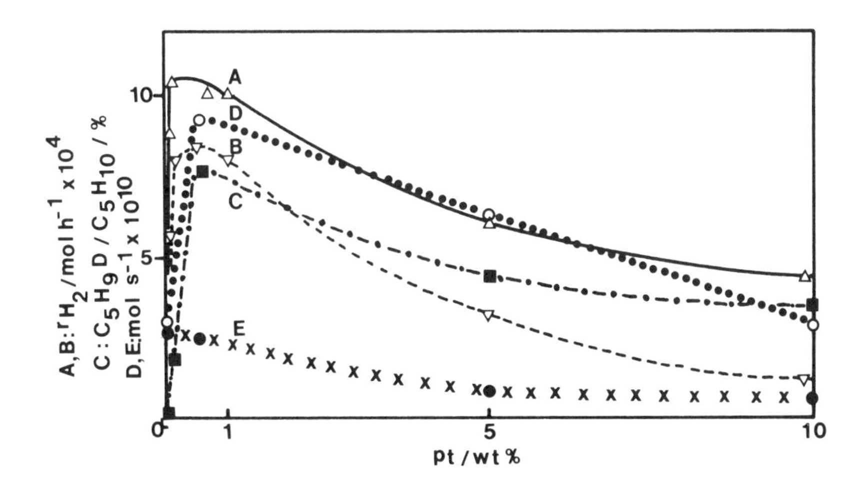

Figure 6. Initial rates vs Pt contents of the Pt/TiO_2 specimens for : liquid methanol (A) or 1-propanol (B) dehydrogenation at 298 K ; cyclopentane-deuterium exchange in gas phase at 263 K (C) ; oxygen isotope heteroexchange at 298 K over non-preoxidized (D) or preoxidized (E) samples.

For neat-liquid it has been established that the dehydrogenation represents almost completely the phenomena observed, since with ethanol and 1-propanol, 1.1 values were found for the ratios H_2/(aldehyde + acetal), which is within the experimental errors, and the formation of CO_2 molecules only amounted to 1 % (ethanol) or 4 % (1-propanol) of the aldehyde + acetal molecules (8). Depending upon the catalyst and the alcohol, quantum yields in the range 0.1-0.8 were found for the following alcohols whose order of reactivity was MeOH > EtOH > 1-PrOH, 2-PrOH, 1-BuOH

However, about three times more of propanal than of acetone was obtained from an equimolecular mixture of 1 and 2-propanol ; this difference might arise from competitive adsorptions ; it seems to indicate that free radicals are not involved under these conditions.

Table I. Mean quantum yields and product distribution for the dehydrogenation of the alcohols indicated (catalyst : 0.5 wt % Pt/TiO_2 ; illumination time : 6 h).

	ϕ	Products (%)		
		Unsat. ald.	Sat. alc.	Sat. ald.
allyl alcohol	0.025	79	4	18
cinnamyl alcohol	0.065	77	17	6
citronellol	0.14	100	0	0
tetrahydrogeraniol	0.14	-	-	100
geraniol	0.35	75 (citral)	24 (citronellol)	1 (citronellol)

The mechanism suggested (8) includes the dissociative adsorption of the alcohol on anatase basic sites :

$$>CHOH \rightarrow >CHO^- + H^+$$

and the intervention of holes (possibly as OH° or $O^{\cdot-}$ species) to abstract a H_α atom (heterolytic breaking) :

$$>CHO^- + p^+ \rightarrow >CO + H^\circ$$

The formation of H_2, which involves electrons, would be favored by the metal crystallites

$$H^\circ + H^+ + e^- \rightarrow H_2$$

This dehydrogenation method was recently extended to various liquid unsaturated alcohols (allyl and cinnamyl alcohols, citronellol, geraniol and, for comparison, tetrahydrogeraniol, chosen as model molecules) (53) (Figure 7). The quantum yields are indicated in Table I. For citronellol (and tetrahydrogeraniol), it was inferred from the H_2/aldehyde ratios, equal to 1 within experimental errors, that the dehydrogenation was by far the dominant phenomenon. In the presence of a conjugated double bond, Table I shows that the corresponding saturated alcohol and the saturated aldehyde (issued from isomerization) were also detected, although the amount of

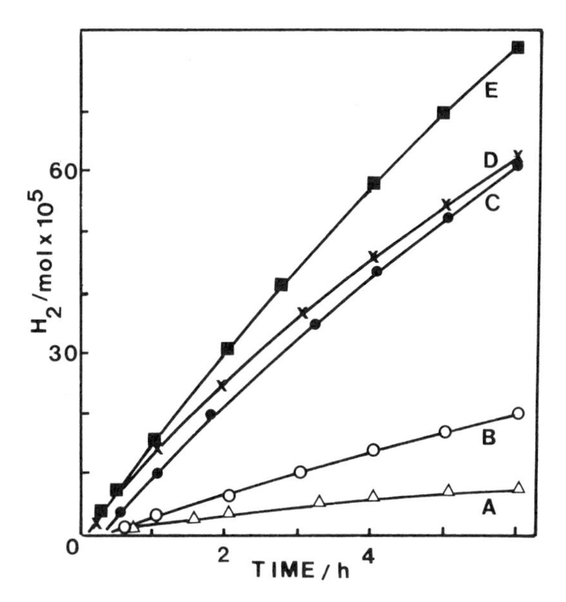

Figure 7. H_2 generated from suspensions of 70 mg 0.5 wt % Pt/TiO_2 in 10 cm^3 (A) allyl alcohol, (B) cinnamyl alcohol, (C) citronellol, (D) tetrahydrogeraniol, (E) geraniol as a function of illumination time.

unsaturated aldehyde predominated markedly. On the contrary, a double bond remote from the alcohol function and in addition hindered with two methyl groups, such as in citronellol or in geraniol, was not reduced. Experiments with naked TiO_2 showed that the Pt deposit was not involved in the isomerization.

Under these conditions involving the absence of oxygen, no significant over-oxidation was detected.

Because of the low incident radiant flux (some 10^{-5} Einstein h^{-1} cm^{-2}), conversions below 1 % were obtained with pure alcohols after illuminating for 10 h, despite reasonable quantum yields. However, high chemical yields could be obtained for comparable numbers of photons and of molecules to transformed, i.e. with diluted solutions. For example, conversions of ~ 50 or ~ 80 % were found after 2 or 6 h, respectively, with 6 x 10^{-4} mol citronellol in 10 cm^3 n-heptane, using a 125 W Hg-lamp.

These alcohol dehydrogenations can be used for in situ hydrogenation of unsaturated compounds. Only a small quantity of H_2 was found to evolve compared with the amount incorporated into the unsaturated molecule. Moreover, when using an alcohol with a conjugated double bond, the presence of a reducible compound allowed the selectivity to the unsaturated aldehyde to be increased (for instance, increase in acrolein percentage for allyl alcohol dehydrogenation in the presence of diphenylacetylene). Unfortunately, at least for the Pt/TiO_2 samples employed, the hydrogenation was not selective ; for instance, both the C-C double bond and the carbonyl group of cinnamaldehyde were saturated by hydrogen abstracted from 2-propanol. Besides the addition of an unsaturated compound did not render catalytic the alcohol dehydrogenation over naked TiO_2.

In conclusion, this dehydrogenation method, beyond its fundamental objectives, could be of interest in organic synthesis.

Interpretation of the optimum metal content for these reactions. As already mentioned an optimum Pt content was found for dehydrogenation of liquid alcohols and cyclopentane-deuterium exchange in gas phase. Also, with Pt/TiO_2 samples which had not been preoxidized and which were accordingly non-stoichiometric according to conductivity measurements, the same optimum content was found for the initial rate of OIE, whereas this rate decreased as a function of Pt content for preoxidized samples (44).

If the metal had only a beneficial catalytic role necessary (i) to evolve H_2 (or HD), and (ii) to regenerate OD groups (C_5H_{10}-D_2 exchange), the reaction rates should increase with increasing metal contents, possibly up to a limit, and no increase should be expected in the case of oxygen isotope exchange.

The maximum rates observed for a Pt content of about 0.5 wt % for instance, regardless of the reaction (Figure 6), show that Pt has also a detrimental effect. This cannot arise from back-reactions, since initial reaction rates have been considered. This cannot result from geometrical features either, since (i) only about 6 % of the anatase surface was occulted by Pt for the highest content (10 wt %), (ii) the TiO_2 areas between the Pt particles remained largely sufficient to allow the adsorption of several C_5H_{10} or alcohol molecules, as shown by TEM (Figure 3), and (iii) the metal particle size did not vary for the contents employed. The OIE experiments show that an optimum Pt content exists only if TiO_2 is sufficiently non-stoichiometric to allow an electron transfer to Pt. This tends to indicate that the electronic role of Pt is at the origin of the optimum content. For low Pt contents, the decreased electron density of TiO_2 in the presence of O_2 can reduce the electron-hole recombination in this material, whereas for high Pt contents (5 and 10 wt %) the recombination at the much more numerous metal particles can significantly compete with the reactions driven by the minority charge carriers. Only this latter effect can occur for preoxidized TiO_2 which has a Fermi level probably very close to that of Pt and this would explain the decline in OIE rate for all the Pt/TiO_2 preoxidized samples.

The fact that only a low amount of metal deposit is required to render catalytic reactions involving hydrogen over illuminated TiO_2 is advantageous to limit back reactions catalyzed by the metal, as well as for economical reasons. Conversely, this result illustrates the limitations encountered when trying to modify the surface of a semiconductor.

Organic acid decarboxylation. Bard et al. (54-57) initiated the investigation of the exergonic decarboxylation of various R-COOH acids (R = CH_3, C_2H_5, C_3H_7, C_4H_9, $C_3H_7C(CH_3)_2$) and of a diacid (hexanedioic) over TiO_2 (or WO_3) with and without photodeposited Pt(1-5 wt %) in various media (aqueous and mixed aqueous/organic solutions (54-57), gas phase (58)). The main reaction was

$$RCO_2H \rightarrow CO_2 + RH,$$

while the formations of the alkane R-R and of H_2 were also detected

$$2RCO_2H \rightarrow 2CO_2 + R - R + H_2$$

In addition, little amounts of intermediate, oxygen-containing compounds, such as ethanal, methanol and ethanol, were produced, which was not unexpected since alkanes are oxidized under these conditions (vide supra). The maximum quantum yield was about 0.1. Much higher conversions were obtained in the presence of O_2. In this case, the Pt deposit seemed to play a minor role, whereas it was more important for deoxygenated solutions.

This beneficial effect was attributed to rate enhancements of reduction processes. With benzoic acid, total degradation of the ring to CO_2 occurred and the detection of salycilic acid suggested the intervention of OH° radicals (56). These radicals were also proposed to explain the oxidation of n-C_x alkanes (x = 6,7,9,10) and of cyclohexane in 1:1 vol. water/hydrocarbon two-phase mixtures over 10 wt % Pt/TiO_2. Traces of alcohols (and of 2-, 4-, 5-decanone with decane) were detected. No transformation occurred without O_2 and in the absence of H_2O the rate was substantially decreased. The role of Pt was attributed to a greater ease of oxygen reduction ; however the oxidation rate was only decreased by a factor of about 1.5 without Pt (59).

Similarly, with the same type of photocatalyst (Pt/TiO_2 or Fe_2O_3) the decomposition of levulinic (4-oxopentanoic) acid in oxygen-free aqueous solution has been investigated in detail (60). In addition to the decarboxylation reaction, oxidative C-C scissions led to propionic and acetic acids (further converted into methane and ethane) and reductive cleavages to acetone and ethanal. The formation of acetone was apparently favoured by higher Pt contents (however product distributions referred to equal illumination durations and not to equal conversions). It was suggested that the variety of products resulted from the presence of two functional groups in levulinic acid. The quantum yield was probably of the order of 5×10^{-3}.

In short, the proposed mechanism was based on the formation of R° radicals (CH_3° was detected by ESR (54) by capture of photoproduced holes ($RCO_2^- + h^+ \rightarrow R^{\circ} + CO_2$), while O_2 was considered to intervene in electron capture, and Pt in the formation of H_2 from H° as in the reactions previously discussed in this text.

Conclusion. As expected, the deposition of a group VIII transition metal onto an appropriate semiconductor presents a very great interest, since it allows one to extend the domain of heterogeneous photocatalysis to reactions involving H_2, such as dehydrogenations and isotopic exchanges, because of the catalytic properties of these metals for either recombining or dissociating H_2. The existence of the spillover phenomenon enables a reversible transfer of hydrogen atoms or protons between both photocatalyst components (61). However, these deposits act as recombination centers for the photoproduced charges. The resulting optimal metal amount, which depends on the metal, on the semiconductor and on their respective particle sizes, should be determined for each case.

Acknowledgments
The author is indebted to his CNRS co-workers Dr J.-M. Herrmann, Mr H. Courbon, Mr J. Disdier, Mrs M.-N. Mozzanega whose contributions appear in the literature cited.

Literature Cited

1. Gerischer, H. Pure & Appl. Chem. 1980, 52, 2649.
2. Pleskov, Yu. V. Sov. Electrochem. 1981, 17, 1.
3. Bard, A.J. J. Phys. Chem. 1982, 86, 172.
4. Wrighton, M.S. In "Inorganic Chemistry : toward the 21st Century" ; Chisholm, M.H., Ed. ; ACS SYMPOSIUM SERIES No. 211, American Chemical Society : Washington, D.C., 1983 ; pp. 59-91.
5. Heller, A. Science 1984, 223, 1141.
6. Bickley, R.I. In "Catalysis", Kemball, C., Ed. ; Specialist Periodical Report, The Chemical Society, London, 1982, Vol. 5, pp. 308-332.
7. Cunningham, J. In "Comprehensive Chemical Kinetics", Bamford, C.H., Tipper, C.F.H., Eds ; Elsevier, Amsterdam, 1984, Vol. 19.
8. Pichat, P. ; Herrmann, J.-M. ; Disdier, J. ; Courbon, H. ; Mozzanega, M.-N. Nouv. J. Chim. 1981, 5, 627.
9. Pichat, P. ; Mozzanega, M.-N. ; Disdier, J. ; Herrmann. J.-M. Nouv. J. Chim. 1982, 6, 559.
10. Prahov, L.T. ; Disdier, J. ; Herrmann, J.-M. ; Pichat, P. Int. J. Hydrogen Energy, 1984, 9, 397.
11. Borgarello, E. ; Pelizzetti, E. Chim. Ind. 1983, 65, 474.
12. Fox, M.A. Acc. Chem. Res. 1983, 16, 314.
13. Frank, S.N. ; Bard, A.J. J. Phys. Chem. 1977, 81, 1484 ; Pruden, A.L. ; Ollis, D.F. J. Catal. 1983, 82, 404 ; Hsiao, C.-Y. ; Lee, C.-L. ; Ollis, D.F. J. Catal. 1983, 82, 418.
14. Clechet, P. ; Martelet, C. ; Martin, J.-R. ; Olier, R. C.R. Acad. Sci. Ser. C 1978, 287, 405 ; Kraeutler, B ; Bard, A.J. J. Am. Chem. Soc. 1978, 100, 4317 ; Hada, H. ; Yonezawa, Y. ; Saikawa, M. Bull. Chem. Soc. Jpn. 1982, 55, 2010.
15. Bickley, R.I. In "Fundamentals and Developments of Photocatalytic and Photoelectrochemical Processes", Proc. of a NATO-ASI, Erice, Italy, 1984 ; Schiavello, M. et al., Eds ; D. Reidel Publ. Co., in press.

16. Curran, J. ; Lamouche, D. J. Phys. Chem. 1983, 87, 5405.
17. Djeghri, N. ; Formenti, M. ; Juillet, F. ; Teichner, S.J. Faraday Disc. Chem. Soc. 1974, 58, 185.
18. Djeghri, N. ; Teichner, S.J. J. Catal. 1980, 62, 99.
19. Mozzanega, M.-N. ; Herrmann, J.-M. ; Pichat, P. Tetrahedron Lett. 1977, 34, 2965, and unpublished results from this group.
20. Pichat, P. ; Herrmann, J.-M. ; Disdier, J. ; Mozzanega, M.-N. J. Phys. Chem. 1979, 83, 3122.
21. Cunningham, J. ; Doyle, B. ; Leahy, E.M. J. Chem. Soc., Faraday Trans. 1 1979, 75, 2000 ; Cunningham, J. ; Hodnett, B.K. J. Chem. Soc., Faraday Trans. 1 1981, 77, 2777 ; Cunningham, J. ; Hodnett, B.K. ; Ilyas, M. ; Leahy, E.M. ; Tobin, J.P. J. Chem. Soc., Faraday Trans. 1 1982, 78, 3297.
22. Bickley, R.I. ; Munuera, G. ; Stone, F.S. J. Catal. 1973, 31, 398 ; Bickley, R.I. ; Jayanty, R.K.M. Faraday Disc. Chem. Soc. 1974, 58, 194.
23. Cundall, R.B. ; Rudham, R. ; Salim, M.S. J. Chem. Soc. Faraday Trans. 1 1976, 72, 1642.
24. Harvey, P.R. ; Rudham, R. ; Ward, S. J. Chem. Soc., Faraday Trans. 1 1983, 79, 1381 and 2975.
25. Walker, A. ; Formenti, M. ; Meriaudeau, P. ; Teichner, S.J. ; J. Catal. 1977, 50, 237.
26. Courbon, H. ; Pichat, P. J. Chem. Soc., Faraday Trans. 1, in press.
27. Pichat, P. ; Herrmann, J.-M. ; Courbon, H. ; Disdier, J. ; Mozzanega, M.-N. Canad. J. Chem. Eng. 1982, 60, 27.
28. Pichat, P. ; Courbon, H. ; Disdier, J. ; Mozzanega, M.-N. ; Herrmann, J.-M. in "New Horizons in Catalysis", Studies in Surf. Sci. Catal., 7A, Proc. 7th Int. Cong. Catal. ; Seiyama, T. ; Tanabe, K., Eds. ; Elsevier, 1981 ; Part B, pp. 1498-1499.
29. Herrmann, J.-M. ; Mozzanega, M.-N. ; Pichat, P. J. Photochem. 1983, 22, 333.
30. Bielanski, A. ; Haber, J. Catal. Rev. 1979, 19, 1 ; Che, M. ; Tench, A.J. Adv. Catal. 1982, 31, 77 ; 1983, 32, 1.
31. Morrison, S.R. "The Chemical Physics of Surfaces" ; Plenum, New-York, 1977 ; Chap. 9.
32. Winter, E.R.S. J. Chem. Soc. 1968, 2889 ; Novakova, J.Catal. Rev. 1970, 4, 77 ; Boreskov, G.K. in "Catalysis" ; Anderson, J.R. ; Boudart, M., Eds. ; Springer-Verlag, 1982 ; Vol. 3 ; Chap. 2.
33. Courbon, H. ; Formenti, M. ; Pichat, P. J. Phys. Chem. 1977, 81, 550.
34. Courbon, H. ; Pichat, P. C.R. Acad. Sci. Ser. C 1977 ; 285, 171.
35. Herrmann, J.-M. ; Disdier, J. ; Pichat, P. Chem. Phys. Lett., 1984, 108, 618.
36. Tanaka, K. J. Phys. Chem. 1974, 78, 555 ; Tanaka, K. ; Miyahara, K. J. Phys. Chem. 1974, 78, 2303.
37. Cunningham, J. ; Goold, E.L. ; Leahy, E.M. J. Chem. Soc., Faraday Trans. 1 1979, 75, 305 ; Cunningham, J. ; Goold, E.L. ; Fierro, J.L.G. J. Chem. Soc., Faraday Trans. 1 1982, 78, 785.
38. Herrmann, J.-M. ; Disdier, J. ; Pichat, P., J. Chem. Soc., Faraday Trans. 1 1981, 77, 2815.
39. Herrmann, J.-M. ; Disdier, J. ; Mozzanega, M.-N. ; Pichat, P. J. Catal. 1979, 60, 369.
40. Boehm, H.P. Chem. Ing. Techn. 1974, 17, 716.
41. Jaeger, C.D. ; Bard, A.J. J. Phys. Chem. 1979, 83, 3146.

42. Völz, H.G. ; Kaempf, G. ; Fitzky, H.G. ; Klaeren, A. In "Photodegradation and Photostabilization of Coatings", Pappas, S.P. ; Winslow, F.H., Eds ; ACS SYMPOSIUM SERIES No. 151, American Chemical Society : Washington, D.C., 1981 ; pp. 163-182 ; Irick, G. J. Appl. Polym. Sci. 1972, 16, 2387.
43. Courbon, H. ; Herrmann, J.-M. ; Pichat, P. J. Catal. 1981, 72, 129.
44. Courbon, H. ; Herrmann, J.-M. ; Pichat, P. J. Phys. Chem., in press.
45. Disdier, J. ; Herrmann, J.-M. ; Pichat, P. J. Chem. Soc., Faraday Trans. 1 1983, 79, 651.
46. Chung, Y.W. ; Tsai, S.C. ; Somorjai, G.A. Surf. Sci. 1977, 64, 588.
47. Bube, R.H. "Photoconductivity of Solids" ; Wiley, 1960 ; Chap. 5.
48. Yamamoto, N. ; Tonomura, S. ; Matsuoka, T. ; Tsubomura, H. Surf. Sci. 1980, 92, 400.
49. Aspnes, D.E. ; Heller, A. J. Phys. Chem. 1983, 87, 4919.
50. Hope, G.A. ; Bard, A.J. J. Phys. Chem. 1983, 87, 1979.
51. Teratani, S. ; Nakamichi, J. ; Taya, K. ; Tanaka, K. Bull. Chem. Soc. Jpn. 1982, 55, 1688 ; Oosawa, Y. Chem. Lett. 1983, 577.
52. Benderskii, V.A. ; Zolovitskii, Ya. M. ; Kogan, Ya.L. ; Khidekel', M.L. ; Shub, D.M. Dokl. Akad. Nauk SSSR 1975, 222, 606 ; Kawai, T. ; Sakata, T. J. Chem. Soc., Chem. Comm. 1980, 695.
53. Pichat, P. ; Disdier, J. ; Mozzanega, M.-N., Herrmann, J.-M. Proc. 8th Int. Cong. Catal., 1984, Verlag Chemie, Dechema : Deerfield Beach, Florida ; Vol. III, pp. 487-498.
54. Kraeutler, B. ; Jaeger, C.D. ; Bard, A.J. J. Am. Chem. Soc. 1978, 100, 4903.
55. Kraeutler, B. . Bard, A.J. J. Am. Chem. Soc. 1978, 100, 5985.
56. Izumi, I. ; Fan,F.-R.F. ; Bard, A.J. J. Phys. Chem. 1981, 85, 218.
57. Yoneyama, H. ; Takao, Y. ; Tamura, H. ; Bard, A.J. J. Phys. Chem. 1983, 87, 1417.
58. Sato, S. J. Phys. Chem. 1983, 87, 3531.
59. Izumi, I. ; Dunn, W.W. ; Wilbourn, K.O. ; Fan F.-R.F. ; Bard, A.J. J. Phys. Chem. 1980, 84, 3207.
60. Chum, H.L. ; Ratcliff, M. ; Posey, F.L. ; Turner, J.A. ; Nozik, A.J. J. Phys. Chem. 1983, 87, 3089.
61. Herrmann, J.-M. ; Pichat, P. In "Spillover of Adsorbed Species", Studies in Surf. Sci. Catal., 17 ; Pajonk, G.M. ; Teichner, S.J. ; Germain ; J.E., Eds. ; Elsevier, 1983, pp. 77-87.

RECEIVED January 10, 1985

3

Semiconductor-Catalyzed Photoreactions of Organic Compounds

KATSUMI TOKUMARU, HIROCHIKA SAKURAGI, TATSUYA KANNO, TAKAHIDE OGUCHI, HIROAKI MISAWA, YASUO SHIMAMURA, and YASUNAO KURIYAMA

Department of Chemistry, University of Tsukuba, Sakura-mura, Ibaraki 305, Japan

The actions of photoexcited semiconductor particles on organic compounds under oxygen is of significant importance from both practical and basic aspects. Semiconductors like titanium dioxide and cadmium sulfide were shown to induce oxidation of olefins and aromatic hydrocarbons under oxygen, and also to sensitize isomerization of unsaturated systems. The mechanisms of these reactions are discussed.

Recently, the photochemical action of semiconductors have been actively investigated. However, this area attracted the interest of many workers nearly thirty years ago (1-5). In those days, Laidler (5), Calvert (6,7), and Markham (8-10) studied the action of zinc oxide under illumination in aqueous alcohols under oxygen producing carbonyl compounds and hydrogen peroxide. The interest in those days can be seen by the appearance of an introductory article by Markham in J. Chem. Ed. in 1955 (11). At nearly the same time, Mashio and Kato worked on the photocatalytic action of titanium dioxide on oxidation of alcohols (12,13) and hydrocarbons (14) in an attempt to elucidate the mechanism by which poly(vinyl acetate) fibres dyed with the aid of titanium dioxide became fragile during use in air under sunshine. They found that the irradiation of titanium dioxide initiates autoxidation of the substrates. In 1960's, when the manufacture of polypropylene by Ziegler-Natta catalyst was started, degradation of the polymer was found to be accelerated by titanium compounds remaining in the polymer, which again attracted attention to the photocatalytic action of titanium dioxide (15-17).

At the end of 1960's, Honda and Fujishima found that in a photochemical cell employing titanium dioxide and platinum electrodes, irradiation of the titanium dioxide electrode resulted in splitting of water into hydrogen and oxygen (18,19). This work has had an extraordinarily strong impact for the research on the photochemical action of various semiconductors inducing evolution of hydrogen from water as well as new catalytic reactions (20).

For the action of semiconductor particles on organic compounds, Bard and his coworker reported that irradiation of titanium dioxide

0097-6156/85/0278-0043$06.00/0

in aqueous acetic acid under oxygen led to the oxidation of acetate anions to methane and ethane (21,22). At nearly the same time, other groups initiated investigations of the photocatalytic action of semiconductor particles on relatively simple organic compounds like alcohols and carboxylic acids, with the hope of photochemically producing hydrogen from water (23-25).

At the end of 1970's we attempted an investigation of the action of semiconductor particles on medium sized organic compounds. In this article, we would like to describe some of the work done in our laboratory on the photocatalytic action of semiconductors on the oxidation of olefins and hydrocarbons, and on the isomerization of unsaturated systems.

Photocatalytic Oxidation of Olefins

We found that excitation of titanium dioxide or cadmium sulfide powder suspended in organic solvents containing olefins like 1,1-diphenylethylene (DPE) under oxygen led to oxidation of the olefins to their epoxides and carbonyl compounds (26). In a typical run, semiconductor powder (ca. 5 mg) was suspended in a solution (ca. 2 ml) containing an olefin (ca. 0.3 mmol), and irradiated under oxygen atmosphere at room temperature with light of longer wavelengths than 350 nm for titanium dioxide and than 430 nm for cadmium sulfide. The products were separated by vpc or tlc for their identification and quantitatively determined by vpc. For example, DPE gave benzophenone and its epoxide, which was converted to 2-methoxy-2,2-diphenylethanol in methanol solution.

$$Ph_2C{=}CR'R'' \xrightarrow[\text{CdS or } TiO_2]{h\nu,\ O_2} Ph_2C{=}O + O{=}CR'R'' + Ph_2C\overset{O}{-}CR'R''$$

$R' = R'' = H, Me$

$R' = H, R'' = OMe$

$$\downarrow MeOH$$

$Ph_2C(OMe)CH_2OH$ or $Ph_2C(OH)CH(OMe)_2$

In aqueous acetonitrile, CdS-sensitized irradiation of DPE gave 1,1-diphenylethan-1,2-diol along with benzophenone, which indicates that the resulting epoxide is hydrolyzed in the solution (27).

Some results are summarized in Table I (27). Figure 1 schematically depicts the reactivity of the olefins examined with their oxidation potentials, and indicates that generally the unreactive olefins are those with higher oxidation potentials, and the reactive olefins are those with lower oxidation potentials. However, among the oxidizable olefins their reactivity is not simply governed by their oxidation potentials as will be discussed later.

The quantum yields for the consumption of DPE and 2-methyl-1,1-diphenylpropene were obtained as at least 0.2 and 0.1, respectively, in TiO_2-sensitized oxygenation without correcting the reflection of the incident light on the semiconductor surfaces.

The above facts indicate that on excitation of the semiconductors the olefins transfer an electron to the photogenerated positive hole in the initiation process to give the olefin radical cation, and concurrently the electron excited to the conduction band is

Table I. Semiconductor Photocatalyzed Oxygenation of Aromatic Olefins (a)

Olefin	Semicon-ductor	Solvent	Time /h	Conver-sion/%	Carbonyl compds/%	Epoxide or its deriv/%
$Ph_2C{=}CH_2$	CdS	MeCN-MeOH (1:1)	6	27	15	26
p-$MeOC_6H_4CH{=}CH_2$	TiO_2	MeCN	6	31	43	trace
$PhMeC{=}CH_2$	TiO_2	MeCN	6	4	77	
$Ph_2C{=}CHOMe$	CdS	MeCN-MeOH (1:1)	7	100	12	69
$Ph_2C{=}CMe_2$	TiO_2	MeCN	6	16	11	9
trans-PhCH=CHPh	CdS	MeCN	12	5	+	5
cis-PhCH=CHPh	CdS	MeCN	12	6	+	2
cyclooctene	TiO_2	MeCN	6	4	-	96
norbornene	TiO_2	MeCN	6	3	-	27
2-cyclohexenone	TiO_2	MeCN	8	No reaction		
PhCH=CHCN	TiO_2	MeCN	8	No reaction		
PhCH=CHCOMe	TiO_2	MeCN	8	No reaction		
$PhCH{=}CHCO_2Et$	TiO_2	MeCN	8	No reaction		

a. Yields are based on the olefin consumed.

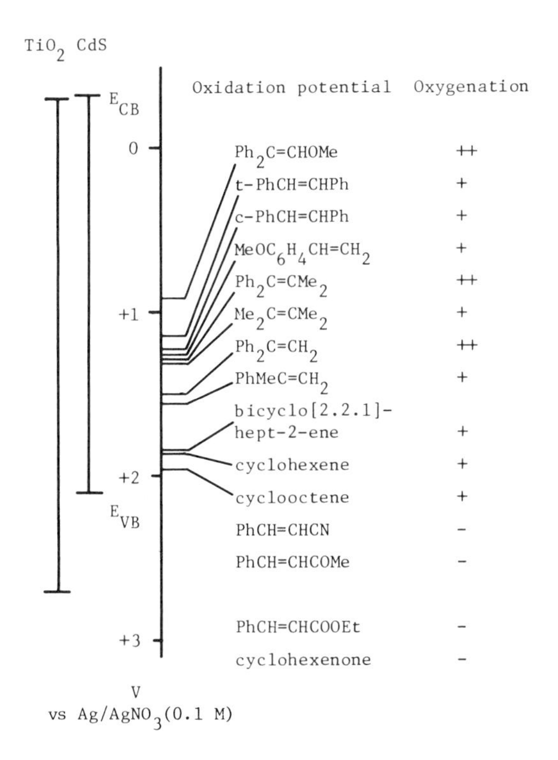

Figure 1. Reactivity of olefins and their oxidation potentials.

transferred to oxygen to give the superoxide anion. These resulting species subsequently initiate free radical chain reactions.

Fox and her coworkers observed transient spectra attributable to radical cations upon laser excitation of powdered semiconductor in the presence of olefins (28-31).

Factors Controlling the Reactivity of Olefins in the Photocatalyzed Oxygenation

To examine the electron transfer from olefins to the excited semiconductor, photoelectrochemical measurements were undertaken. In a cell comprising of two compartments divided by a glass frit, a titanium dioxide working electrode was immersed in an acetonitrile solution of an olefin (0.2 M) in one compartment, and a platinum counter electrode was put in acetonitrile in the other compartment; tetraethylammonium perchlorate (0.1 M) was used as a supporting electrolyte. After bubbling argon into the working cell solution, the titanium dioxide electrode was illuminated ($\lambda > 350$ nm) under voltages varying between -0.5 and 0.8 V vs $Ag/AgNO_3$ (0.1 M). Some olefins exhibited an increase of photocurrent with increase of the applied potential to give a nearly constant current around 0.5 V; however, some olefins did not show any photocurrent at these potentials (32).

Figure 2 shows that the photocurrent measured at 0.5 V increases with decreasing the oxidation potentials of the olefins examined; however, benzylideneacetone, etc. with higher oxidation potentials did not show any photocurrent (32).

Although the absolute amount of the photocurrents is governed by various factors such as the oxidation potentials of olefins and the extent of adsorption of olefins on the electrode, the above findings show that the reactive olefins in the photocatalytic oxygenation exhibit photocurrents and the olefins which do not exhibit photocurrents are unreactive in the photocatalytic oxygenation. On the other hand, the olefins which exhibit photocurrents are not always reactive. For example, stilbene shows a higher photocurrent than DPE, but is not so reactive as DPE. The electron transfer to the excited semiconductor takes place more efficiently from stilbene than from DPE due to the lower oxidation potential of the former, but in the subsequent free radical reactions, stilbene is less reactive than DPE (33).

Therefore, it can be concluded that for the photocatalytic oxygenation to occur, the electron transfer from the olefin to the positive hole has to take place, but the overall reactivity of the olefins is governed by the efficiency of free radical processes as exemplified in Table II.

In view of the above results, in photocatalytic oxidation of a series of 4-substituted diphenylethylenes, an increase in reactivity with decreasing Hammett's sigma constants (31) seems to arise not only from the lowering of the oxidation potentials of the olefins in this sequence but also from the general trend of the increase in the reactivity of olefins toward peroxyl radicals with increasing the electron donating ability of olefins (33).

To get insight into the reactivity of olefin radical cations toward oxygen, anodic oxidation of olefins under oxygen was attempted. DPE was electrolyzed at 1.5 V vs SCE in a mixture of

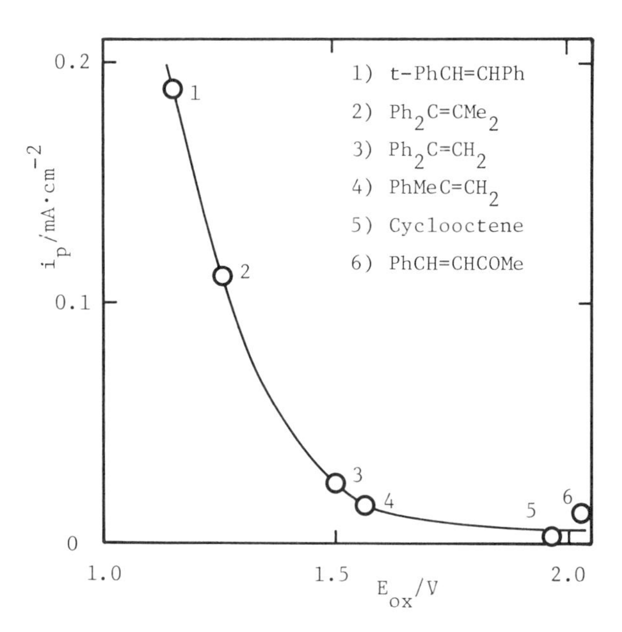

Figure 2. Correlation of photocurrents with oxidation potentials.

Table II. Reactivity of Olefins in Photocatalyzed Oxygenation

Olefin	Efficiency		
	Electron transfer	Radical chain oxidation	Overall
PhCH=CHPh	good	poor	medium
$Ph_2C=CH_2$	medium	good	good
PhCH=CHCOMe	poor	poor	poor

acetonitrile and methanol (1:3 by volume) in the presence of tetraethylammonium perchlorate as a supporting electrolyte under oxygen. The electrolysis carried out to nearly 70% conversion of the olefin afforded a 1,2-dimethoxylated product as the main product along with a small amount of benzophenone, but 2,2-diphenyl-2-methoxyethanol, which had been expected to arise from oxygenation of the radical cation in methanol, was formed in a very low yield (27). These results indicate that the DPE radical cation is not highly reactive with oxygen.

For generation of olefin radical cations, recently several attempts have been made to use 9,10-dicyanoanthracene (DCA) as an electron accepting sensitizer under oxygen. On excitation of DCA, the resulting DCA singlet excited state accepts an electron from an olefin to give the radical anion of DCA and the radical cation of the olefin, and the former anion can transfer an electron to oxygen to give the superoxide anion. For example, DCA-sensitized oxygenation of DPE was reported to give a mixture of various oxidation products involving benzophenone as the main product along with the corresponding epoxide and 1,1-diphenyl-2-methoxyethanol, etc., and the products were proposed to arise from the reaction between the olefin radical cation and superoxide anion (34-37).

Among olefin radical cations studied only adamantylideneadamantane (ADA) is established to react facilely with oxygen to give an oxygen adduct which subsequently accepts an electron from ADA to give the corresponding dioxetane and to regenerate the ADA radical cation, thus accomplishing a chain reaction (38).

Therefore, it seems reasonable to suppose that in the semiconductor catalyzed oxidation of DPE, the resulting radical cation of the olefin could react with the superoxide anion to give a peroxyethyl-1,4-diradical which subsequently reacts with the olefin and oxygen through the free radical chain processes.

Autoxidation of aromatic olefins initiated by a radical chain initiator like azobisisobutyronitrile (AIBN) or by photoirradiation under 1 atm oxygen usually gives carbonyl compounds as the main products along with epoxides, and the production of the epoxides tends to increase with decrease of oxygen pressure (33,39). Accordingly, the production of a considerable amount of the epoxide accompanied by benzophenone in the semiconductor photocatalyzed oxygenation of DPE under 1 atm oxygen shows that in the semiconductor catalysis the reaction might proceed through a mechanism slightly different from the autoxidation, possibly due to the participation of the DPE radical cation and superoxide anion in place of neutral radicals and oxygen in the chain initiation step.

Photocatalyzed Isomerization of Unsaturated Systems

The sensitizing action of semiconductor affects isomerization of olefins. On excitation of CdS or TiO_2 in the presence of trans- or cis-stilbene under argon, the isomerization occurred only from the cis- to trans-isomer; the trans-isomer scarcely underwent isomerization into the cis-isomer. Irradiation under oxygen resulted in the oxygenation to give benzaldehyde and trans-stilbene oxide, irrespective of the configuration of the starting olefin, along with the isomerization from the cis- to trans-isomer (40).

The isomerization taking place only from cis- to trans-stilbene is consistent with the isomerization through the olefin radical cation. Figure 3 depicts the energies of the radical cations of cis- and trans-stilbene. In the ground state, cis-stilbene is nearly 3 kcal/mol (ca. 0.13 eV) higher in energy than the trans-isomer (41). Since the oxidation potentials of trans- and cis-stilbene are 1.15 and 1.23 V vs Ag/$AgNO_3$ (0.1 M), respectively, the cation radical of cis-stilbene lies 0.21 eV higher in energy than that of trans-stilbene. The inefficient production of the cis-isomer from the trans-isomer rules out a possibility for the participation of the triplet state of stilbene (2.1 eV over the trans-isomer). This means that the pair of the trans radical cation with the electron in the conduction band of TiO_2 (ca. -0.6 V vs Ag/$AgNO_3$) or CdS (ca. -1.1 V) lies in the energies (ca. 1.8 and 2.3 eV, respectively over the trans-isomer) not much exceeding the triplet state.

Accordingly, it is reasonable to conclude that in the photocatalyzed isomerization of olefins, the cis radical cation twists to the more stable trans or trans-like twisted radical cation (42-46), which subsequently receives an electron from the excited semiconductor to give the trans-isomer. Similar isomerization using CdS was reported recently (47). On the other hand, the "trans" radical cation resulting from oxidation of the trans-isomer cannot isomerize to the less stable cis radical cation. The formation of trans-stilbene oxide irrespective of the configuration of the starting isomer can be attributed to the reaction of the "trans" radical cation with oxygen giving a peroxyl radical cation which subsequently collapses to the trans-epoxide.

The behavior of stilbene radical cations in the semiconductor catalysis is in keeping with the result of photoisomerization of other olefins like β-methylstyrene sensitized by electron acceptors like chloranil in polar solvents (48). The semiconductor photocatalyzed isomerization of strained cyclobutanes to strained dienes (isomerization of quadricyclene to norbornadiene and similar reactions of complex cage compounds (49)) is related to the olefin isomerization discussed above.

Photocatalytic Oxygenation of Hydrocarbons

Irradiation of powdered titanium dioxide suspended in solutions containing aromatic compounds and water under oxygen has recently been shown to induce hydroxylation of aromatic nuclei giving phenolic compounds and oxidation of side chains of the aromatic compounds (50-55). These reactions have been assumed to proceed through hydroxyl and other radical intermediates, but the mechanism for their generation, whether reactive free radicals result from oxidation of water, from reduction of oxygen, or from oxidation of the substrates on the surfaces of the excited titanium dioxide, has not been clear.

An attempt was made to reveal the mechanism for the formation of free radicals upon irradiation of titanium dioxide in the presence of benzene and toluene. Careful examination of the effects of oxygen and water showed that the presence of oxygen is essential for the reaction, and that under oxygen the oxidation of water contributes to the aromatic hydroxylation and the oxidation of toluene as a substrate leads to oxidation of its side chain (56).

Powdered titanium dioxide (30 mg) was suspended in an aceto-

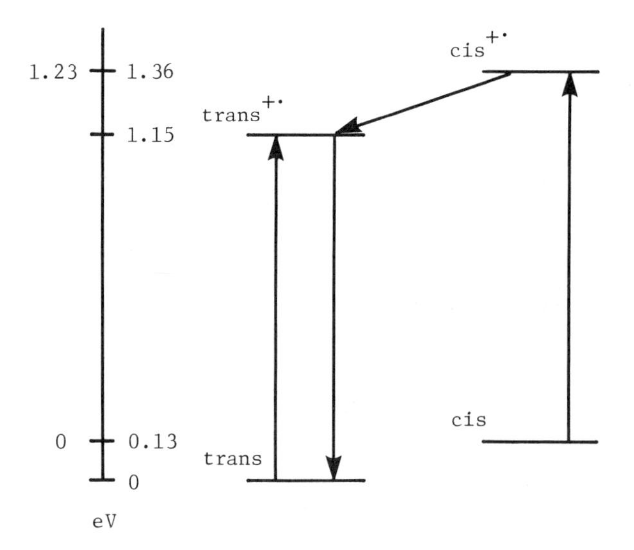

Figure 3. Energies of stilbene radical cations.

nitrile solution (2 ml) containing benzene or toluene (16 vol%) and varying amounts of water (0-8 vol%), and irradiated with light longer than 350 nm for 15 h at 20°C with and without oxygen.

Figure 4 depicts typical results of the oxidation of benzene under oxygen and without oxygen with varying concentrations of water. As Fig. 4 clearly shows, only a trace amount of phenol was produced, even in the presence of water without oxygen, while under oxygen the yield of phenol is still low in the absence of water, but increases with increasing concentration of water. For the reaction of toluene under careful removal of oxygen, only trace amounts of oxidation products were produced even in the presence of water, as was found for benzene.

It is noticeable that in the oxidation of toluene under oxygen (Fig. 5) o-, m-, and p-cresols resulting from nuclear hydroxylation of the substrate are produced in very low yields in the absence of water, but tend to increase with increasing concentration of water; on the contrary, benzaldehyde and benzyl alcohol arising from oxidation of the side chain are produced in much higher yields than cresols in the absence of water, but their yields do not vary with water concentration.

The above results with benzene and toluene show that the presence of oxygen is necessary for the formation of oxidation products, but the presence of water alone is not enough to induce the oxidation of the substrates. These facts indicate that water can be oxidized to hydroxyl radicals by the photogenerated positive holes with concurrent removal of electrons in the conduction band by oxygen, presumably adsorbed on the semiconductor surfaces. The resulting hydroxyl radicals will hydroxylate the aromatic nuclei.

$$H_2O \xrightarrow[-H^+]{p^+} HO\cdot \xrightarrow[X = H \text{ or } CH_3]{C_6H_5X} (\text{X-}C_6H_5(\cdot)(OH)(H)) \longrightarrow \text{X-}C_6H_4\text{-OH}$$

$$C_6H_5CH_3 \xrightarrow{p^+} (C_6H_5CH_3)^{+\cdot} \xrightarrow{-H^+} C_6H_5CH_2\cdot$$

$$\xrightarrow{O_2} \longrightarrow C_6H_5CH_2OH,\ C_6H_5CHO$$

$$O_2 \xrightarrow{e^-} O_2^{\overline{\cdot}} \xrightarrow{H^+} HO_2\cdot$$

Under oxygen in the absence of water, toluene will transfer an electron to the positive hole, concurrently with electron transfer from the conduction band to oxygen, to give a toluene radical cation. On the other hand, in the presence of water, both toluene and water will transfer an electron to the positive holes. The resulting toluene radical cation may subsequently lose a proton affording a benzyl radical, which will be oxidized with oxygen or the superoxide anion to benzyl alcohol and benzaldehyde, as proposed for the reactions of Fenton's reagent with toluene (57).

Therefore, it is reasonable to conclude that upon irradiation of titanium dioxide under oxygen, the electron transfer from water to the positive hole mainly results in hydroxylation of the aromatic

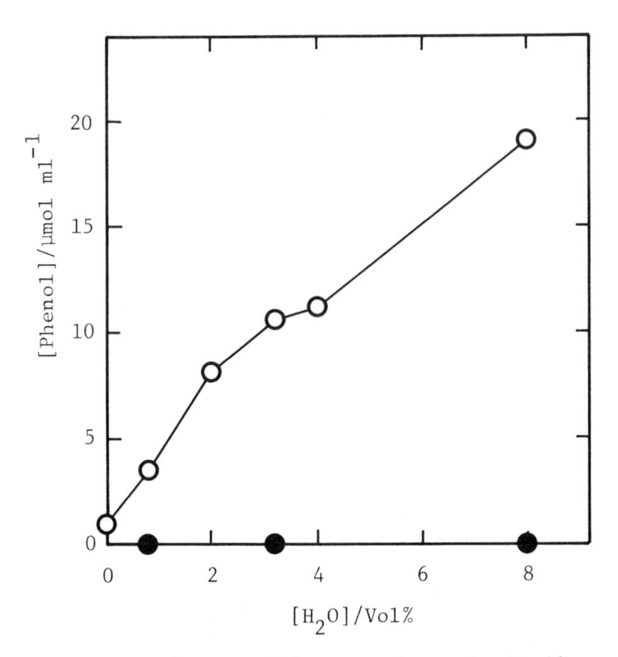

Figure 4. Hydroxylation of benzene in acetonitrile under degassing (●) and under oxygen (○).

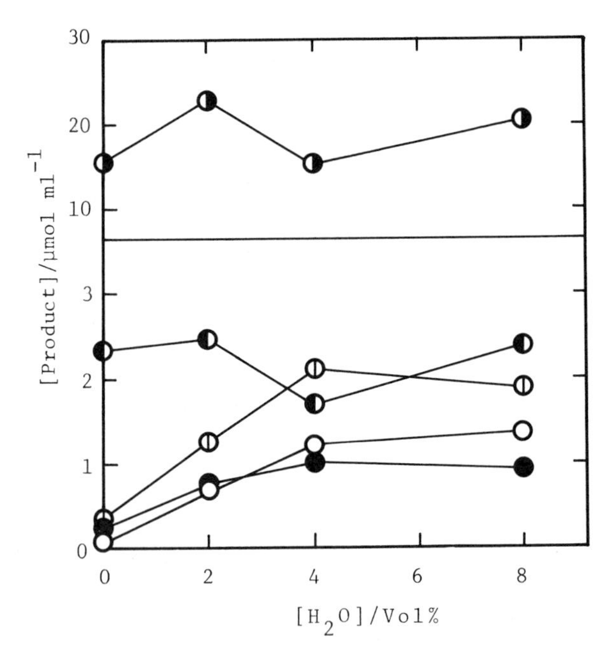

Figure 5. Oxidation of toluene in acetonitrile under oxygen; products: benzaldehyde (◑), benzyl alcohol (◐), o-cresol (○), m-cresol (○), p-cresol (⦶).

nuclei and the electron transfer from toluene mainly leads to oxidation of its side chain, and that oxygen contributes to these reactions by concurrently accepting an electron from the conduction band of semiconductor.

Titanium dioxide photocatalyzed oxidation of neat tetralin was previously reported to give its hydroperoxide (14). Reinvestigation showed that tetralol and tetralone are also formed in acetonitrile possibly through electron transfer from tetralin to the positive holes (27).

Concluding Remarks

Semiconductors act as photocatalysts on various organic compounds. The essential features of the mechanism are becoming clear. However, how the initiation steps proceed, particularly how the species on the semiconductor surface react with water and oxygen, is not yet well understood. Moreover, another area to be investigated is the mechanisms through which the resulting radical cations of the substrates undergo transformation to products.

Acknowledgments

The authors express their thanks to Ministry of Education, Science, and Culture in Japan for the special project research on energy under grant in aid of scientific research and to Japan-US Cooperative Research on Photoconversion and Photosynthesis for enabling discussions with the scientists concerned.

Literature Cited

1. Freund, T; Gommes, W. P. Catal. Rev. 1970, 3, 1.
2. Egerton, G. S. J. Text. Inst. 1948, 39, 293, 305.
3. Treiber, E. Kolloid Z. 1953, 130, 39.
4. Jacobsen, E. Ind. Eng. Chem. 1949, 41, 523.
5. Markham, M. C.; Laidler, K. J. J. Phys. Chem. 1953, 57, 363.
6. Rubin, T. R.; Calvert, J. G.; Rankin, G. T.; MacNevin, W. M. J. Am. Chem. Soc. 1953, 75, 2850.
7. Calvert, J. G.; Theurer, K.; Rankin, G. T.; MacNevin, W. M. J. Am. Chem. Soc. 1954, 76, 2575.
8. Markham, M. C.; Hannan, M. C.; Paternostro, R. M.; Rose, C. B. J. Am. Chem. Soc. 1958, 80, 5394.
9. Kuriacose, J. C.; Markham, M. C. J. Catalysis 1962, 1, 498.
10. Markham, M. C.; Upreti, M. C. J. Catalysis 1965, 4, 229.
11. Markham, M. C. J. Chem. Ed. 1955, 32, 540.
12. Mashio, F.; Kato, S. Japanese Patent 252 007, 1959.
13. Mashio, F.; Kato, S. U.S. Patent 2 910 415, 1959.
14. Kato, S.; Mashio, F. Kogyo Kagaku Zasshi 1964, 67, 1136.
15. Allen, N. S.; Kellar, J. F.; Phillips, G. O.; Wood, D. G. M. J. Polym. Sci. 1974, B 12, 241.
16. Allen, N. S.; Kellar, J. F.; Phillips, G. O.; Chapman, C. B. J. Polym. Sci. 1974, B 12, 723.
17. For chalking, for example, Pappas, S. P.; Fischer, R. M. J. Paint Tech. 1974, 46, 65.
18. Fujishima, A.; Honda, K.; Kikuchi, S. Kogyo Kagaku Zasshi 1969, 72, 108.

19. Fujishima, A.; Honda, K. Nature 1972, 238, 37.
20. For example, Watanabe, T.; Takizawa, T.; Honda, K. Shokubai 1978, 20, 370.
21. Kraeutler, B.; Bard, A. J. J. Am. Chem. Soc. 1977, 99, 7729.
22. Kraeutler, B.; Bard, A. J. J. Am. Chem. Soc. 1978, 100, 2239, 5985.
23. Miyake, M.; Yonezawa, H.; Tamura, H. Chem. Lett. 1976, 635.
24. Sakata, T.; Kawai, T. Yukigosei Kagaku Kyokaishi 1981, 39, 589, and references cited therein.
25. Kawai, T.; Sakata, T. Nature 1980, 288, 474.
26. Kanno, T.; Oguchi, T.; Sakuragi, H.; Tokumaru, K. Tetrahedron Lett. 1980, 21, 467.
27. Oguchi, T. Thesis, University of Tsukuba, 1982.
28. Fox, M. A. Acc. Chem. Res. 1983, 16, 314.
29. Fox, M. A.; Chen, C. C. J. Am. Chem. Soc. 1981, 103, 6757.
30. Fox, M. A.; Lindig, B. A.; Chen, C. C. J. Am. Chem. Soc. 1982, 104, 5828.
31. Fox, M. A.; Chen, C. C. Tetrahedron Lett. 1983, 24, 547.
32. Misawa, H.; Kanno, T.; Sakuragi, H.; Tokumaru, K. Denki Kagaku 1983, 51, 81.
33. Mayo, F. R. Acc. Chem. Res. 1968, 1, 193.
34. Eriksen, J.; Foote, C. S. J. Am. Chem. Soc. 1980, 102, 6083.
35. Araki, Y.; Dobrowolski, D. C.; Goyne, T. E.; Hanson, D. C.; Jiang, Z. R.; Lee, K. J.; Foote, C. S. J. Am. Chem. Soc. 1984, 106, 4570, and references cited therein.
36. Schaap, A. P.; Zaklika, K. A.; Kaskar, B.; Fung, L. W.-M. J. Am. Chem. Soc. 1980, 102, 389.
37. Mettes, S. L.; Farid, S. In "Organic Photochemistry"; Padwa, A., Ed.; Marcel Dekker: New York, 1983; Vol. 6, pp. 233-326.
38. Nelsen, S. F.; Akaba, R. J. Am. Chem. Soc. 1981, 103, 2096.
39. Kanno, T.; Hisaoka, M.; Sakuragi, H.; Tokumaru, K. Bull. Chem. Soc. Jpn. 1981, 54, 2330.
40. Oguchi, T.; Kuriyama, Y.; Sakuragi, H.; Tokumaru, K., to be published.
41. Taylor, T. W. J.; Murray, A. R. J. Chem. Soc. 1938, 2078.
42. Haselbach, E.; Bally, T. Pure Appl. Chem. 1984, 56, 1203.
43. Shida, T.; Haselbach, E.; Bally, T. Acc. Chem. Res. 1984, 17, 180.
44. Suzuki, H.; Koyano, K.; Shida, T.; Kira, A. Bull. Chem. Soc. Jpn. 1982, 55, 3690.
45. Suzuki, H.; Ogawa, K.; Shida, T.; Kira, A. Bull. Chem. Soc. Jpn. 1983, 56, 66.
46. Shida, T.; Hamill, W. H. J. Chem. Phys. 1966, 44, 4372.
47. Al-Ekabi, H.; de Mayo, P. J. Chem. Soc., Chem. Commun. 1984, 1231.
48. Roth, H.; Schilling, M. L. M. J. Am. Chem. Soc. 1980, 102, 4303.
49. Okada, K.; Hisamitsu, K.; Mukai, T. J. Chem. Soc., Chem. Commun. 1980, 941.
50. Izumi, I.; Fan, F.-R. F.; Bard, A. J. J. Phys. Chem. 1981, 85, 218.
51. Fujihira, M.; Satoh, Y.; Osa, T. Nature 1981, 293, 206.
52. Fujihira, M.; Satoh, Y.; Osa, T. Chem. Lett. 1981, 1053.
53. Fujihira, M.; Satoh, Y.; Osa, T. J. Electroanal. Chem. 1981, 126, 277.

54. Fujihira, M.; Satoh, Y.; Osa, T. Bull. Chem. Soc. Jpn. 1982, 55, 666.
55. Teratani, S.; Okuse, F.; Ikuo, A.; Choi, S.; Takagi, Y.; Tanaka, K. 45th Annual Meeting Chem. Soc. Jpn., Tokyo, April 1982, Abstracts I, p. 106.
56. Shimamura, Y.; Misawa, H.; Oguchi, T.; Kanno, T.; Sakuragi, H.; Tokumaru, K. Chem. Lett. 1983, 1691.
57. Walling, C. Acc. Chem. Res. 1975, 8, 125.

RECEIVED February 21, 1985

4

Single Potential Step Electrogenerated Chemiluminescence

A Nonradiative Method for the Production of Excited States

WILLIAM G. BECKER

Department of Chemistry, Portland State University, Portland, OR 97207

Two methods of electrogenerating excited states by single potential steps of a working electrode are discussed. Both anodic and cathodic mechanisms are initiated by a heterogeneous electron transfer. Electrogenerated chemiluminescence (ECL) is observed during the reaction of aromatic radical ions ($R^{-}\cdot$) with the dissociative ion peroxydisulfate ($S_2O_8^{2-}$). The ECL spectra that arise from these reactions agree with the fluorescence spectra of the corresponding aromatic compounds. The cyclic voltammetric reduction waves in the presence of $S_2O_8^{2-}$ were generally of the catalytic type, with R regenerated by the following chemical reaction with peroxydisulfate. Formation of R* and ECL is primarily caused by the $R^{-}\cdot/R^{+}\cdot$ reaction, with $R^{+}\cdot$ generated via oxidation of R by $SO_4^{-}\cdot$ (a product of the reduction of $S_2O_8^{2-}$). The relative ECL efficiencies qualitatively depend upon the stability of the aromatic radical cation. A tertiary reactant system further illustrates the importance of radical cations to the reductive single potential step ECL mechanism.

Since its discovery in 1964 electrogenerated chemiluminescence (ECL) has been an active area of chemical research (1). The light producing mechanism common to all ECL systems consists of an oxidation and reduction, followed by charge annihilation to produce the excited state of an emitting compound.

$$D \rightarrow D^{+\cdot} + e^{-} \quad \text{(anode oxidation)} \tag{1}$$

$$A + e^{-} \rightarrow A^{-\cdot} \quad \text{(cathode reduction)} \tag{2}$$

$$A^{-\cdot} + D^{+\cdot} \rightarrow A^{*} + D \text{ or } A + D^{*} \quad \text{(charge annihilation)} \tag{3}$$

The reagents A and D can be, but are necessarily, the same compound. The excited states produced by this methanism can have either singlet

0097-6156/85/0278-0057$06.00/0

or triplet spin multiplicity depending on the energetics of the charge annihilation step. These excited states will undergo the same decay reactions as photoexcited states. Most of the early ECL reactions were studied in aprotic solvents and involved sequential cathodic and anodic potential steps to produce radical anions and cations as intermediates.

Recently, the dissociative oxalate ion has been utilized as a powerful reducing reagent in single potential step ECL mechanisms. These reactions have been shown to occur in both aqueous and non-aqueous media (2,3). In this scheme the oxidized form of D reacts with an oxalate ion to produce the unstable radical anion $C_2O_4^{-\cdot}$. Decomposition of this intermediate yields carbon dioxide and the strong reducing reagent $CO_2^{-\cdot}$. The $CO_2^{-\cdot}$ can transfer an electron to A thus generating the reduced form of A in the vicinity of the anode where oxidation of D is taking place. The charge annihilation reaction then yields an excited state which can emit its characteristic fluorescence. Alternately, electron transfer from $CO_2^{-\cdot}$ to $D^{+\cdot}$ can be sufficiently energetic to generate D* directly. The term "oxidative reduction" has been used to describe the mechanisms by which these electron transfers take place.

$$D \rightarrow D^{+\cdot} + e^- \quad (4)$$

$$D^{+\cdot} + C_2O_4^{2-} \rightarrow D + C_2O_4^{-\cdot} \quad (5)$$

$$C_2O_4^{-\cdot} \rightarrow CO_2 + CO_2^{-\cdot} \quad (6)$$

$$CO_2^{-\cdot} + A \rightarrow CO_2 + A^{-\cdot} \quad (7)$$

$$A^{-\cdot} + D^{+\cdot} \rightarrow A^* + D \text{ or } A + D^* \quad (8)$$

$$CO_2^{-\cdot} + D^{+\cdot} \rightarrow CO_2 + D^* \quad (9)$$

The most efficient ECL from this mechanism occurred when A and D are the same compound. In non-aqueous media polyaromatic hydrocarbons such as rubrene and 9,10-diphenylanthracene have yielded ECL by this mechanism (3). In aqueous media, $Ru(bpy)_3^{2+}$ (bpy = 2,2' bipyridine) has been used (2). Analytical application of anodic single potential step ECL has been suggested for determination of oxalate concentration in urine samples (4).

This report concerns the analogous "reductive oxidation" ECL mechanism which has now also been examined. The principal reagent in this scheme is the dissociative anion peroxydisulfate ($S_2O_8^{2-}$). This anion is a two electron oxidant whose reduction potentials (vs. SCE) in aqueous solution have been estimated to be (5):

$$S_2O_8^{2-} + e^- \rightarrow SO_4^{2-} + SO_4^{-\cdot} \qquad E^\circ < 0.35V \quad (10)$$

$$SO_4^{-\cdot} + e^- \rightarrow SO_4^{2-} \qquad E^\circ > 3.15V \quad (11)$$

The intermediate $SO_4^{-\cdot}$, formed during the reduction of $S_2O_8^{2-}$, is therefore an exceedingly strong oxidizing agent. Initial reductive oxidation ECL studies employed polypyridine-type complexes of Cr(III), Ru(II) and Os(II) as light emitting reagents (6,7). In this scheme the overall reaction process is the two electron reduction of $S_2O_8^{2-}$ mediated by a transition metal complex. The by-product of this process is the characteristic light emission from the excited state transition metal complex.

$$S_2O_8^{2-} + 2e^- \xrightarrow{M} 2SO_4^{2-} + h\nu \tag{12}$$

ECL is obtained from these systems only when the potential of the working electrode is sufficiently negative that reduction of the transition metal complex occurs. In all cases peroxydisulfate will also be reduced at this potential. However, the $S_2O_8^{2-}$ which reacts at the electrode does not participate in the ECL mechanism. Rather, it is the reaction of the reduced form of M with $S_2O_8^{2-}$ that initiates the luminescence process. The first two steps of this reductive oxidation ECL mechanism have therefore been proposed to be (6):

$$M + e^- \rightarrow M^- \tag{13}$$

$$M^- + S_2O_8^{2-} \rightarrow SO_4^{2-} + SO_4^{-\cdot} \tag{14}$$

The strongly oxidizing $SO_4^{-\cdot}$ species can now generate the metal complex excited state via two different paths. Both paths are energetically feasible. In the first case, reaction of $SO_4^{-\cdot}$ with M will give M^+. The highly energetic electron transfer from M^- to M^+ can then produce M*.

$$SO_4^{-\cdot} + M \rightarrow SO_4^{2-} + M^+ \tag{15}$$

$$M^- + M^+ \rightarrow M + M^* \tag{16}$$

Alternately, reaction of $SO_4^{-\cdot}$ with M^- produced at the electrode can yield M* directly.

$$SO_4^{-\cdot} + M^- \rightarrow SO_4^{2-} + M^* \tag{17}$$

In both of these mechanisms the intensity of ECL is dependent on the concentrations of metal complex and peroxydisulfate. However, in the case of the $Ru(bpy)_3^{2+}$-$S_2O_8^{2-}$ system, the metal complex excited state is also quenched by peroxydisulfate. The quantum efficiency of this quenching has been reported to equal two (8). Upon further examination, steady-state and lifetime measurements of this quenching process revealed ground state ion-pairing between $Ru(bpy)_3^{2+}$ and $S_2O_8^{2-}$ in CH_3CN/H_2O solvents (9). The photoexcited state of this ion-pair possesses an unusually long lifetime which shows a direct dependence on solvent composition. As a result of this quenching process there exists an optimum peroxydisulfate concentration and solvent composition for which the reductive oxidation ECL intensity is at a maximum. A 1:1 (v/v) mixture of CH_3CN/H_2O containing 1 mM of $Ru(bpy)_3^{2+}$ has a maximum ECL intensity at 18 mM $S_2O_8^{2-}$. The estimated maximum efficiency of ECL by reductive oxidation can be obtained by examination of the overall reaction (equation 12). The maximum coulombic efficiency of this reaction would be two cathodic electrons consumed per excited state generated. The optimized reductive oxidation ECL from $Ru(bpy)_3^{2+}$ has a coulombic yield near unity (7).

Further examination of "reductive oxidation" ECL using polyaromatic compounds in non-aqueous media has revealed three significant features of the luminescence mechanism (10). First, the cyclic voltammograms for the reduction of the polyaromatic compounds in the presence of $S_2O_8^{2-}$ were of a highly catalytic type. Second, the efficiency of ECL was qualitatively dependent on the stability of the aromatic radical cation rather than of the aromatic radical anion. Third, the importance of the aromatic radical cation ion in the mechanism for the formation of excited states was illustrated using a tertiary reactant system. The results of these studies are summarized below.

The cyclic voltammogram of rubrene (1 mM in 2:1 v/v CH_3CN/benzene, 0.1 M $TBABF_4$) displayed reversible one electron waves for both oxidation and reduction at a platinum electrode (Figure 1). In contrast, the electrochemical reduction of the tetra-n-butylammonium peroxydisulfate (20 mM) consists of a broad, drawn-out wave beginning at -0.3 V (vs. SCE). The contribution of the current measured from this process to the total current was negligible at the rubrene concentrations employed. Addition of rubrene (1 mM) to a solution containing 20 mM $S_2O_8^{2-}$ and 0.1 M $TBABF_4$ resulted in drastically different electrochemical behavior. The reduction wave of rubrene became irreversible and had a peak current that was 18 times larger than the peak current in the absence of $S_2O_8^{2-}$ (Figure 2b). The rubrene reduction peak potential showed a scan rate dependence, shifting to more negative values with increasing scan rate. Upon scan reversal, an anodic wave was not observed at scan rates up to 10 V s^{-1}. Furthermore, cyclic voltammograms which were run at scan rates of 2 mV s^{-1} did not show the "S" shaped curve that would be expected from a simple catalytic system. Changing the relative concentration of $S_2O_8^{2-}$ to achieve a 100:1 excess with respect to rubrene had no effect on the shape of the catalytic wave and therefore ruled out any explanation based on depletion of $S_2O_8^{2-}$ near the electrode. This behavior in the presence of $S_2O_8^{2-}$ is generally characteristic of a catalytic process in which there is a fast following irreversible reaction.

$$RUB + e^- \rightarrow RUB^{-\cdot} \quad (18)$$

$$RUB^{-\cdot} + S_2O_8^{2-} \rightarrow RUB + S_2O_8^{3-} \quad (19)$$

The presence of a scan rate dependence under such highly catalytic conditions, however, is contrary to a mechanism in which the following reaction is a single electron transfer (11). Instead, a mechanism in which the following reaction is a two electron oxidation by $S_2O_8^{2-}$ to generate $RUB^{+\cdot}$ and two SO_4^{2-} near the cathode would better explain the observed cyclic voltammetry. In this mechanism the catalytic current would be larger (two catalytic electrons for each initial RUB reduction) and the persistent scan rate dependence might be attributed to the rate determining following reaction (eq. 19).

$$RUB^{-\cdot} + S_2O_8^{2-} \rightarrow RUB^{+\cdot} + 2\ SO_4^{2-} \quad (20)$$

$$RUB^{+\cdot} + 2\ e^- \rightarrow RUB^{-\cdot} \quad (21)$$

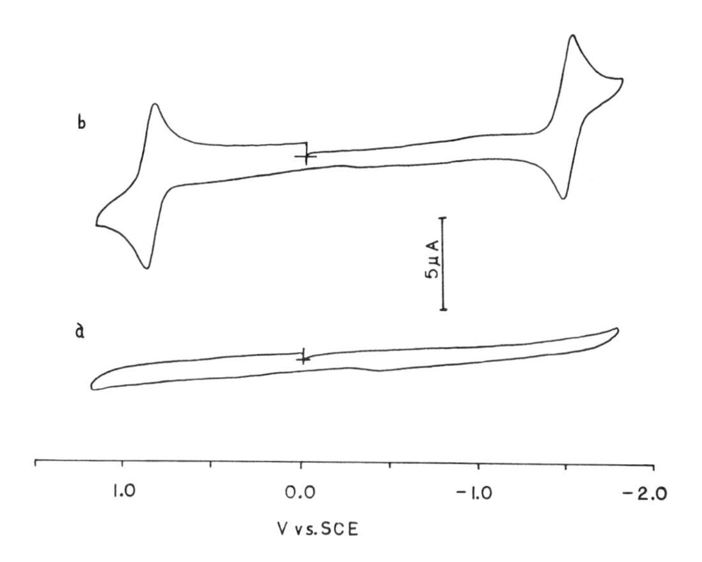

Figure 1. Cyclic voltammograms (0.1 V s^{-1}) at a Pt disk electrode in 2:1 CH_3CN-benzene (v/v) containing: (a) supporting electrolyte (0.1 M $TBABF_4$), (b) 1.0 mM RUB and 0.1 M $TBABF_4$.

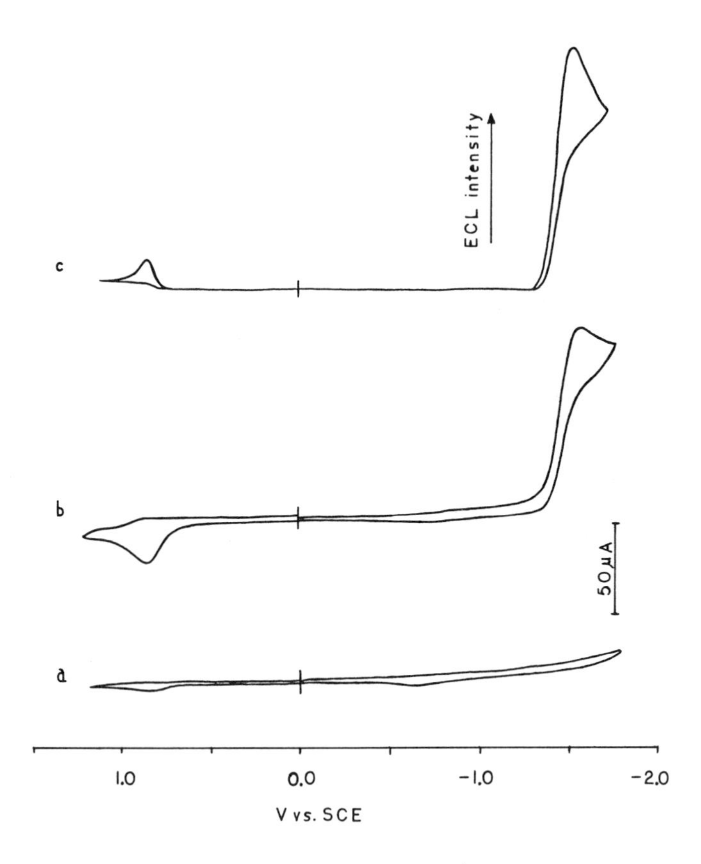

Figure 2. Cyclic voltammograms (0.1 V s^{-1}) at a Pt disk electrode in 2:1 CH_3CN-benzene (v/v) containing: (a) 0.1 M $TBABF_4$ and 20 mM $(TBA)_2S_2O_8$, (b) solution (a) with 1.0 mM RUB, (c) relative ECL intensity vs. potential for solution (b).

Repeated cycling through the RUB reduction wave resulted in a decrease in size of the catalytic current. This occurred even when the solution was stirred between cycles. This behavior implies that a blocking or filming of the electrode occurred during the reduction process. Repeated cycling over the oxidation wave removed the film and reactivated the electrode. The electrochemical reduction of 9,10-diphenylanthracene (DPA), 1,3,6,8-tetraphenylpyrene (TPP), anthracene (ANT), fluoranthene (FLU) and 2,5-diphenyl-1,3,4-oxadiazole (PPD) in the presence of $S_2O_8^{2-}$ all showed similar cathodic waves. As in the case of $Ru(bpy)_3^{2+}$, luminescence was not observed at potentials more positive than the fluorophore reduction wave where only the small amount of direct reduction of $S_2O_8^{2-}$ appears (Figure 2c) (12). The similarity between the intensity-potential profile and the reduction voltammogram indicates that the mechanism which leads to luminescence is intimately associated with the electroreduction of the fluorophore.

Single potential step ECL was observed upon electrochemical reduction of RUB, DPA, TPP, ANT, FLU and PPD in the presence of a 20:1 excess of $S_2O_8^{2-}$. The ECL spectra agreed with the photoexcited fluorescence spectra in all cases. The relative ECL intensities were determined by integrating the luminescence intensity from a platinum cathode for potential steps of 15 s duration. These results are summarized in Table 1. It is interesting to note that relative ECL intensities do not correlate with the size of the catalytic current or the exothermicity of electron transfer. Rather it appears that the stability of the organic radical cation is the important factor in determining the efficiency of ECL (13). RUB, the compound which gave the highest ECL intensity, was the only fluorophore to show a reversible oxidation in the presence of $(TBA)_2S_2O_8$. DPA and TPP had reversible oxidations in pure electrolyte solution, but when $(TBA)_2S_2O_8$ was added the oxidation voltammograms became irreversible. The reason for the instability of these radical cations in the presence of $(TBA)_2S_2O_8$ is not known at this time. Accordingly, the ECL intensities from these fluorophores were significantly diminished. ANT and FLU did not give reversible electrochemical oxidations in the solution employed. The ECL intensities from these fluorophores were also correspondingly small. Finally, in the case of PPD, oxidation occurred on the edge of the solvent limit (14). The single potential step ECL intensity for this fluorophore was extremely low.

The tertiary reactant system of PPD, thianthrene (TH), and $S_2O_8^{2-}$ in CH_3CN demonstrates the importance of the radical cation to the reductive oxidation ECL mechanism. This system is interesting in that both PPD and TH have accessible excited states. Also, PPD can only be reduced to a stable radical anion and TH can only be oxidized to a stable radical cation within the electrochemical limits of the solvent. The cyclic voltammogram of a CH_3CN solution containing PPD, TH and electrolyte is shown in Figure 3. The E_o's of these couples are separated by 3.42 V, making ECL "energy sufficient" only for TH ($E_{o,o}$ = 3.31 eV for TH singlets and $E_{o,o}$ = 4.00 eV for PPD singlets). Upon addition of $(TBA)_2S_2O_8$ (30 mM), the cyclic voltammogram of this system showed a catalytic wave for PPD reduction that was identical to that observed in the absence of TH (Figure 4). A cathodic potential step of this tertiary system resulted in ECL that

Table I. Experimental Results from $S_2O_8^{2-}$ Catalyzed Reductions.

	$E^0(R/R^{+\cdot})$[a] (V)	$E^0(R/R^{-\cdot})$[a] (V)	E_{oo}[b] (eV)	i_c/i_d[c]	I_{ecl}[d]
RUB	+0.89	-1.47	2.28	18.0	10.2
DPA	+1.28	-1.88	3.04	20.0	0.16
TPP	+1.19	-1.83	3.00	24.8	0.14
ANT	+1.2[e]	-1.96	3.21	16.0	0.03
FLU	+1.6[e]	-1.77	3.01	15.8	0.10
PPD[f]	-	-2.13	4.00	12.4	trace
TH[f]	+1.29	-	3.31	-	-

a E^0 = average of peak potential of the cathodic and anodic waves measured at a Pt disk versus SCE in 2:1 CH_3CN-benzene containing 0.1 M TBA(BF_4), v = 50 mV s^{-1}.

b Energy of the lowest singlet excited state measured from the overlap of the photoabsorption and photoemission spectra.

c Peak catalytic reduction current from solutions containing 20 mM $S_2O_8^{2-}$ and 1 mM fluorophore divided by peak diffusional current without $S_2O_8^{2-}$.

d Relative integrated intensity for a 15 s cathodic step to 100 mV negative of the peak potential.

e Onset of irreversible anodic wave.

f Values measured in CH_3CN.

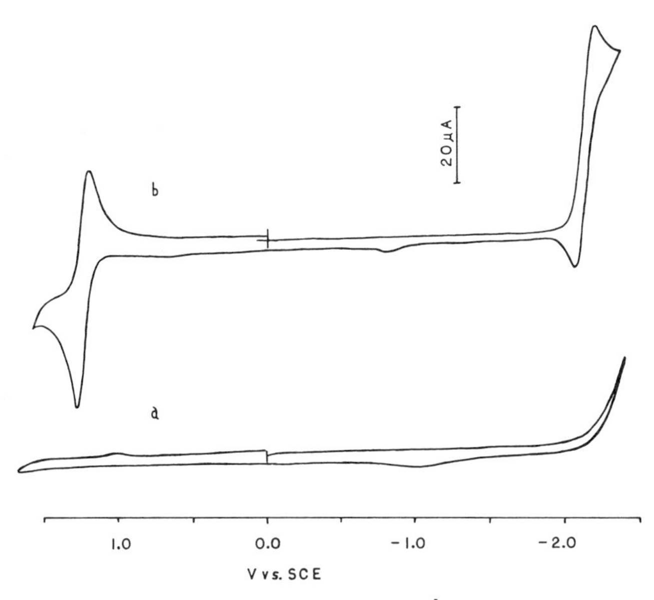

Figure 3. Cyclic voltammograms (0.1 V s^{-1}) at Pt disk electrode in CH_3CN containing: (a) supporting electrolyte (0.1 M $TBABF_4$), (b) 3.0 mM PPD, 3.0 mM TH and 0.1 M $TBABF_4$.

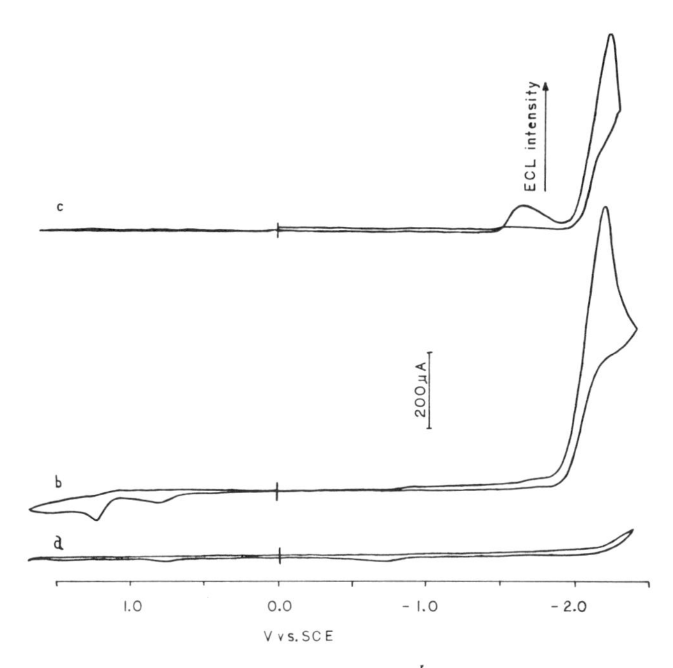

Figure 4. Cyclic voltammograms (0.1 V s^{-1}) at a Pt disk electrode in CH_3CN containing: (a) supporting electrolyte (0.1 M $TBABF_4$), (b) 3.0 mM PPD, 3.0 mM TH, 20 mM $(TBA)_2S_2O_8$ and 0.1 M $TBABF_4$, (c) relative ECL intensity vs. potential for solution (b).

had an intensity in excess of 35 times that measured in the absence of TH. Furthermore, the ECL spectrum was that of TH and not PPD fluorescence. The photoluminescence of this solution was dominated by PPD fluorescence (λ_{ex} = 300 nm), as would be expected from the larger fluorescence quantum yield of PPD (Φ = 0.89 for PPD and Φ = 0.036 for TH) (15). Therefore, the ECL mechanism for this system can only be explained by the formation of $TH^{+\cdot}$ and cannot be attributed to TH quenching of PPD luminescence. The only species in solution capable of oxidizing TH was $SO_4^{-\cdot}$. Thus the mechanism for this system would be:

$$PPD + e^- \rightarrow PPD^{-\cdot} \quad (22)$$

$$PPD^{-\cdot} + S_2O_8^{2-} \rightarrow PPD + SO_4^{2-} + SO_4^{-\cdot} \quad (23)$$

$$TH + SO_4^{-\cdot} \rightarrow TH^{+\cdot} + SO_4^{2-} \quad (24)$$

$$TH^{+\cdot} + PPD^{-\cdot} \rightarrow {}^1TH^* + PPD \quad (25)$$

Direct formation of excited states by reaction of $SO_4^{-\cdot}$ with radical anions generated at the electrode does not appear to be an important step in the reductive oxidation ECL of polyaromatic fluorophores. From the above considerations, the general mechanism which best describes reductive oxidation ECL in the presence of $S_2O_8^{2-}$ would be:

$$A + e^- \rightarrow A^{-\cdot} \quad (26)$$

$$A^{-\cdot} + S_2O_8^{2-} \rightarrow A + SO_4^{2-} + SO_4^{-\cdot} \quad (27)$$

$$D + SO_4^{-\cdot} \rightarrow D^{+\cdot} + SO_4^{2-} \quad (28)$$

$$A^{-\cdot} + D^{+\cdot} \rightarrow A^* + D \text{ or } A + D^* \quad (29)$$

The reagents A and D can be, but are not necessarily, the same compound.

Currently, research in this area involves the use of single potential step ECL in analytical applications. In particular, oxidative reduction ECL from the $Ru(bpy)_3^{2+}$-$S_2O_8^{2-}$ system appears to be sensitive and specific enough to use in competitive protein binding reactions (16). The ECL intensity shows a linear response for aqueous concentrations of $Ru(bpy)_3^{2+}$ in the region between 10^{-13} and 10^{-7} M. Studies involving single potential step ECL from polyaromatic compounds in multiphase systems are also in progress. Envisioned analytical applications of these schemes include the detection of trace amounts of carcinogenic fluorophores in water.

Acknowledgments

The majority of the findings discussed in this report arose from research performed at the University of Texas at Austin in the laboratories of Allen J. Bard. The author wishes to recognize the contribution of Henry S. White, Deniz Ege and Hyanjune S. Seung.

Literature Cited

1. Faulkner, L.R.; Bard, A.J. in "Electroanalytical Chemistry"; Bard, A.J., Ed.; Marcel Dekker: New York, 1977; Vol. 10, chpt. 1.
2. Rubinstein, I.; Bard, A.J. J. Am. Chem. Soc. 1981, 103, 512.
3. Chang, M-M.; Saji, T.; Bard, A.J. J. Am. Chem. Soc. 1977, 99, 5399.
4. Rubinstein, I.; Martin, C.; Bard, A.J. Anal. Chem. 1983, 55, 1580.
5. Memming, R. J. Electrochem. Soc. 1969, 116, 785.
6. Bolletta, F.; Balzani, M.V.; Serpone, N. Inorg. Chim. Acta 1982, 62, 207.
7. White, H.S.; Bard, A.J. J. Am. Chem. Soc. 1982, 104, 6891.
8. Boletta, F.; Juris, A.; Maestri, M.; Sandrini, D. Inorg. Chim. Acta 1980, 44, L175.
9. White, H.S.; Becker, W.G.; Bard, A.J. J. Phys. Chem. 1984, 88, 1840.
10. Becker, W.G.; Seung, H.S.; Bard, A.J. J. Electroanal. Chem. 1984, 167, 127.
11. (a) Nicholson, R.S.; Shain, I. Anal. Chem. 1964, 706; (b) Saveant, J.M.; Vianello, E. Electrochim. Acta 1965, 10, 905.
12. Berlman, I.B. in "Handbook of Fluorescence Spectra of Aromatic Molecules"; Academic: New York, 1971.
13. Mann, C.K.; Barnes, K.K. in "Electrochemical Reactions in Nonaqueous Systems"; Marcel Dekker: New York, 1970; chpt. 3.
14. Kesztheyli, C.P.; Tachikawa, H.; Bard, A.J. J. Am. Chem. Soc. 1972, 94, 1522.
15. The small amount of light observed at 0.8V occurred only during the initial anodic scan at filmed electrodes in RUB solutions. This luminescence was therefore attributed to the presence of some reduced material in the film generated during the cathodic scan.
16. Ege, D.; Becker, W.G.; Bard, A.J. submitted for publication.

RECEIVED January 10, 1985

5

Controlled Organic Redox Reactivity on Irradiated Semiconductor Surfaces

MARYE ANNE FOX, CHIA-CHUNG CHEN, KOON-HA PARK, and JANET N. YOUNATHAN

Department of Chemistry, University of Texas at Austin, Austin, TX 78712

Photoexcitation of n-type semiconductors renders the surface highly activated toward electron transfer reactions. Capture of the photogenerated oxidizing equivalent (hole) by an adsorbed oxidizable organic molecule initiates a redox sequence which ultimately produces unique oxidation products. Furthermore, specific one electron routes can be observed on such irradiated surfaces. The irradiated semiconductor employed as a single crystalline electrode, as an amorphous powder, or as an optically transparent colloid, thus acts as both a reaction template and as a directed electron acceptor. Recent examples from our laboratory will be presented to illustrate the control of oxidative cleavage reactions which can be achieved with these heterogeneous photocatalysts.

The recent recognition that the surfaces of n-type semiconductors become effective redox catalysts when irradiated with light has captured the attention of a wide range of scientists interested in solar energy conversion and has spawned innumerable studies by electrochemists, physical chemists, and solid state physicists in the last decade.[1] Most of these investigations have concentrated on detailed depictions of the semiconductor or the interface formed as the semiconductor is brought into contact with a metal, with another semiconductor, or with a liquid phase electrolyte.

Somewhat less attention has been directed toward characterizing chemical redox reactions which can be stimulated by the initial surface photoexcitation and most of these have been directed at a

0097-6156/85/0278-0069$06.00/0

single fundamentally important reaction, namely water splitting to form hydrogen (a combustible fuel) and oxygen. By comparison, until recently, the mechanistic investigation of organic reactions initiated in parallel fashion has been almost completely overlooked. For the last several years we have studied the semiconductor-mediated photochemically induced oxidation of a variety of organic compounds.[2] Even our preliminary results have convinced us that the excited semiconductor surface provides a unique environment for controlling the chemical fate of intermediates formed by photoinduced electron transfer. We present in this article a summary of several recent reactions which illustrate how a desired reaction pathway from an oxidized intermediate can be specified on the surface of the irradiated semiconductor.

Basic Principles of Charge Separation

A semiconductor is characterized by band structure, Figure 1. There exist two sets of closely spaced energy levels or bands, one (the valence band) which is electronically fully occupied by the constituent electrons associated with each atom of the material and one (the conduction band) which is electronically vacant. The energy difference between the top of the valence band and the bottom of the conduction band, called the band gap, defines the wavelength sensitivity of the material. When such a semiconductor, as an electrochemical half cell, is brought into contact with a liquid phase electrolyte containing a redox couple, electronic equilibration occurs, establishing a common occupancy (Fermi) level. Ordinarily, the Fermi level lies just below the conduction band in the bulk of a negatively-doped (n-type) semiconductor and the equilibration process results in a bending of the bands from the

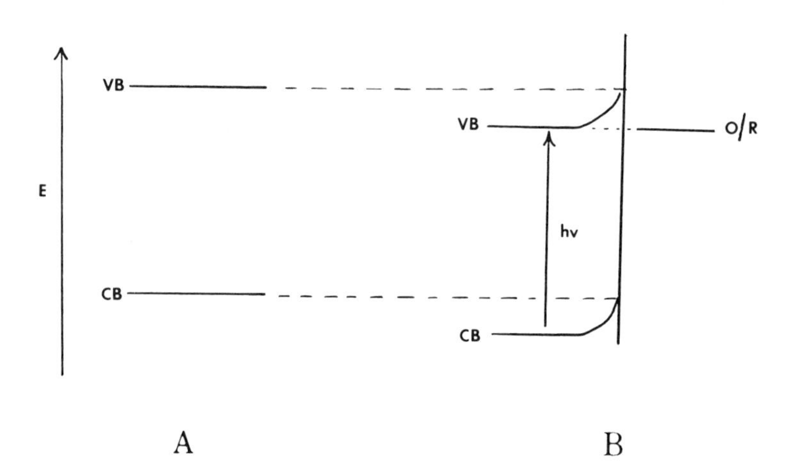

Figure 1. Band Structure in a n-type Semiconductor A. Solid State. B. In contact with a liquid phase redox couple (O/R). E_V=energy of the conduction band. Vertical line indicates solid-liquid interface. CB= conduction band; VB = valence band.

surface toward the bulk. In an n-type material, this band bending causes a sharp energy level rise as one moves along a single energy level from the bulk toward the surface and hence toward the interfacial region between the solid and liquid phases.

When the semiconductor is excited with a photon of energy greater than or equal to its band gap, an electron is promoted from the valence band to the conduction band. The vacancy, or hole, created by the excitation is a strong oxidant whose chemical behavior can be reasonably approximated by an oxidation couple poised at the valence band edge potential. Intrinsic band bending of the semiconductor causes electrons in the interfacial region to move toward the bulk to fill the hole, thus causing a migration of the oxidizing hole to the surface of the semiconductor. At the interface, an oxidizable substrate can then fill the photogenerated hole and become a surface adsorbed oxidized species.

The electronic excitation also promotes an electron to the conduction band, where it can function effectively as a reductant at a potenial governed by the position of the conduction band edge. In parallel fashion, reduced products can accumulate at a metal counterelectrode which has collected the photogenerated electrons from the conduction band.

A similar situation also is encountered in a miniaturized photoelectrochemical cell, i.e., on a metallized semiconductor powder, Figure 2. Here, the individual particle can be thought of as two electrochemical half cells which have eventually collapsed onto each other as the conductive wire connecting them became shorter and shorter. The oxidizing and reducing sites are thus found in close spatial proximity and the potential for subsequent chemical reaction between the initial oxidation and reduction products is excellent. In fact, so long as the respective rates of the oxidation and reduction half reactions differ appreciably, it may be unnecessary to metallize the semiconductor powder in order to

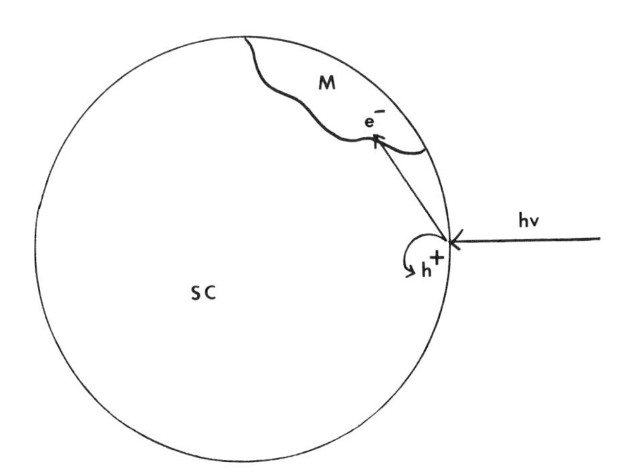

Figure 2. Electron-hole separation on a metallized (M) semiconductor (SC) powder.

initiate chemical reactivity. The powder suspension thus becomes a "reagent" which can be handled conveniently in laboratories not equipped for detailed electrochemical investigations.

Selectivity in Photoelectrochemical Reactions

By virtue of the relatively high positive potentials associated with the valence band edges of commonly encountered semiconductors, substantial oxidizing power is created upon photoexcitation of common semiconductor suspensions.[3] A wide variety of organic functional groups may potentially function as redox participants on irradiated semiconductor powders, and studies have begun to appear in the last few years which indicate that indeed a wide range of hydrocarbons and heteroatom-containing substrates can be oxidized in unusual ways on these photoactivated supports.

Although it is clear that photoinduced redox exchange can occur efficiently at the surface of an irradiated semiconductor powder, this redox chemistry will not find extensive use unless it provides access to new chemical transformations which are inaccessible with conventional reagents or to an improved selectivity in multifunctional molecules or in mixtures of reagents.

The heterogeneous surface of an irradiated semiconductor powder provides several potential ways by which selectivity can be influenced. First, since electron exchange occurs only with those molecules found at the semiconductor-electrolyte interface, directed oxidation (or reduction) will be governed by adsorption characteristics, with preferential reactivity occurring for those substrates or that functional group which is preferentially adsorbed. Second, the semiconductor surface provides a reaction milieu in which the oxidized and reduced products of the photoinduced electron exchange are generated as adsorbed species with minimal physical separation. Their bimolecular chemical reaction may therefore be expected to occur with greater facility than in homogeneous solution, where the directive template effect of the surface is absent. Finally, since the semiconductor interface becomes an effective electron donor or acceptor only when photoexcited, it may prove possible to constrain electron exchange at low light flux to single electron processes. That is, after donating or accepting a single electron, the individual semiconductor particle (excited by a single photon) loses completely its reducing or oxidizing power, rendering it unable to complete multielectron exchanges unless the redox product is sufficiently chemically and physically persistent on the surface to await a subsequent photoexcitation.

Adsorption Control

It has long been recognized by electrochemists that controlling the chemical composition of the double layer formed at the surface of a poised electrode should allow specific electrochemical reactions of molecules contained therein. On semiconductor powders suspended in nonaqueous solutions of a substrate of interest, a substantial adsorption/solubilization differential will occur as the dielectric of the liquid phase is altered. For example, when titanium dioxide powder is suspended in acetonitrile containing diphenylethylene, and the resulting suspension is irradiated with long wavelength ultraviolet light, an impressively clean oxidative cleavage is observed, eqn (1).[4]

$$Ph_2C{=}CH_2 \xrightarrow[CH_3CN]{TiO_2,\ O_2} Ph_2C{=}O \quad (100\%) \qquad (1)$$

If, however, the same reaction is attempted in methanolic solution, no olefin oxidative cleavage is detected, and solvent oxidation dominates the observed chemical process. Presumably, the polar solvent preferentially binds to the oxide surface, effectively negating the adsorption of the less polar hydrocarbon. The observed reactivity is then restricted to molecules held at the surface, i.e., to solvent oxidation.

In similar fashion, it has recently been reported that a chemically significant difference in oxidizability in primary and secondary alcohols is observed on irradiated TiO_2.[5] For example, eqn (2), excellent yields of aldehyde are obtained from primary alcohol:

$$n\text{-}C_7H_{15}CH_2OH \xrightarrow[N_2,\ C_6H_6]{TiO_2^*/Pt} n\text{-}C_7H_{15}CHO \qquad (2)$$

Since gas phase ionization potentials indicate that secondary alcohols should be more easily oxidized than primary alcohols, the relative reactivity cannot be governed by thermodynamic factors associated with primary electron exchange. A logical explanation for the ordering of the reactivity may relate to the preferential adsorption of the primary alcohol on the metal oxide surface.

The Semiconductor Surface as a Reaction Template

In principle, the arrangement of reactive intermediates generated by electron - hole pair capture by two redox couples on the semiconductor surface may allow for divergent reaction paths when the same reactive intermediates are generated on the irradiated surface and in an isotropic environment. If a particular reactive intermediate is quite stable, the overall chemistry observed may be identical on the surface of this heterogeneous photocatalyst and in homogeneous solution. For example, the oxidative cleavage of 1-methoxynaphthalene, eqn 3,

OCH_3 ; TiO_2^*, O_2 / CH_3CN ; NC–⟨◯⟩–CN / O_2, CH_3CN ; CO_2CH_3 ; CO_2H (3)

gives identical products when the radical cation formed by single electron oxidation is generated on irradiated TiO_2 or by electron transfer to a dissolved electron acceptor (p-dicyanobenzene).[6] The similarity in reactivity may in part be attributed to the relative stability of the radical ion, which persists either in solution or on the surface until it is captured by oxygen or superoxide, initiating a sequence which ultimately leads to carbonyl compounds.

In contrast, the analogous 1-methylnaphthalene gives a completely different product if the reaction is conducted in homogeneous solution compared with that found on the irradiated solid photocatalyst. Exclusive side chain oxidation is observed, eqn 4,

CH_3 ; TiO_2^*, O_2 / CH_3CN ; $COCH_3$; CO_2H ; NC–⟨◯⟩–CN / O_2, CH_3CN ; CHO (4)

in solution, whereas a ring cleavage parallel to that observed with the methoxylated analogue is found on irradiated TiO_2.[6] This result is understandable if the solvent acts as a weak base enveloping the highly acidic radical cation formed by single electron exchange in solution. The alpha-methylnaphthyl radical thus generated by deprotonation can be efficiently scavenged by oxygen, initiating a radical chain process in which the alpha-hydrogen is specifically involved. On the semiconductor, however, the faster reaction with adsorbed oxygen or with photogenerated superoxide apparently dominates the observed chemistry, leading to oxidative cleavage.

A particularly graphic example of the divergent chemical reactivity of common radical cationic intermediates generated under different experimental conditions is found in the chemistry of diphenylethylene. With this system, eqn 5,

TiO_2^* / O_2, CH_3CN ; + ; O (5)

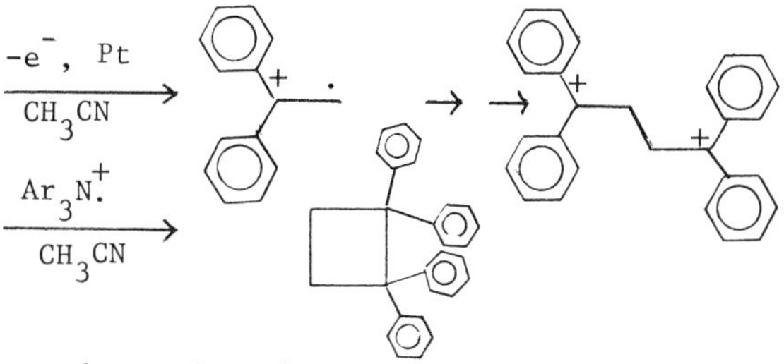

$$\xrightarrow[CH_3CN]{-e^-,\ Pt} \qquad \xrightarrow[CH_3CN]{Ar_3N^{+\cdot}} \qquad (5)$$

oxidative cleavage is observed on irradiated TiO_2 in the presence of oxygen, oxidative dimerization is attained at a poised metal electrode, and cyclodimerization is found with a homogeneously dispersed single electron oxidant, the triarylaminium ion. All three pathways are thought to involve the same intermediate, i.e., the radical cation formed by one-electron oxidation of diphenylethylene. It is clear that the control afforded by conducting redox reactions on irradiated semiconductor surfaces can provide a new dimension in mechanistic chemistry.

Surface Control of the Number of Electrons in a Redox Reaction

Semiconductor powders whose valence and conduction band edges straddle the oxidation and/or reduction potentials of a redox couple of interest can neither accept or donate electrons to the substrate in the ground state. Upon photoexcitation, however, these surfaces become highly effective sites for electron exchange. After the primary electron transfer, however, they resume their original poor catalytic redox reactivity. Since a wide range of organic substrates are known to react via multielectron exchange in conventional preparative electrochemical cells, it might be possible to induce novel reaction pathways from the initially formed one-electron redox product.

A case in point involves the electrochemical oxidation of vicinal diacids. A standard synthetic method for the preparation of carbon-carbon double bonds occurs by the bis-decarboxylation of such diacids. Even relatively strained, synthetically inaccessible double bonds have been introduced in this way, e.g., eqn 6.[7]

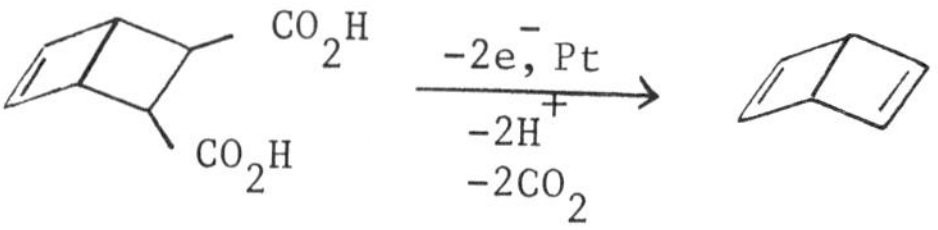

$$CO_2H,\ CO_2H \xrightarrow[-2H^+,\ -2CO_2]{-2e^-,\ Pt} \qquad (6)$$

A parallel electrooxidation of <u>cis</u>- or <u>trans</u>-cyclohexene-4,5-dicarboxylic acids would thus be expected to produce 1,4-cyclohexadiene, eqn 7.

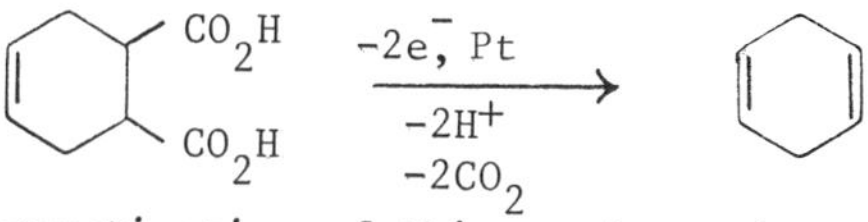

$$CO_2H,\ CO_2H \xrightarrow[-2H^+,\ -2CO_2]{-2e^-,\ Pt} \qquad (7)$$

Oxidative aromatization of this product might reasonably lead to benzene. When the oxidation was conducted by irradiating an aerated suspension of titanium dioxide powder, however, the bis-decarboxylation product could not be detected. Instead, the monodecarboxylation product was obtained in reasonable chemical yield as the sole isolable product, eqn 8.[8]

$$\text{cyclohexene-4,5-}(CO_2H)_2 \xrightarrow[\text{1 M aq. } HNO_3,\ O_2]{TiO_2{}^*} \text{cyclohexene-4-}CO_2H \quad (8)$$

This reaction course can be understood if we consider the mechanisms for the two different products formed in eqns 7 and 8. In the anodic oxidation, the top route in eqn 9,

$$\text{cyclohexene-}(CO_2H)_2 \xrightarrow[-CO_2,\,H^+]{-e^-} \text{radical}(CO_2H) \xrightarrow{-e^-} \text{cation}^+(CO_2H) \xrightarrow{-CO_2} \text{cyclohexadiene}$$

$$\text{cyclohexene-}(CO_2H)_2 \xrightarrow[-CO_2,\ -H^+]{TiO_2{}^*} \text{radical}(CO_2H) \xrightarrow{TiO_2{}^{\cdot-}} \text{anion}^-(CO_2H) \xrightarrow{H^+} \text{cyclohexene-}CO_2H \quad (9)$$

the first electron is removed to produce a carboxy radical, a species well-known from the Kolbe electrolysis to rapidly lose CO_2. This bond cleavage thus provides a beta-carboxy radical. This species is more readily oxidized than is its precursor, since a second oxidation can convert it to an anchimerically stabilized cation. Thus, the second step shown in the top mechanism of equation 9 occurs rapidly compared with the first. The first two steps may be considered therefore to occur simultaneously, i.e., by a two electron oxidation. It is not possible to avoid the second oxidation at a poised electrode because the applied potential cannot be switched off at a rate comparable to that of the second electron transfer. The cation formed by the two electron oxidation then rearranges electronically, expelling the second molecule of CO_2 and forming the observed olefinic product.

With the semiconductor oxidation catalyst, however, the surface becomes activated only upon photoexcitation. At low light intensities, the possibility that many holes are formed in the valence band is remote, so that the irradiated semiconductor powder becomes an effective one-electron oxidant. Now if the same chemistry ensues on the photochemically activated TiO_2 surface, then the reaction will proceed as in the bottom route of eqn 9. Thus, the carboxy radical is formed, producing an alkyl radical after loss of carbon dioxide. Since the semiconductor cannot continue the oxidation after the first step, the radical persists, eventually recapturing the conduction band electron, either directly or through the intervention of an intermediate relay, perhaps superoxide. The resulting anion would be rapidly protonated to product. Preferential adsorption of the dicarboxylate starting material (compared with the mono-carboxylate) ensures the isolability of this potentially oxidizable product, at least at low fractional conversion. Note that this route does not constitute a net change in oxidation level, but is rather a redox-catalyzed dismutation.

Reactions in which an electron transfer (E) is followed sequentially by a chemical reaction (C) and a second electron transfer (E), are called ECE processes. We can now see that both the top and bottom routes of eqn 9 involve ECE pathways. The top route, however, involves an oxidation as the second electrochemical step, while the bottom route is characterized by a reduction in the second stage of the electron exchange sequence.

The role of oxygen in this chemical transformation can be addressed if the reaction shown in Figure 8 is conducted as a deoxygenated aqueous acidic suspension. Under these conditions, monodecarboxylation can still be detected, eqn 10,

$$\text{(cyclohexene-dicarboxylic acid: } CO_2H, CO_2H) \xrightarrow[\underline{1M}\ HNO_3,\ N_2]{TiO_2^*} \text{(cyclohexane-}CO_2H\text{) major} \qquad (10)$$

but it is accompanied by extensive hydrogenation of the C-C double bond of the reactant and product. Since platinum is an excellent hydrogenation catalyst in the presence of gaseous hydrogen, this product argues for the photoelectrochemcial formation of hydrogen by concomitant proton reduction. Since no hydrogenation can be detected in the presence of oxygen, it is clear that oxygen is necessary for the cathodic half reaction in the photoelectrochemical conversion.

If a sacrificial electron donor replaces the oxidizable dicarboxylate in the reaction mixture of eqn 10, the irradiated powders can serve as effective hydrogenation catalysts, providing an effective, safe source of hydrogen gas at low, but chemically effective, concentrations.

The example shown in eqn 8 is but the first of many potential photoelectrochemical conversions which might differ substantively from analogous conventional anodic oxidations. Further investigations of this area are currently being undertaken in our research group.

Summary

We have shown how the band structure of photoexcited semiconductor particles makes them effective oxidation catalysts. Because of the heterogeneous nature of the photoactivation, selective chemistry can ensue from preferential adsorption, from directed reactivity between adsorbed reactive intermediates, and from the restriction of ECE processes to one electron routes. The extension of these experiments to catalyze chemical reductions and to address heterogeneous redox reactions of biologically important molecules should be straightforward. In fact, the use of surface-modified powders coated with chiral polymers has recently been reported to cause asymmetric induction at prochiral redox centers.[10] As more semiconductor powders become routinely available, the importance of these photocatalysts to organic chemistry is bound to increase.

Literature Cited

1. Bard, A.J. Science 1980, 207, 139.

2. Fox, M.A. Accts. Chem. Res. 1983, 16, 314.

3. For a listing of typical band positions for a variety of semiconductors, see Nozik, A.J. Ann. Rev. Phys. Chem. 1983, 16, 314.

4. Fox, M.A.; Chen, C.C J. Amer. Chem. Soc. 1981, 103, 6757.

5. Hussein, F.H.; Pattenden, G.; Rudham, R.; Russell, J.J. Tetrahedron Lett. 1984, 25, 3363.

6. Fox, M.A.; Chen, C.C.; Younathan, J.N.; J. Org. Chem. 1984, 49, 1969.

7. For example, see Radlick, P.; Klem, R.; Spurlock, S.; Sims, J.J. van Tamelen, E.E; Whitesides, T. Tetrahedron Lett. 1968, 49, 5117.

8. Fox, M.A.; Park, K. J. Amer. Chem. Soc. 1985, submitted for publication.

9. Fox, M.A.; Kamat, P.V.; Hohman, J.R. Can. J. Chem. 1983, 61, 888.

10. Kawai, T. Proceedings, Fifth International Conference on Photochemical Conversion and Storage of Solar Eneregy, Osaka, 1984.

RECEIVED January 21, 1985

Electron and Energy Transfer from Phenothiazine Triplets

A. M. BRAUN, M.-A. GILSON, M. KRIEG, M.-T. MAURETTE[1], P. MURASECCO, and E. OLIVEROS[1]

Institut de Chimie Physique, Ecole Polytechnique Fédérale de Lausanne, Ecublens, CH-1015 Lausanne, Switzerland

Phenothiazine derivatives have found many applications in todays chemical industry; they have been and are used primarily as dyestuffs, antioxidants and in pharmaceutical preparations. This application is due to the discovery of their neuroleptic activity (1) (e.g. Chlorpromazine), but allergic skin reactions and ocular opacity are known to occur during therapy.

Assuming that these side effects might be primarily initiated by light induced electron or energy transfers, investigations on the conditions where such transfers take place with high efficiency and with some specificity are of fundamental interest. Phenothiazine and its N-alkylated derivatives have also been of interest as electron donating substrates in early investigations of photochemically induced charge separation (2-5), necessary for a successfull photochemical energy conversion. Besides phototoxicity, which might involve oxygen, and energy conversion, just two examples of model investigations, phenothiazine derivatives are convenient probes in micelles and mixed aggregates, model systems themselves for preparative applications (6) as well as biomimetic environments (7).

Phenothiazine

Phenothiazine is readily oxidized when irradiated in solution with chlorinated hydrocarbones (8). The reaction has been shown to be an electron transfer generating the phenothiazine (PTH) radical cation ($PTH^{+\cdot}$) and the halogen anion as shown in equation 1.

$$PTH + RCl \xrightarrow{h\nu} PTH^{+\cdot} + R^{\cdot} + Cl^{-} \qquad (1)$$

Photooxidation may lead to a similar electron transfer yielding $PTH^{+\cdot}$ and superoxide,

$$PTH + O_2 \xrightarrow{h\nu} PTH^{+\cdot} + O_2^{-\cdot} \qquad (2)$$

[1]Current address: Laboratoire IMRCP, ERA 264, Univ. Paul Sabatier, F-31062 Toulouse, France

0097-6156/85/0278-0079$06.00/0

in fact, the neutral phenothiazinyl radical ($PT^{\cdot}$), produced by subsequent deprotonation of $PTH^{+\cdot}$ (Equation 3), has been identified by ESR spectroscopy (9).

$$PTH^{+\cdot} \longrightarrow PT^{\cdot} + H^{+} \qquad (3)$$

The corresponding ESR signal has also been assigned to the nitroxide derivative of phenothiazine (10) which is thought to be generated by insertion of singlet oxygen ($^{1}O_2$) (Equation 4) into the N-H bond (Equation 5) and subsequent homolysis of the hydroperoxide (Equation 6).

$$PTH + O_2 \xrightarrow{h\nu} PTH + {}^{1}O_2 \qquad (4)$$

$$PTH + {}^{1}O_2 \longrightarrow \text{N-OOH phenothiazine} \qquad (5)$$

$$\text{N-OOH phenothiazine} \longrightarrow \text{N-O}^{\cdot}\text{ phenothiazine} + HO^{\cdot} \qquad (6)$$

However, the results of the sensitized oxygenation do not support such a sequence of reactions (vide infra).

In addition to the possibilities of electron transfer (Equation 2) and energy transfer (Equation 4), electron transfer to singlet oxygen (Equation 7) and subsequent deprotonation (Equation 3), or hydrogen transfer to singlet oxygen (Equation 8) (8) are to be taken into consideration and make an attempt of a differentiation between those postulated mechanisms extremely difficult.

$$PTH + {}^{1}O_2 \longrightarrow PTH^{+\cdot} + O_2^{-\cdot} \qquad (7)$$

$$PTH + {}^{1}O_2 \longrightarrow PT^{\cdot} + HO_2^{\cdot} \qquad (8)$$

The identification of the major products of the photooxidation of phenothiazine upon its direct excitation does not support the idea of a N-hydroperoxy intermediate (Equation 5) (11). In fact, phenothiazine-5-oxide (A) which has been first identified by comparing R_f values in thin layer chromatography of isolated and separately prepared samples (11) represents the best evidence of a singlet oxygen reaction (Equation 9) (12).

$$2\ PTH + {}^{1}O_2 \longrightarrow 2\ \text{A} \qquad (9)$$

A

Other isolated products consist of the 3H-phenothiazine-3-one derivatives B (13), C and D (11), the formation of which may be explained by a radical hydroxylation of PT$^{\bullet}$.

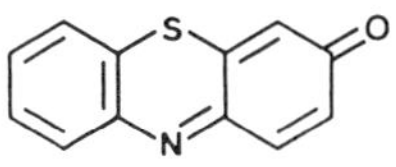

B

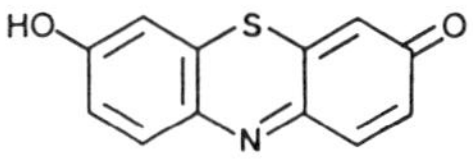

C

Since superoxide reduces readily quinoid structures (e.g. B) (14), product C (7-hydroxy-3H-phenothiazine-3-one) may have a similar genesis, and 7-(10-phenothiazinyl)-3H-phenothiazine-3-one (D) represents one of the many possible recombination products starting from PT$^{\bullet}$.

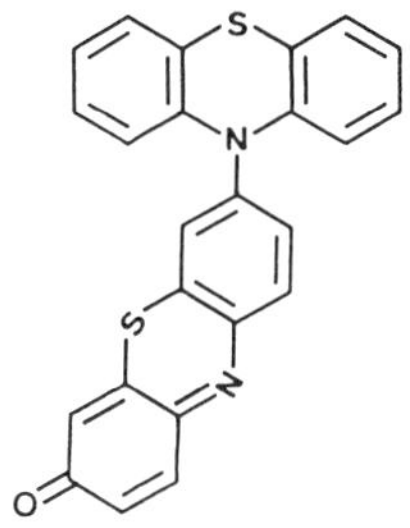

D

Dye sensitized photooxidation of phenothiazine should reveal whether in addition to reaction 9 any competitive singlet oxygen reaction (Equation 5) or electron transfer process (Equation 7) is occuring. Rosenthal and Poupko (15) reported that radical formation could be observed by ESR spectroscopy during a methylene blue sensitized photooxidation of phenothiazine; the signal was suppressed, however, upon addition of DABCO. Since this quencher does not effect ESR spectra of several commercially available nitroxides, thus, indicating that a nitroxide intermediate (Equation 6) does not exist (16), it was claimed that singlet oxygen undergoes an electron transfer reaction (Equation 7), the product of which (PT$^{\bullet}$ after reaction 3) is no longer detectable when the singlet oxygen is quenched.

Benzophenone (BP) sensitized photooxidation of phenothiazine leads mainly to the formation of products A and B (17). In view of the many data available from the literature and from our own experiments the published interpretation needs, however, some revision: provided 3H-phenothiazine-3-ones are not products of a singlet oxygen reaction with ground state phenothiazine (c.f. methylphenothiazine), formation of products A and B implies two different oxidation mechanisms. Since singlet oxygen sensitization by benzophenone (Equations 10 to 12) is known to be very inefficient, reaction sequence (10), (13), (14), and (9)

$$ {}^{1}_{0}BP \xrightarrow{h\nu} {}^{1}_{1}BP \tag{10}$$

$$ {}^{1}_{1}BP \xrightarrow{ISC} {}^{3}_{1}BP \tag{11}$$

$$ {}^{3}_{1}BP + O_2 \longrightarrow {}^{1}_{0}BP + {}^{1}O_2 \tag{12}$$

$$ {}^{3}_{1}BP + {}^{1}_{0}PTH \longrightarrow {}^{1}_{0}BP + {}^{3}_{1}PTH \tag{13}$$

$$ {}^{3}_{1}PTH + O_2 \longrightarrow {}^{1}_{0}PTH + {}^{1}O_2 \tag{14}$$

seems logical. For product B, either reaction 13 could be paralleled by reaction 15

$$ {}^{3}_{1}BP + PTH \longrightarrow BP^{-\bullet} + PTH^{+\bullet} \tag{15}$$

leading in oxygen saturated benzene to superoxide (Equation 16), or reaction 7 is successfully competing with reaction 9.

$$ BP^{-\bullet} + O_2 \longrightarrow BP + O_2^{-\bullet} \tag{16}$$

Considering reactions 3 and 17, both ways are leading finally to product B by radical hydroxylation.

$$ O_2^{-\bullet} + H^{+} \rightleftharpoons HO_2^{\bullet} \tag{17}$$

It is evident that, for the reaction sequences proposed, the quantitative analysis of product B as a function of added DABCO cannot be used as a decisive argument in favour of or against a singlet oxygen mechanism. Moreover, the increase of the rate of phenothiazinone formation with the concentration of protic solvent where singlet oxygen is quenched efficiently supports the given interpretation.

In conclusion, no differentiation regarding reactions 2 and 7 can be drawn from the available data. Many experimental difficulties arising from the rather great number of products found due to the deprotonation of $PTH^{+\bullet}$ (reaction 3) can, however, be reduced in taking simple N-alkylated phenothiazine derivatives as model compounds. In addition, the use of microheterogeneous systems (micelles, microemulsions) as a reaction environment, favouring charge separation, might be advantageous for the differentiation between electron and energy transfer reactions.

N-Methyl-phenothiazine

Triplets of N-methyl-phenothiazine ($^3_1$MPT) are found to reduce almost quantitatively Cu^{2+} but to transfer their energy to Ni^{2+} and Co^{2+} (18). Experiments in solution (ethanol/water:1/2) and in aqueous $Cu(LS)_2$ micelles show that the electron transfer in functionalized organized media is extremely efficient. The rate constant of electron transfer (Equation 18) in solution has been measured

$$^3_1\text{MPT} + Cu^{2+} \longrightarrow \text{MPT}^{+\cdot} + Cu^{+} \quad (18)$$

to be $1.0(\pm 0.1) \times 10^9\ M^{-1}s^{-1}$ (18) and is, thus, ~40 times slower than that estimated in $Cu(LS)_2$ micelles. Energy transfer to Ni^{2+} and Co^{2+} (Equation 19), respectively, is slower by about a factor of 100; table I shows the corresponding quenching rate constants, indicating

$$^3_1\text{MPT} + Co^{2+} \longrightarrow {}^1_0\text{MPT} + Co^{2+} \quad (19)$$

a faster triplet quenching in appropriately functionalized micelles by about a factor of 600.

Table I. Rate constants of the energy transfer from $^3_1$MPT (~10^{-6} M) to Ni^{2+} and Co^{2+} in solution (ethanol/water:1/2) and aqueous lauryl sulphate micelles (18).

Ni^{2+} (5×10^{-3}M)	$7.7(\pm 0.1) \times 10^6\ M^{-1}s^{-1}$	
$Ni(LS)_2$ (2×10^{-2}M)		$2.1(\pm 0.1) \times 10^7\ s^{-1}$
Co^{2+} (5×10^{-3}M)	$1.4(\pm 0.1) \times 10^7\ M^{-1}s^{-1}$	
$Co(LS)_2$ (2×10^{-2}M)		$4.2(\pm 0.1) \times 10^7\ s^{-1}$

$^3_1$PTH reduces Eu^{3+} (E^o = -0,4 V) at a rate comparable to that of reaction 19, but only triplet quenching is observed when Mn^{2+} is used as a quencher. The rate constants of the latter reaction in solution and aqueous micellar systems correspond to energy transfer rate constants measured with Ni^{2+} and Co^{2+} (Equation 19, Table I).

In contrast to the electron transfer reactions in question, the efficiency of which is primarily determined by the difference of redox potentials of the excited donor molecule and the acceptor and, hence, by the variation in free energy (21), efficiencies of exothermic energy transfers depend solely on the local concentration of an appropriate quencher.

If MPT is solubilized in aqueous CTAB micelles together with a hydrophobic quencher, statistical and proximity effects will influence the efficiency and rate of the energy transfer in the same direction. Thus reaction 20 is enhanced in the confined space of a CTAB micelle by a factor of ~10 (19).

$$^{3}_{1}\text{MPT} + \text{trans-stilbene} \longrightarrow {}^{1}_{0}\text{MPT} + {}^{3}_{1}\text{trans-stilbene} \qquad (20)$$

In both cases, where the quencher is identical with the counterion or where the quencher is hydrophobic, confinement of the reaction space to the micelle size induces both an enhancement of the observed reaction as well as a reduction of the reaction order from two to one.

The observed reversible energy transfer (Equation 21) places the energy of ${}^{3}_{1}$MPT at ~255 kJ.mol^{-1} (19).

$$^{3}_{1}\text{MPT} + {}^{1}_{0}\text{naphthalene} \rightleftharpoons {}^{1}_{0}\text{MPT} + {}^{3}_{1}\text{naphthalene} \qquad (21)$$

Since triplet N-methyl-phenothiazine reduces NO_3^-, the redox potential of ${}^{3}_{1}$MPT must be <-1 V and has been estimated to be -1.8 V (20). The electron transfer rate observed for reaction 22 in aqueous DTAC micellar systems is 3.5×10^5 $M^{-1}s^{-1}$ and, thus, under the experimental conditions reported, about a factor of 10^4 slower than reaction 18.

$$^{3}_{1}\text{MPT} + NO_3^- \longrightarrow \text{MPT}^{+\cdot} + NO_3^{2-} \qquad (22)$$

This is in agreement with the predictions based on the Marcus theory (21), that the rate of reaction should be considerably below the diffusion-controlled limit due to the relatively small decrease in free energy in the reduction of NO_3^-.

It is interesting to note that the rate of reaction 22 is enhanced in cationic micellar systems with respect to anionic micellar systems but remains slower by a factor of 10 than in homogeneous solution (water-ethanol). This has been interpreted in terms of reaction energetics attributing to ${}^{1}_{0}$MPT and ${}^{3}_{1}$MPT solubilized in micellar systems higher redox potentials than in homogeneous solution (20). Table II shows, however, that the ground state redox potential of N-methyl-phenothiazine is found smaller in micellar aggregates (SLS) and the substance hence a better reductant under these conditions (22).

Table II. Standard redox potentials of MPT in solution and in micellar aggregates.

[MPT]	System	E^o	Remarks
10^{-3}	$EtOH/H_2O$ (2/1)	0.62	$[LiClO_4]$: 4.8×10^{-2}M
10^{-4}	10^{-1}M SLS in H_2O	0.43	

The difference of these redox potentials is too big to be explained by a complexation with the surfactant aggregate (23). A detailed study of the dependence of the redox potential of MPT from the nature and the concentration of the surfactant leads to the conclusion that $\text{MPT}^{+\cdot}$ is interacting with the anionic surfactant system, interaction most pronounced with the monomeric surfactant entities (24).

Experimental evidence for this interpretation is given by ESR spectroscopy, and the UV spectra indicate that $MPT^{+\cdot}$ is localized in a highly polar region of the aggregate, that is now the anionic head groups of the surfactant.

Similar triplet energies (18) and ground state donor characteristics of MPT and PTH in solution imply that triplet redox potentials of the two substances must also be close.

Based on these results of model electron and energy transfer reactions ^3_1MPT must produce superoxide by both direct (Equation 23), as well as proton assisted transfer (Equation 24).

$$^3_1MPT + O_2 \longrightarrow MPT^{+\cdot} + O_2^{-\cdot} \quad (23)$$

$$E^o_{O_2/O_2^{-\cdot}} = -0.33 \ (25)$$

$$^3_1MPT + O_2 + H^+ \longrightarrow MPT^{+\cdot} + HO_2^{\cdot} \quad (24)$$

$$E^o_{O_2/HO_2^{\cdot}} = -0.05 \ (25)$$

Indeed, photooxidation of MPT in water/ethanol (1/2) leads, as expected from similar experiments with PTH (v. supra), to a complex mixture of products. The main components of the mixture have been separated and identified (22). 3-Hydroxy-10-methyl-phenothiazine (E) is not formed in a singlet oxygen reaction (26); in analogy to the products from the photooxidation of directly excited phenothiazine, the formation of E may be explained by a recombination of the products of reaction 23,

(25)

followed by proton shift and reduction of the intermediate hydroperoxide (Equation 25). The chemical yields of E and F (10-methyl-phenothiazine-5-oxide) are practically the same (23% at 60% conversion) indicating the high efficiency of electron transfer when compared with singlet oxygen production (Equations 26 and 27).

$$ {}^{3}_{1}MPT + O_2 \longrightarrow {}^{1}_{0}MPT + {}^{1}O_2 \quad (26) $$

$$ 2\ MPT + {}^{1}O_2 \longrightarrow 2\ F \quad (27) $$

Hovey also found the ESR signal corresponding to $MPT^{+\cdot}$ when irradiating MPT solubilized in aqueous SDS micelles (27). He deduces from reaction 28 yielding 5% of benzylsulfoxide that singlet oxygen is

$$ MPT^{+\cdot}ClO_4^- + PhCH_2SCH_2Ph + KO_2 \longrightarrow \quad (28) $$
$$ \longrightarrow MPT + PhCH_2SOCH_2Ph + KClO_4 $$

formed from the energy of recombination of $MPT^{+\cdot}$ and $O_2^{-\cdot}$ (Equation 29).

$$ MPT^{+\cdot} + O_2^{-\cdot} \longrightarrow MPT + {}^{1}O_2 \quad (29) $$

However, no recombination products according Equation 25 have been looked for and, thus, sulfoxide production due to sulfide oxidation by an intermediate 10-methyl-phenothiazine-3-hydroperoxide cannot be excluded.

In addition to products E and F, 16% (at 60% conversion) of 9-methyl-carbazole G has been isolated. Carbazole formation implies a loss of SO_2 from sulfone H and subsequent ring closure (Equation 30).

$$ H \longrightarrow G + SO_2 \quad (30) $$

Sulfone production itself is not a singlet oxygen reaction (26); the reaction requires most probably an electronically excited phenothiazine intermediate since superoxide and MPT do not react in model reactions using KO_2 and crown ethers in pyridine.

Our hypothesis is based on the observation that the photooxidation upon direct excitation of F yields G as well as a hydroxylated sulfone, too (28). Again, this reaction must involve oxidants other than singlet oxygen, for no reaction has been observed between the latter and ground state 10-methyl-phenothiazine-5-oxide (F).

Superoxide production cannot be achieved by reaction 31, a hypothesis similar to that formulated with reaction 7:

$$ MPT + {}^{1}O_2 \longrightarrow MPT^{+\cdot} + O_2^{-\cdot} \quad (31) $$

We have confirmed earlier results (26) where singlet oxygen, produced by Rose Bengal sensitization, reacts with ground state MPT only to yield the corresponding sulfoxide F (Equation 27). Since redox potentials of PTH and MPT are about the same, electron transfer (Equation 7) does not seem a plausible explanation of the ESR signals observed (15); investigations in micellar systems give evidence of a prominent electron transfer reaction to the Rose Bengal triplets oxidizing the phenothiazine derivative present (vide infra). Electron transfer would then imply a similar reaction sequence as postulated for the benzophenone sensitized oxidation of phenothiazine (reactions 10, 11, 15, 16), however, laser photolysis experiments necessary for the detection of a similar electron transfer reaction between methylene blue triplets and 10-methyl-phenothiazine have not been reported so far.

$$\text{MPT} + {}^1O_2 \longrightarrow \text{MPT} + O_2 \qquad (32)$$

MPT reacts somewhat more slowly with singlet oxygen than PTH. The combined rate constants of physical quenching (e.g. Equation 32) and chemical reaction (e.g. Equation 27) have been calculated from photooxidation experiments in bromobenzene/methanol (2/1) to be $1.2\text{x}10^6$ $M^{-1}s^{-1}$ for MPT and $4.2\text{x}10^7$ $M^{-1}s^{-1}$ for PTH (26). Hovey has calculated a rate constant of physical quenching for MPT in chloroform of $2.9\text{x}10^6$ $M^{-1}s^{-1}$ (27). Experiments in aqueous methanol or aqueous micellar systems show slower rate constants of the chemical reaction, most probably due to the more efficient deactivation of singlet oxygen by the solvent ($k_{d(H_2O)} \simeq 2.3\text{x}10^5$ s^{-1}).

Photooxidation of MPT is, however, accelerated in O/W microemulsions; this might be due to local concentration and organization effects (24) accelerating dismutation as proposed by Foote and Peters (12) for the formation of sulfoxides (Equation 27).

Possibilities in the application of microheterogeneous systems for the enhancement of electron transfer reactions and the stabilization of charged intermediates suggest a vast domain of investigations using ionic surfactants. No enhancement of charge separation has been found subsequent to reaction 23, when anionic micelles are used. This indicates that the back transfer of the electron is faster than the expulsion of the superoxide anion from the aggregate. The stability of $MPT^{+\cdot}$ towards the addition of charged nucleophiles present in the aqueous bulk phase increases from cationic to neutral to anionic micellar systems; the rate constants change from first to second order and imply a high local concentration of charged mucleophile in the Stern and in the Gouy-Chapman space of cationic aggregates (29). A similar nucleophilic addition of superoxide passing the micellar interface is practically impossible as found by electron transfer experiments using hydrophilic and hydrophobic acceptors (14).

Chlorpromazine

Rosenthal, Bercovici and Frimer (26) report that the dye sensitized photooxidation of Chlorpromazine (I) leads to 2-chloro-phenothiazine-5-oxide (J); since upon shorter irradiation times 2-chloro-phenothiazine (K) has been isolated as a minor product of photolysis, K has been proposed as an intermediate (Scheme 33).

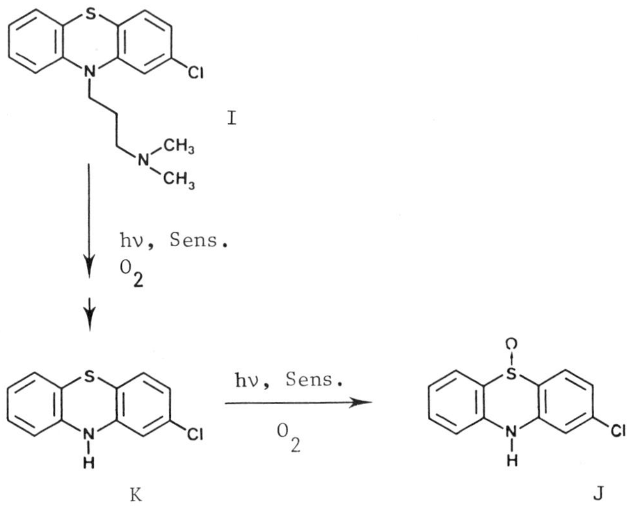

I

hν, Sens.
O_2

K

hν, Sens.
O_2

J

(33)

Whereas the formation of J fits into the interpretation given for the reaction of $^1O_2$ with PTH and with MPT, the first step implying dealkylation of the 10-position is open to mechanistic speculations.

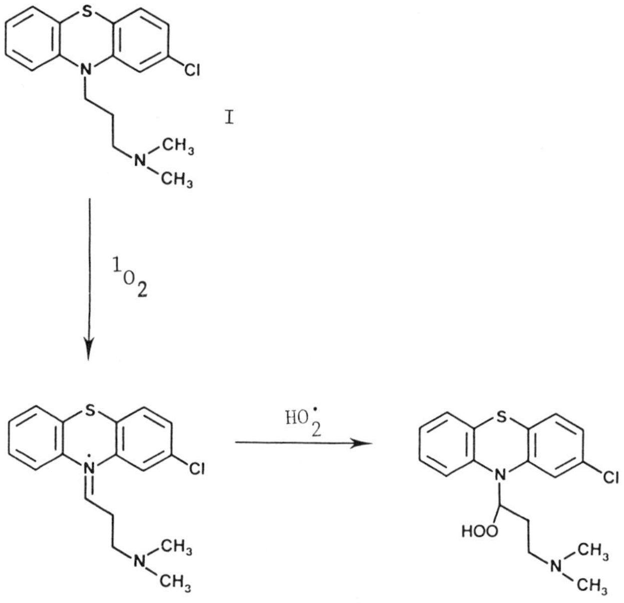

I

$^1O_2$

$HO_2^{\cdot}$

(34)

The facts that singlet oxygen is needed for this dealkylation process and that the reaction is about 10 times faster then the sulfoxide production do not support the postulated hydroxylation in the α-position (26). Hydrogen abstraction at that position would produce a highly stabilized radical intermediate (Scheme 34).

Peroxylation involving a N-peroxide intermediate (30) (Scheme 35) would lead to the dealkylated derivative K. Experimental conditions used (26) would allow the detection of the highly stabilized immonium ion by spectrophotometric analysis. However, no such intermediate has yet been reported.

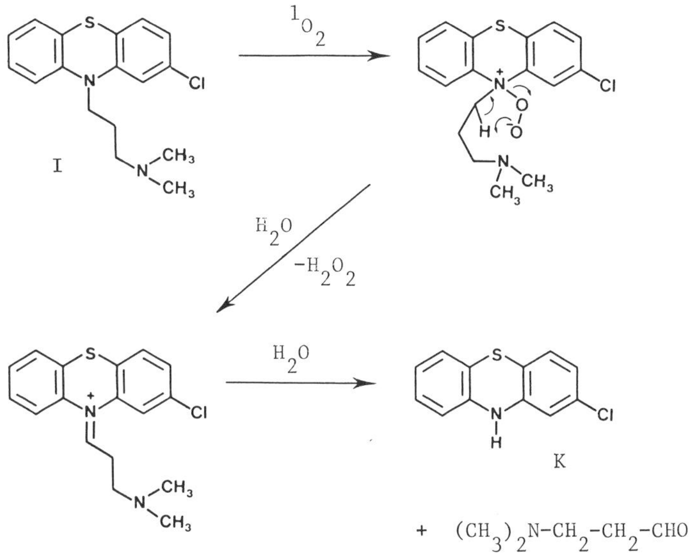

I $\xrightarrow{^1O_2}$; $\xrightarrow[-H_2O_2]{H_2O}$; $\xrightarrow{H_2O}$ K + $(CH_3)_2N\text{-}CH_2\text{-}CH_2\text{-}CHO$ (35)

Neglecting semantic differences for the description of an intermediate (Structures 36) not yet characterized, we may assume that the physical quenching of singlet oxygen by phenothiazine proceeds via

a b (36)

some charge transfer interaction (31,32) producing a charge deficiency at the phenothiazine moiety (Equation 37).

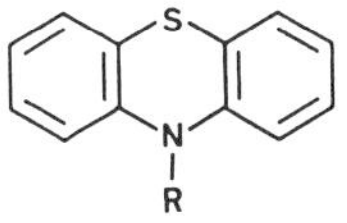

$+ \ ^1O_2 \longrightarrow (36)$ (37)

Since no dealkylation is observed in the reaction of MPT with singlet oxygen, an intermediate leading to a reaction sequence as postulated in Scheme 35 (30,33) does not seem realistic.

Efficient physical quenching of singlet oxygen by phenothiazine derivatives might then involve a charge transfer complex (structure 36a), and depending on substituents linked to an alkyl chain in 10-position, anchimeric effects may lead to stabilization of the charge deficiency at the phenothiazine moiety by irreversible fragmentation (34). Thus, dealkylation of chlorpromazine might follow Scheme 38, and the fact that corresponding β-amine derivatives do not dealkylate supports this interpretation.

I $\xrightarrow{^1O_2}$ (δ+ N···O_2 δ−) $\xrightarrow{H_2O}$ K + O_2 + ... (38)

Unfortunately, preparative experiments of Iwaoka and Kondo (35) are of no direct relation to mechanistic investigations; the use of a low pressure mercury lamp provides no selectivity as far as excitation of substrate or products is concerned. However, the fact that photolysis in strong acid solution decreased the bleaching rate would indicate the absence of an anchimeric effect and the results of their investigations by flash photolysis are in agreement with the electron (Equation 2) and energy transfer (Equation 4) reactions upon direct excitation.

Pursuing the idea that physical quenching efficiency and dealkylation might depend on mesomeric and anchimeric charge stabilisation, a series of phenothiazine derivatives is currently under investigation in order to determine rate constants of physical quenching and chemical reaction.

Iwaoka and Kondo were also interested in the transients of the phenothiazine derivatives K to P in solution and aqueous micelles (36).
From NMR spectral patterns of the aromatic protons they deduce that L, O and P are incorporated into SDS aggregates, whereas M and N are attached to the double layer. Their flash photolysis results as well

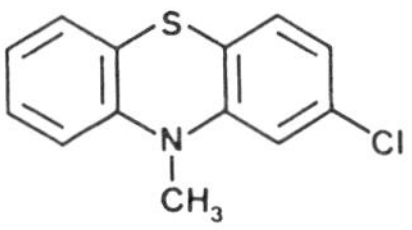

L

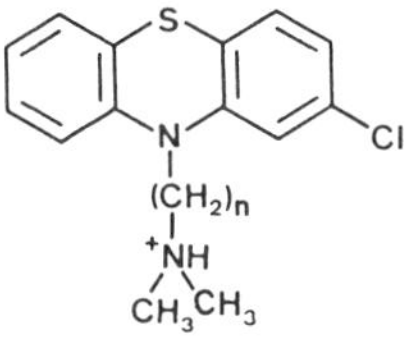

n = 2 : M
3 : N
6 : O
10 : P

as those of Hoffmann, Tagesson and Ulbricht (37) do, however, not support this interpretation, since no difference in radical cation stability has been found. Furthermore, high yields of photoinduced radical cation formation are explained for all compounds by efficient electron ejection from the aggregate.

12-(10'-Phenothiazinyl)-Dodecyl-1-Sulfonate

In contrast to the ammonium compounds cited above, the 12-(10'-phenothiazinyl)-dodecyl-1-sulfonate (Q) is a stable surfactant molecule forming aggregates over a wide pH range by itself or with SLS. Its CMC has been determined by different methods to be approx. $5x10^{-4}$ M with an aggregation number of $4.2x10^{3}$ (38). Various experiments imply that aggregate formation begins at concentrations far below the given CMC. The redox potential (0.45 V) is practically independent of concentration for a range from $5x10^{-5}$ to $5x10^{-3}$ M and differs insignificantly from that of MPT in aqueous SLS micelles.

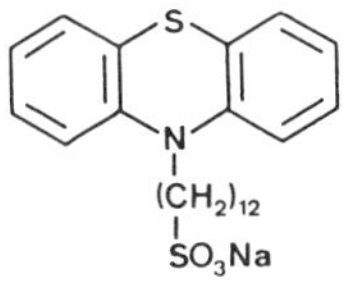

Q

From Table III, the triplet lifetime is seen to increase strongly upon increasing the concentration of Q, especially above the CMC value. The higher viscosity of the micelle medium impedes the quenching of triplets by impurities and ions; the triplet lifetime is then governed by the slow exit of Q^T from the micelle and is seen in their primarily first-order decay.

Upon increasing the laser intensity, the triplet decay takes on a new form with the appearance of a short-lived component superimposed on the long-lived decay. When the excitation pulse intensity was varied in the range of 5-100 mJ, the slow decay process was found to be independent of the intensity, and the amount of triplet, 2μs after the laser pulse, was also invariant once saturation had been achieved.

An additional feature observed upon micellization is an increase in the relative yield of radical cation production. From Table IV can be seen that at low concentration of Q, the amount of radical cation produced is essentially nil and that triplet production is unity. Above CMC the radical cation production becomes comparable to the

Table III. Lifetimes and yields of triplets of Q in aqueous solutions of different concentrations of Q (38).

[Q]	$\tau(\mu s)$ a)	Φ_{isc}
$2.55x10^{-5}$	68	1.0 b)
$6.37x10^{-5}$	70	1.0 b)
$1.10x10^{-4}$	82	0.78
$2.22x10^{-4}$	88	0.40
$5.00x10^{-4}$	322	0.15
$9.10x10^{-4}$	448	0.093
$1.15x10^{-3}$	535	0.066
$2.39x10^{-3}$	686	0.058

a) Lifetimes determined for long-lived component at low laser intensity

b) Triplet quantum yield assumed as unity at this concentrations (38)

triplet production. A relative enhancement in radical cation production has been observed by Iwaoka and Kondo (35,36) in aerated micellar systems, conditions totally different from those reported here. The concentration of radical cations produced by photoionization was found to increase nearly linearly with laser intensity, indicative

Table IV. Lifetimes and yields of the radical cation of Q in aqueous solution of different concentrations of Q (38)

[Q]	$\tau_{1/2}$	Φ_{CAT} a)
$2.55x10^{-5}$	0.8 ms	$8x10^{-4}$
$6.37x10^{-5}$		$3x10^{-4}$
$1.10x10^{-4}$	900 ms	0.048
$2.20x10^{-4}$	$5x10^{2}$ s	0.17
$5.00x10^{-4}$		0.33
$9.10x10^{-4}$	$1.19x10^{5}$ s	0.34
$1.15x10^{-4}$		0.25
$2.385x10^{-3}$	$1.80x10^{5}$ s	0.1

a) Determined from relative yields of triplet and cationic species.

of a monophotonic process. Under dilute conditions this process is biphotonic in nature. The net effect of micelle formation is then to enhance the radical cation yield with a subsequent decrease in the yield of triplets produced.

The fate of the radical cation produced is also interesting. As has been observed previously, the stability of the radical cation is greatly enhanced in the micellar environment (18,29). This system demonstrates this clearly by considering the variation in half-life of the radical cation with concentration. The increase in lifetime above CMC exemplifies the stabilizing features of the micelle. The e^-_{aq} decay process is experimentally observed. The loss of the radical cation of Q must then be controlled by the rate of its exit to the surface of the aggregate, as has been observed for $MPT^{+\cdot}$ in other amphiphilic environments (29): even at pH 10, the micellar effect is strong enough to prevent fast degradation by hydroxide ions.

It is proposed that each micelle can support only a certain number of radical cations before the forces holding the micelle together are negated by the repulsive forces between cations. On the basis of this assumption, it would seem the maximum number of radical cations per micelle is 22 in a system containing a micelle concentration of 4.4×10^{-7} M (29).

Direct photolysis of aqueous aggregates of Q in the presence of oxygen produces about 10 times more radical cations than triplets. Nevertheless, preparative photolysis yields only minute traces of secondary products attribuable to electron transfer and subsequent reactions. This means then that the back transfer of the electron from superoxide to the radical cation of Q is much faster than the ejection of superoxide from the anionic aggregate, and energy transfer to produce singlet oxygen and its subsequent reactions are at the origin of the formation of the corresponding sulfoxide R as the major product.

Figure 1 shows a plot of $\Phi^{-1}_{ox} = f([Q]^{-1})$ as derived from Equation 39,

$$\frac{1}{\Phi_{ox}} = \frac{k_r + k_q}{\Phi_{^1O_2} \cdot k_r} + \frac{\beta(k_r + k_q)}{\Phi_{^1O_2} \cdot k_r} \cdot \frac{1}{[Q]} \qquad (39)$$

where Φ_{ox} is the quantum yield of oxygen consumption, $\Phi_{^1O_2}$ the quantum yield of singlet oxygen production,

$$\beta = \frac{k_d}{k_r + k_q}, \text{ and} \qquad (40)$$

k_d, k_r and k_q are the rate constants of the deactivation by the solvent, of the chemical reaction and the physical quenching of singlet oxygen by a given acceptor, respectively.

Assuming that physical quenching is negligible compared to the chemical reaction of Q with singlet oxygen ($k_q \ll k_r$), the intercept of the slope points to an evaluation of the quantum yield of singlet oxygen production ($\Phi_{^1O_2} = 0.095$), whereas the rate constant of the chemical reaction can be calculated from the slope ($k_r = 1.4 \times 10^8$ $M^{-1}s^{-1}$).

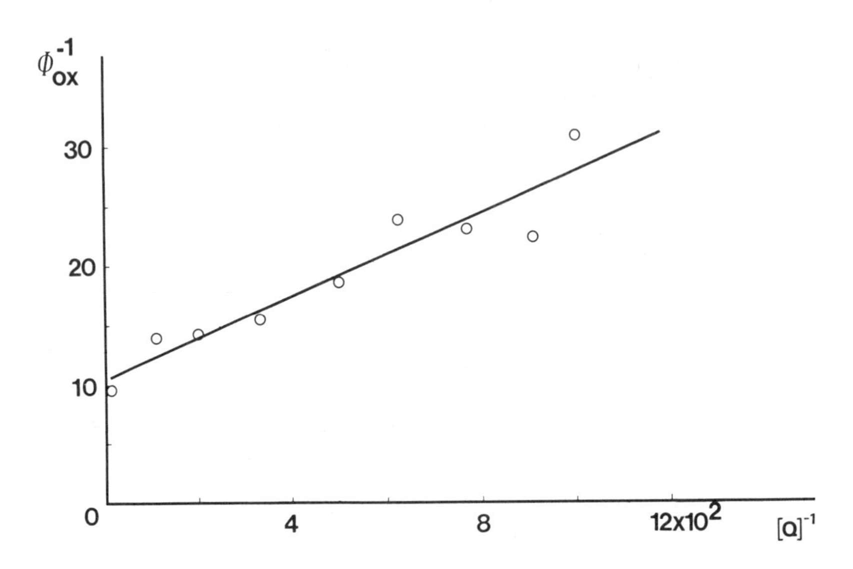

Figure 1. Quantum yields of photooxidation upon direct excitation of Q shown in a plot derived from Equation 39.

The Rose Bengal sensitized photooxidation of Q produces the corresponding sulfoxide R with almost quantitative yield and can be easily analyzed by Equation 39 for concentrations not exceeding 5×10^{-3} M of Q (Figure 2). The quantum yield of oxidation extrapolated from the linear part of the plot is 0.25 and, thus, only 1/3 of the quantum yield of singlet oxygen production by Rose Bengal in aqueous solution (39). The calculated rate constant of the chemical reaction is 8.1×10^6 $M^{-1}s^{-1}$; it is a factor of 15 smaller than found upon direct excitation, the difference might be due to an enhanced deactivation of singlet oxygen when passing the micellar double layer ($k_q = 1.6 \times 10^7$ $M^{-1}s^{-1}$).

O
S
N
$(CH_2)_{12}$
SO_3Na R

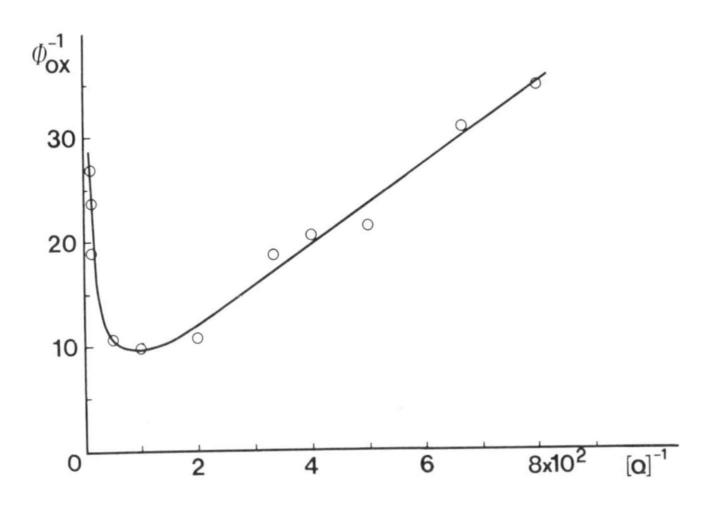

Figure 2. Quantum yields of the Rose Bengal sensitized photo-oxidation of Q shown in a plot derived from Equation 39.

At concentrations of Q greater than $5x10^{-3}$ M the plot (Figure 2) deviates strongly towards small Φ_{ox}. At these surfactant concentrations, Rose Bengal might be integrated into the organizates, and charge transfer interactions with the phenothiazine moiety might compete successfully with the singlet oxygen producing energy transfer. In fact, laser photolysis experiments with deoxygenated samples yield only small concentrations of Rose Bengal triplets, and no signal of the radical cation of the sensitizer can be detected.

In conclusion, the study of energy and electron transfer processes involving phenothiazine derivatives provides a good opportunity to show the usefullness of microheterogeneous systems for mechanistic investigations. The chromophore might also serve as a probe for the characterization of microheterogeneous media (polarity, viscosity, solubility of oxygen ...), and suitable derivatives are currently used for investigations on mixed aggregates and adsorption effects on corresponding interfaces.

Acknowledgments

The authors thank the Swiss National Science Foundation for financial assistance (project nrs. 4.397-0.80.04 and 2.824-0.83) and the French Centre National de la Recherche Scientifique for a grant from its ATP International Program.

Literature Cited

1. "Phenothiazines and Structurally Related Drugs". In "Advances in Biochemical Psychopharmacology"; Forrest, I. S.; Carr, C. J.; Usdin, E., Eds.; Raven Press: New York, 1974; Vol. 9.
2. Alkaitis, S. A.; Beck, G.; Grätzel, M. J. Am. Chem. Soc. 1975, 97, 5723.
3. Alkaitis, S. A.; Grätzel, M. J. Am. Chem. Soc. 1976, 98, 3549.
4. Maestri, M.; Infelta, P. P.; Grätzel, M. J. Chem. Phys. 1978, 69, 1522.
5. Turro, N. J.; Grätzel, M.; Braun, A. M. Angew. Chem. Int. Ed. 1980, 19, 675.
6. Farques, R.; Maurette, M. T.; Oliveros, E.; Rivière, M.; Lattes, A. Nouv. J. Chim. 1979, 3, 487
7. Barbaric, S.; Luisi, P. L. J. Am. Chem. Soc. 1981, 103, 4239
8. Burrows, H. D.; Dias da Silva, J.; Ventura Batista, M. I. "Charge Transfer Interactions in the Photooxidation of Phenothiazine". In "Excited States Biol. Mol."; Birks, J. B., Ed.; Proc. Int. Conf., 1974; Wiley: Chichester, 1976; p. 116.
9. Shine, H. S.; Veneziani, C. V.; Mach, E. E. J. Org. Chem. 1966, 31, 3395.
10. Jackson, C; Patel, N. K. B. Tetrahedron Letters 1967, 2255.
11. Roseboom, H.; Fresen, J. A. Pharm. Acta Helv. 1978, 50, 55.
12. Foote, C. S.; Peters, J. W. J. Am. Chem. Soc. 1971, 93, 3795.
13. Ram, N.; Bansal, W. R.; Uppal, S. S.; Sidhu, K. S. Proc. Symp. Singlet Mol. Oxygen, 1976, p. 141.
14. Maurette, M.-T.; Oliveros, E.; Infelta, P. P.; Ramsteiner, K.; Braun, A. M. Helv. 1983, 66, 722.
15. Rosenthal, I.; Poupko, R. Tetrahedron 1975, 31, 2103.
16. Borzo, M.; Heubeck, J. C.; Potenza, J. A.; Wagner, B. E. J. Coll. Interf. Sci. 1984, 97, 428.
17. Ram, N.; Sidhu, K. S. J. Photochem. 1979, 10, 329.
18. Moroi, Y.; Braun, A. M.; Grätzel, M. J. Am. Chem. Soc. 1979, 101, 567.
19. Rothenberger, G.; Infelta, P. P.; Grätzel, M. J. Phys. Chem. 1979, 83, 1871.
20. Frank, A. J.; Grätzel, M. Inorg. Chem. 1982, 21, 3834.
21. Marcus, R. A. Ann. Rev. Phys. Chem. 1964, 15, 155.
22. Krieg, M. Thesis, EPFL, Lausanne, 1981.
23. Tomlinson, E.; Davis, S. S.; Mukhayer, G. I. In "Solution Chemistry of Surfactants"; Mittal, K. L., Ed.; Plenum Press: New York, 1979 ; Vol. 2, p. 889.
24. Mc Intire, G. L.; Blount, H. N. In "Solution Behaviour of Surfactants"; Mittal, K. L.; Fendler, E. J., Eds.; Plenum Press: New York, 1982; Vol. 2, p. 1101.
25. Allen, A.O.; Bielski, B. H. J. In "Superoxide Dismutase"; Oberley, L. W.; CRC Press: Boca Raten, 1982; Chap. 6, p. 125.
26. Rosenthal, I.; Bercovici, T.; Frimer, A. J. Heterocycl. Chem. 1977, 14, 355.

27. Hovey, M. C. J. Am. Chem. Soc. 1982, 104, 4196.
28. Maurette, M.-T.; Oliveros, E.; Krieg, M.; Braun, A. M., to be published.
29. Lardet, D.; Laurent, E.; Thomalla, M. Nouv. J. Chim. 1982, 6, 349.
30. Bellus, D. Adv. Photochem. 1979, 11, 105.
31. Young, R. H.; Martin, R. L.; Feriozi, D.; Brewer, D.; Kayser, R. Photochem. Photobiol. 1973, 17, 233.
32. Ogryzlo, E. A.; Tang, C. W. J. Am. Chem. Soc. 1970, 92, 5034.
33. Gollnik, K.; Lindner, J. H. E. Tetrahedron Letters 1973, 1903.
34. Grob, C. A. Angew. Chem. 1969, 81, 543.
35. Iwaoka, T.; Kondo, M. Bull. Chem. Soc. Japan 1974, 47, 980.
36. Iwaoka, T.; Kondo, M. Chem. Lett. 1978, 731.
37. Hoffmann, H.; Tagesson, B.; Ulbricht, W. Ber. Bunsenges. Phys. Chem. 1979, 81, 148.
38. Humphry-Baker, R.; Braun, A. M.; Grätzel, M. Helv. 1981, 64, 2036.
39. Gassmann, E.; Braun, A. M., to be published.

RECEIVED January 10, 1985

7

Photosensitized Water Reduction Mediated by Semiconductors Dispersed in Membrane Mimetic Systems

YVES-M. TRICOT, RAFAEL RAFAELOFF, ÅSA EMEREN, and JANOS H. FENDLER

Department of Chemistry, Institute of Colloid and Surface Science, Clarkson University, Potsdam, NY 13676

Colloidal semiconductor particles were in situ generated and coated by catalysts in reversed micelles, surfactant vesicles and polymerized surfactant vesicles.

Sodium bis-2-ethylhexylsulfosuccinate (AOT) was used to form reversed micelle-entrapped water pools in isooctane. Platinized CdS, in situ generated in those water pools, sensitized water photoreduction by thiolphenol dissolved in the organic phase.

Colloidal 30-50 Å diameter CdS particles were also in situ generated in 800-1000 Å diameter single bilayer vesicles prepared from dihexadecylphosphate (DHP), dioctadecyl dimethylammonium chloride (DODAC) and from the polymerizable surfactant $[C_{15}H_{31}CO_2(CH_2)_2]_2N^+(CH_3)CH_2C_6H_4CH{=}CH_2$, Cl^- (1). Band gap excitation of the vesicle-embedded CdS resulted in weak fluorescence which could be quenched by electron donors and acceptors. Electron transfer was also examined by laser flash photolysis. Visible-light excitation of in situ rhodium-coated DHP-, DODAC- and 1-entrapped CdS particles led to sustained hydrogen production at the expense of sacrificial electron donors. Generation of CdS in polymerized 1 offered a number of advantages. Optimization of these systems and their potential in artificial photosynthesis are described.

The goal of artificial photosynthesis is to lay the foundation for efficient and economically viable conversion of sunlight to chemical energy. Considerable advances towards this goal have been made by different laboratories around the world (1-4). One general approach has involved the separate studies of sacrificial reduction and oxidation half cells. Hydrogen has been generated in reduction half cells, following light harvesting by a sensitizer (S), electron transfer through relays (R), and catalytic water reduction at the expense of a sacrificial electron donor (D).

Sacrificial photosensitized water reduction has been extensively investigated both in homogeneous solutions and in the presence of organized assemblies (micelles, microemulsions, polyelectrolytes, polymers, vesicles and

0097–6156/85/0278–0099$06.00/0

even clay particles) (5,6). Organized assemblies mimic the functions of the thylakoid membrane. They provide compartments of potentially controllable microenvironments for the sensitizers, relays, electron donors and catalysts and allow vectorial charge separations. Typically, ruthenium complexes and porphyrins have been used as sensitizers, viologens as relays (Figure 1A), ascorbic acid, thiols and EDTA as electron donors, and colloidal platinum and palladium as catalysts.

Although this approach continues to be used and improved upon, replacement of the sensitizer and the relay by dispersed colloidal semiconductors has gained widespread popularity (Figure 1B) (4). Colloidal semiconductors, like their more robust solid-state analogs, are characterized by their band gap, the energy difference between the filled valence band (VB) and the vacant conduction band (CB). Photoexcitation of the semiconductor (SC) at a wavelength corresponding to the band gap results in the promotion of an electron from the valence to the conduction band, and hence in charge separation:

$$SC \xrightarrow{h\nu} e^- + h^+ \quad (1)$$

Electron (e^-) and hole (h^+) recombinations, in the absence of quenching, result in radiationless transition (i.e., emission of heat) and, in some cases, in fluorescence (i.e., emission of light):

$$e^- + h^+ \rightarrow \Delta \quad ; \quad e^- + h^+ \rightarrow h\nu \quad (2)$$

Charge carriers escaping to the surface of the semiconductors can transfer an electron to an acceptor (A) or accept one from a donor (D):

$$e^- + A \longrightarrow A^- \quad (3)$$

$$h^+ + D \longrightarrow D^+ \quad (4)$$

at the semiconductor interface. Advantage has been taken of these interfacial electron transfers for mediating catalytic photosensitized water reduction (Figure 1B). Colloidal dispersed TiO_2, CdS, CdSe, Fe_2O_3, $SrTiO_3$ and their mixtures have been used as semiconductors either in the absence of in the presence of sensitizers. Sacrificial electron donors used have included EDTA, thiols and alcohols and Pt, Pd and Rh have been used as catalysts.

There are a number of advantages of using colloidal semiconductors in artificial photosynthesis. They are relatively inexpensive. They have broad absorption spectra and high extinction coefficients at appropriate band gap energies. Nevertheless, they can be made optically transparent enough to allow direct flash photolytic investigations of electron transfers. They can be modified by derivatization or sensitizer adsorption. Importantly, electrons produced by band gap excitation can be used directly without relays for catalytic water reduction (Figure 1B).

Unfortunately, colloidal semiconductors also suffer from a number of disadvantages. They are notoriously difficult to form reproducibly as small (smaller than 200 Å in diameter) monodispersed particles. Such small particles are needed for obtaining high surface areas and achieving efficient electron transfer to catalysts and/or other molecules adsorbed at the semiconductor interfaces. However, a minimal size is necessary to obtain semiconductor properties and spatial separation of the electron from the hole. A size range of 40 - 60 Å has been suggested to be ideal for the use of colloidal semiconductors in photochemical solar energy conversion (7). Semiconductors are equally difficult to maintain in solution for extended times in the absence of stabilizers which bound to affect their photoelectrical behavior unpredictably. Their modification, and their coating by catalysts are, at present, more of an art than science. Furthermore, the lifetime of electron-hole pairs in semiconductors is orders of magnitude shorter than the excited state lifetime of typical organic sensitizers. This is due to a very fast undesirable electron-hole recombination. Quantum yields for charge separations in colloidal semiconductors are, therefore, disappointingly low.

Incorporation of colloidal semiconductors into polyurethane films (8) and Nafion membranes (9) were reported to have overcome some of these disadvantages. Research in our laboratories is focussed upon the utilization of membrane mimetic systems for realizing the full potential of colloidal semiconductors. We have developed methologies for the in situ formation of small (ca. 40 Å diameter) and uniform catalyst-coated colloidal semiconductors in reversed micelles (10), surfactant vesicles (11) and polymerized surfactant vesicles (12). Utilization of these surfactant aggregate-stabilized semiconductor particles in photosensitized charge separation and hydrogen generation is the subject of the present report.

Semiconductors in Reversed Micelles

Colloidal CdS was in situ generated in reversed micelles formed by sodium bis-2-ethylhexyl sulfosuccinate (AOT). Aqueous $CdCl_2$ or $Cd(NO_3)_2$ was added to isooctane solutions of the surfactant to obtain a water to AOT ratio of 20 (10). Exposure to controlled amounts of gaseous H_2S resulted in CdS formation. Excess H_2S was removed by Argon bubbling. The onset of absorption of this optically clear CdS colloidal dispersion was approximately 500 nm (Figure 2), which is slightly blue-shifted compared to the band gap of 520 nm (2.40 eV) of bulk crystalline CdS (4). Such shifts in band gap have been shown to occur in CdS particles smaller than 100 Å in diameter (13-15). At 400 nm, the extinction coefficient was 900 ± 100 M^{-1} cm^{-1}, based on Cd^{2+} atomic absorption determinations. Dynamic light-scattering measurement prior and subsequent to H_2S exposure revealed only a slight increase in the apparent radius of the water pool from 60 to 75 Å.

AOT reversed micelle-entrapped, colloidal TiO_2 particles were prepared by mixing anhydrous isooctane solutions of Ti-tetraisopropoxide ($Ti(OC_3H_7)_4$) with AOT/H_2O dispersions in isooctane having a ratio of 10 H_2O per AOT molecule. Fast hydrolysis resulted in TiO_2 formation. Particular care had to be taken to avoid uncontrolled growth of the TiO_2 particles. Due to the dynamic exchange between aggregates, initially dispersed TiO_2 could form bigger particles by encountering other small particles, until they reached a point where they had a very low probability of colliding with each other. In 0.1 M AOT solution in isooctane and 1.0 M H_2O, 2 x 10^{-4} M colloidal TiO_2 remained stable and optically clear for several days. At higher TiO_2 concentrations, the turbidity increased with time and TiO_2 particles eventually precipitated.

Reversed micelle-entrapped, colloidal CdS showed the characteristic weak fluorescence emission (Figure 2), previously observed in homogeneous solutions (16-19). However, the maximum emission intensity corresponded to full band gap emission (approximately 500 nm) and was not red-shifted, as observed in homogeneous solution (17). This discrepancy might arise from the mode of preparation (H_2S instead of Na_2S), or from the specific effect of surfactant aggregates. Alternatively, this can be the result of a size

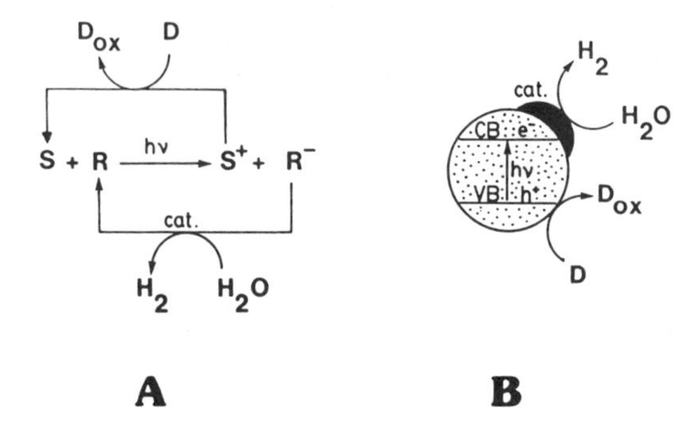

Figure 1. Basic features of sacrificial water reduction systems. (A) Homogeneous solution, with sensitizer, S, electron relay, R, sacrificial electron donor, D, and metal catalyst. (B) Catalyst-coated colloidal semiconductor dispersion, obviating the need for electron relay.

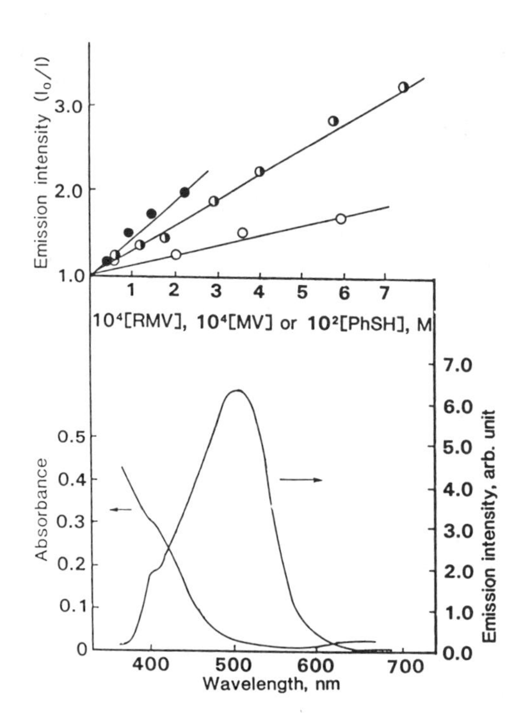

Figure 2. (Top) Stern-Volmer plots for the quenching of the fluorescence of colloidal CdS in AOT-entrapped water pools in isooctane by RMV^{2+} (●), MV^{2+} (◐), and PhSH (○). (Bottom) Absorption and emission spectra of colloidal CdS in AOT-entrapped water pools in isooctane. The shoulder observed at 400 nm is due to a spectrometer artifact.

distribution of the colloidal CdS particles, which could produce different band gap energies and hence, different fluorescence spectra (13-15). Replacing $CdCl_2$ by $Cd(NO_3)_2$ did not change the fluorescence characteristics of colloidal CdS. Fluorescence intensities of these solutions decayed biexponentially with lifetimes of 2.4 and 28.5 nanoseconds. Addition of 2.3 x 10^{-4} M methylviologen, MV^{2+}, decreased the fluorescence lifetime of the long-lived component to 18.7 nanoseconds. Similarly, addition of MV^{2+} and a surface-active viologen, $CH_2=C(CH_3)COO(CH_2)_{11}(C_6H_4N^+)_2CH_3$, Br^-, I^- (RMV^{2+}), as well as PhSH, quenched the emission intensity (Figure 2). Apparent Stern-Volmer constants for quenching the fluorescence of colloidal CdS by MV^{2+}, RMV^{2+} and PhSH are 2.6 x 10^3 M^{-1}, 4.6 x 10^3 M^{-1}, and 11.6 M^{-1}, respectively. The low quenching constant of PhSH arises from local concentration effects. Polar quenchers, such as MV^{2+} and RMV^{2+}, are locally concentrated in the water pool / AOT interface, whereas PhSH is homogeneously dissolved in isooctane. Due to the very fast electron-hole recombination, the quenching mechanism must be, at least in part, of static nature and dependant upon the absorbed concentration of quencher at the CdS particle surface. The Stern-Volmer appearance of the quenching data in Figure 2 may be, therefore, coincidental. In fact, a different behavior is observed in surfactant vesicles (vide infra).

To be active in H_2-production, a catalyst had to be incorporated in the system and deposited on the semiconductor surface. Platinization was carried out by adding aqueous K_2PtCl_4 solutions to the reversed micelle entrapped, colloidal CdS and irradiating it by a 450 W Xenon lamp under Ar bubbling for 30 minutes. Platinization was monitored absorption spectrophotometrically.

Irradiation of degassed, reversed micelle-entrapped, platinized CdS by visible light (450 W Xenon lamp, λ < 350 nm) resulted in hydrogen formation upon addition of 1.0 x 10^{-3} M PhSH, and it could be sustained for 12 h. This was the consequence of electron transfer from PhSH to the positive holes in the colloidal CdS, which diminished electron-hole recombinations (Figure 3).

The reversed micelles had the specific advantage of providing a means for charge separation by continuously removing the product (PhSSPh) from the semiconductor, located in the water pool, to the organic solvent. However, due to water pool exchanges, the semiconductor particles could interact with each other, which limited the concentration of semiconductor tolerable by the system. A better insulation of inner compartment was provided by aqueous surfactant vesicles.

Semiconductors in unpolymerized and polymerized surfactant vesicles

CdS particles were in situ generated from Cd^{2+} and H_2S in vesicles prepared from negatively charged dihexadecylphosphate, DHP, positively charged dioctadecyldimethylammonium chloride, DODAC, and positively charged polymerizable $[C_{15}H_{31}CO_2(CH_2)_2]N^+(CH_3)CH_2C_6H_4CH=CH_2$, Cl^-, 1. Sizes and sites of CdS particles were controlled by the amount of Cd^{2+} adsorbed and by the method of preparation. Positively charged Cd^{2+} ions were readily attracted, of course, to the negative surfaces of anionic DHP vesicles. Adsorption of Cd^{2+} ions to cationic vesicles (DODAC and 1) was achieved by complexation with EDTA. In anionic DHP vesicles, up to one Cd^{2+} ion per three surfactant molecules could be adsorbed. For cationic vesicles, up to one Cd/EDTA complex per four surfactant molecules could be adsorbed, the difference arising presumably from steric hindrance.

CdS particles were generated selectively inside, or outside, or on both sides of anionic DHP vesicles. After nucleation, CdS particles grew by encounter with neighboring Cd^{2+} ions or with other CdS particles. In most experiments, a ratio of one Cd^{2+} ion per ten surfactant molecules was maintained. Electron micrographs showed CdS particles to be in the 30 - 50 Å size range (11a).

Band gap excitation (λ < 500 nm) of colloidal CdS in anionic DHP vesicles resulted in characteristic weak fluorescence, due to electron-hole recombination. The quantum yield of fluorescence was estimated to be ca. 10^{-4}. Excitation at 330 nm gave fluorescence maximum at approximately 500 nm. Most surprisingly, detectable fluorescence was produced only from colloidal CdS which was generated inside of DHP vesicles (prepared by sonicating DHP with Cd^{2+} and by removing Cd^{2+} from the outer vesicle surfaces by ion exchange chromatography prior to forming CdS) and also when the sonication pH was higher than 6. No fluorescence was observed on CdS generated from Cd^{2+} added to already formed vesicles, regardless of the pH and of the added Cd^{2+} concentration. It is believed that colloidal CdS on the outside of vesicles can "age" either by aggregation with other colloidal CdS particles during vesicle-vesicle collision, or by dissolution/recrystallization from Cd^{2+} and S^{2-} ions in the outside aqueous solution (13). This aging could result in a decrease of the fluorescence quantum yield, although the mechanism of this effect is not clear. Colloidal CdS particles, when formed inside of the vesicles, are isolated from the outside bulk and hence are protected against these aging processes. Their fluorescence was very reproducible and independent of the anions (which were excluded from the vesicles). The quenching of CdS fluorescence by methylviologen, MV^{2+}, a membrane-impermeable electron acceptor, is shown in Figure 4. The highest quenching efficiency was obtained when both MV^{2+} and CdS were located inside of the vesicles. Lower quenching occurred when MV^{2+} and CdS were symmetrically distributed on both sides of the anionic vesicles, and no quenching was observed when MV^{2+} was externally added, regardless of the location of CdS. This confirmed that only "inside" CdS fluoresced and, that MV^{2+} did not diffuse through the DHP membrane. Observation of a time-dependent fluorescence quenching indicated a gradual relocation of CdS particles deeper inside the hydrocarbon part of the membrane.

Fluorescence quenching by MV^{2+} is the consequence of electron transfer from the conduction band of excited CdS to MV^{2+}, with formation of radical cation $MV^{+\bullet}$ (Figure 5). Due to the very short (< 1 ns) lifetime of the electron - hole pair in CdS, only adsorbed MV^{2+} on the CdS surface was able to capture an electron. This is reflected in the non Stern-Volmer behavior of the quenching of colloidal CdS by co-entrapped MV^{2+}. In other configurations, the quenching appears to be Stern-Volmer, but it is again misleading since not all the MV^{2+} participates in the quenching process. Direct evidence of this electron transfer was substantiated in anionic vesicles by laser flash photolysis (Figure 5).

CdS fluorescence was also observed in cationic DODAC and 1 vesicles, albeit with much less intensity than that found in DHP vesicles. The preparation pH influenced the CdS fluorescence in all vesicles. Interestingly, CdS fluorescence was somewhat stronger in unpolymerized than polymerized vesicles prepared from 1. This was not a consequence of the uv irradiation applied to polymerize the vesicles, since such irradiation had no effect on CdS fluorescence in nonpolymerizable surfactant vesicles (such as DHP or DODAC). However, this effect correlated with the ability of EDTA to act as a sacrificial electron donor in hydrogen production experiments (vide infra).

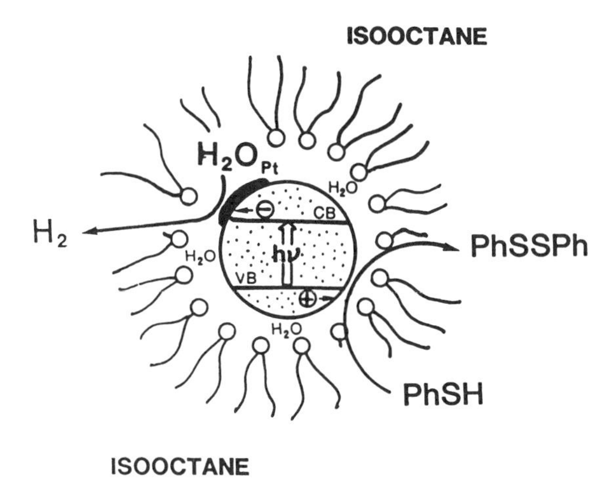

Figure 3. An idealized model for the CdS-sensitized water photoreduction by PhSH in AOT reversed micelles in isooctane. VB = valence band, CB = conduction band.

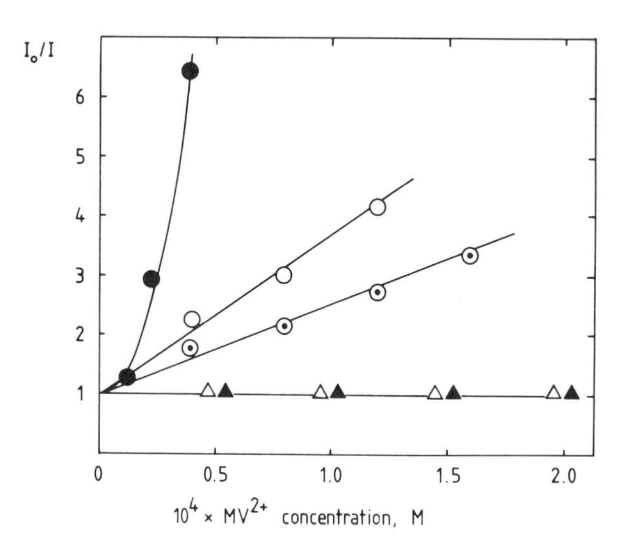

Figure 4. Quenching of CdS (2×10^{-4} M) fluorescence in DHP (2×10^{-3} M) vesicles by MV^{2+} in different locations: (●) : CdS and MV^{2+} only inside; (○) : CdS and MV^{2+} are on both sides; 30 min. after CdS formation; (⊙) : same samples, 8 days after CdS formation; (△) : CdS inside only, MV^{2+} outside only; (▲) : CdS on both sides, MV^{2+} outside only. λ_{ex} = 330 nm, λ_{em} = 500 nm.

In the presence of a suitable electron donor, catalyst-coated, vesicle stabilized, in situ-generated colloidal CdS particles produced hydrogen in deaerated solutions under visible-light irradiation (λ < 350 nm). In anionic DHP vesicles, rhodium was used as catalyst, whose precursor, Rh^{3+}, was co-adsorbed with Cd^{2+} on the surface of the vesicles. Reduction of Rh^{3+} to Rh° could be achieved either by uv irradiation in deaerated solutions prior to addition of the electron donor, or by visible-light irradiation in the presence of the electron donor. A slight induction period was observed in the latter case, but after 2 h of irradiation the same amount of hydrogen had been produced, and the rate was the same. Figure 6 shows the hydrogen production rate from PhSH as sacrificial electron donor in DHP vesicle-stabilized, rhodium-coated, colloidal CdS. Various catalyst concentrations were used, and the maximum efficiency was obtained with 8 x 10^{-5} M Rh^{3+}, with 2 x 10^{-4} M CdS generated on both sides of 2 x 10^{-3} M DHP vesicles. Several blanks demonstrated that the presence of all components of the system were necessary to observe hydrogen production. Hydrogen production could be sustained for about 15 h, until at least 90% of the PhSH had been consumed. Room temperature irradiation (25°C) was found to be optimal. The site of CdS generation (formed from Cd^{2+} on the inside or at the outside surface) was not very critical for hydrogen production. However, lower rates of hydrogen production with poorer reproducibility were usually obtained on irradiating CdS particles generated from Cd^{2+} adsorbed only on the outside surfaces of DHP vesicles. A reproducibility better than 10% was obtained in systems prepared by cosonicating Cd^{2+}, Rh^{3+} and DHP. A perfectly homogeneous mixing of Cd^{2+} and Rh^{3+} on the surfaces of the vesicles was easily obtained when those ions were cosonicated with the surfactant. Addition of EDTA to cationic vesicles containing already (non-adsorbed) Cd^{2+} and Rh^{3+} resulted in the formation of Rh-coated CdS exclusively on the outer surface. The hydrogen production efficiency in this system was the same as in systems prepared by cosonicating Cd^{2+}, Rh^{3+}, EDTA, and the cationic surfactants (DODAC or 1).

Hydrogen production from PhSH in colloidal, rhodium-coated CdS located selectively in the inside of DHP vesicles, was reproducible and had about the same efficiency that symmetrically distributed CdS particles, at equivalent CdS concentration. It seems, therefore, that homogeneously distributed CdS particles contributed to hydrogen production even if they did not exhibit a detectable fluorescence.

Figure 7 illustrates the mechanism of photosensitized H_2 formation from PhSH in rhodium-coated colloidal CdS in nonpolymerized cationic vesicles. The proposed position of the CdS particle (partially buried in the vesicle bilayer) is supported by the following observations, the last two made specifically in DHP vesicles:

(a) CdS particles generated from externally adsorbed Cd^{2+} ions did not precipitate, even after months; therefore, they had to remain bound to the vesicle interface.

(b) CdS fluorescence was efficiently quenched by PhSH, which was located in the hydrophobic membrane; therefore, the colloidal CdS particles had a direct contact with the inner part of the membrane.

(c) The CdS particle retained access to the surface where it originated, since entrapped polar electron acceptors such as MV^{2+}, while unable to penetrate the DHP membrane, could also quench the fluorescence of inside-generated CdS particles.

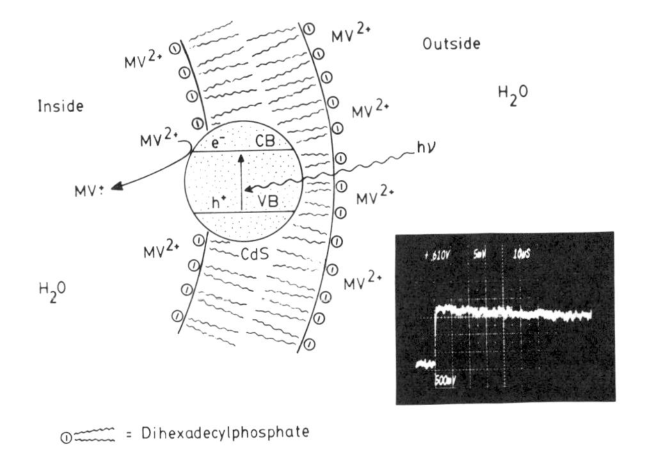

Figure 5. An idealized mechanism of photoinduced electron transfer from CdS conduction band to methylviologen (MV^{2+}), resulting in formation of methylviologen radical cation ($MV^{\cdot+}$). The colloidal CdS particle, as represented, was generated at the inside surface of the DHP vesicle. Its exact location is based on fluorescence quenching experiments (Figure 5). Insert: oscilloscope trace showing the formation of $MV^{\cdot+}$ by the absorbance change at 396 nm, after a laser pulse at 355 nm.

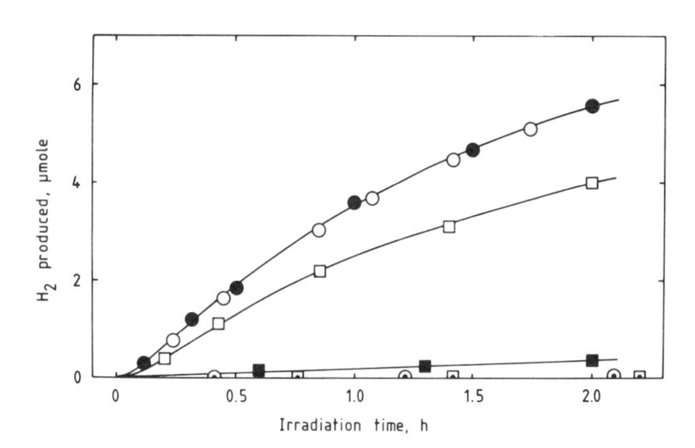

Figure 6. Hydrogen production at 25°C in deaerated solutions as a function of catalyst (rhodium) concentration during the first two hours of irradiation using 350 nm cut-off and water filters. Plotted are the amount of hydrogen produced in 25 ml DHP vesicle solution and measured in the gas phase (16 ml) by GC : 2×10^{-3} M DHP, 2×10^{-4} M CdS symmetrically distributed on both sides of the vesicles, and 10^{-3} M PhSH as electron donor; pH approximately 7 at sonication and during photolysis. Concentrations of the catalyst, reduced by uv irradiation prior to visible light photolysis: (●) : Rh 1.2×10^{-4} M; (○) : Rh 0.8×10^{-4} M; (□) : Rh 0.4×10^{-4} M; (■) no Rh; (⊡) no Rh and no CdS (only PhSH); (⊙) : Rh^{3+} (unreduced) and no CdS (with PhSH).

However, this quenching decreased with time, showing a gradual penetration of the CdS towards the middle of the bilayer.

(d) CdS particles generated at the vesicle interiors remained on their original side of the membrane, since externally added quencher such as MV^{2+} and Rh^{3+}, while adsorbed on the outside surface of DHP vesicles, did not quench "inside" CdS fluorescence, even after several weeks.

A different situation was encountered on using polymerized vesicles prepared from 1 as hosts for the catalyst-coated colloidal semiconductors. Polymerization of 1 vesicles was shown to result in pulling together some 20 styrene headgroups at the vesicle surfaces, thereby creating clefts of some 15 Å diameter (20). These relatively aqueous clefts were proposed to be the site of CdS particles (Figure 8). The observed behavior of EDTA in the photolysis of Rh-coated CdS particles embedded in 1 vesicles were in accord with this interpretation. In nonpolymerized 1 vesicles, EDTA was unable to act as a sacrificial electron donor; no photosensitized hydrogen formation was observed (11b). Conversely, in polymerized vesicles prepared from 1, efficient photosensitized hydrogen formation was observed from EDTA (12). This effect is rationalized in terms of improved access of EDTA to the CdS surface in polymerized vesicles, due to surface irregularities (Figure 8). The observed increase in fluorescence intensity of CdS in polymerized vesicles as compared to that in their nonpolymerized counterparts also supports this rationalization. Controlled degrees of vesicle surface polymerization provide, therefore, a means for improving the efficiencies of electron donors. It should be pointed out, however, that EDTA is not an ideal electron donor since in high concentrations it destabilized the vesicles.

In an attempt to overcome the problem of accumulation of the oxidized electron donor, we have incorporated a recyclable surface-active electron donor in DODAC vesicles (12). This electron donor contains a sulfide moiety which dimerizes upon light-induced oxidation. Simultaneously, hydrogen is evolved via vesicle-stabilized, catalyst-coated, colloidal CdS particles. The dimer could be chemically reduced for additional hydrogen formation. Figure 9 is an idealized view of this cyclic process (12).

Conclusion

Our initial efforts to develop the use of surfactant aggregate-stabilized, catalyst-coated, colloidal semiconductors in artificial photosynthesis are promising. In situ generation of the particles provides an unprecedented control over their morphologies. Colloidal semiconductors embedded in reversed micelles, vesicles or polymerized vesicles show remarkable long-term stabilities and diminish electron-hole recombination rates. These systems have the potential of decreasing photocorrosion and separating charges. Subsequent to optimization of the hydrogen generating systems, sacrificial nitrogen and carbon dioxide reductions can be accomplished.

Our ultimate goal is to combine oxidation and reduction half cells in a bilayer vesicle which would efficiently split water upon solar irradiation. Figure 10 represents a highly idealized system. It would require the immobilization of CdS (or other suitable semiconductors) particles in polymerized vesicles in such a way that they span the bilayer. Coating the opposite sides of the vesicle-embedded semiconductor by catalysts would result in cyclic water splitting. The hydrogen and oxygen formed could then be harvested in compartments separated by the bilayer (Figure 10).

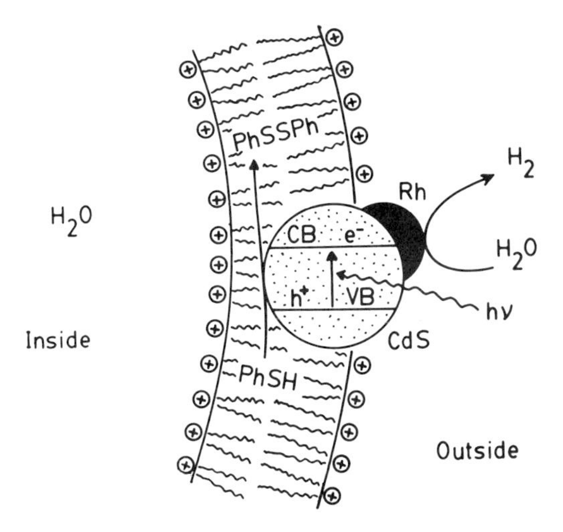

Figure 7. An idealized model for CdS sensitized photoreduction of water by PhSH in aqueous DODAC or DODAB vesicles. The exact position of the colloid represented here as generated on the outside surface of the vesicles, is based on fluorescence quenching experiments performed in anionic DHP vesicles, assuming similar interactions of the CdS particles with both types of vesicles.

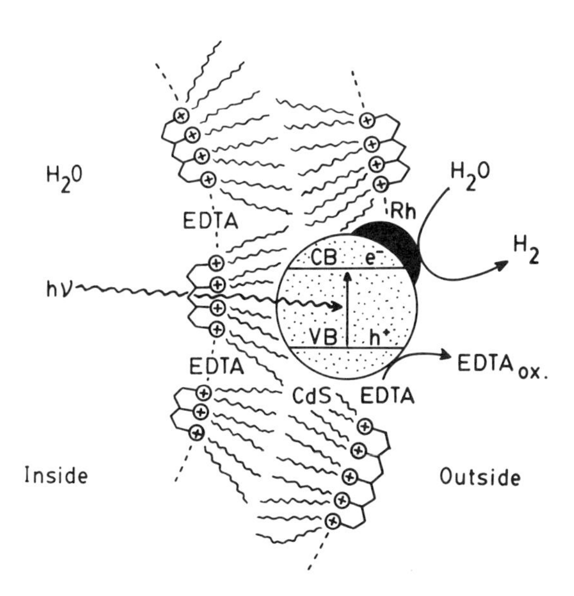

Figure 8. An idealized model for CdS sensitized photoreduction of water by EDTA in polymerized 1 vesicles. The location of the particle is represented as similar to that in unpolymerized vesicles, but should not be taken for granted in this case, however.

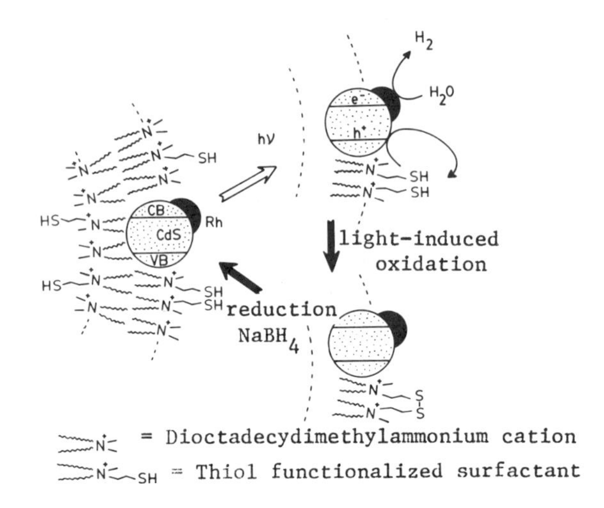

Figure 9. An idealized model for the cyclic oxidation-reduction process using thiol-functionalized surfactant, incorporated into DODAC vesicles together with Rh-coated colloidal CdS, for sustained hydrogen generation under visible light irradiation in one part of the cycle.

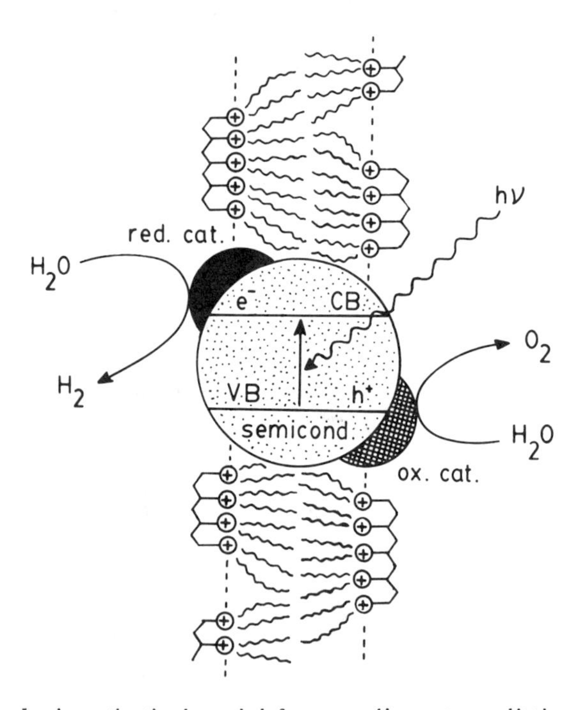

Figure 10. An hypothetical model for a cyclic water splitting system, based on a semiconductor particle immobilized in a polymerized membrane and having access to both aqueous solutions on each side of the membrane. Specific and selective coating by catalysts leads to simultaneous and separate hydrogen and oxygen generation on each side of the polymerized membrane.

Acknowledgment

Support of this work by the Department of Energy is gratefully acknowledged.

Literature Cited

1. Calvin, M. Acc. Chem. Res., 1978, 11, 369 - 374.
2. Porter, G. Proc. R. Soc. London Ser A., 1978, 362, 281 - 303.
3. Fendler, J. H. J. Phys. Chem., 1980, 84, 1485 - 1491.
4. Gratzel, M. "Energy Resources through Photochemistry and Catalysis", Academic Press, New York, 1983.
5. Fendler, J. H. "Membrane Mimetic Chemistry", Wiley-Interscience, New York, 1982.
6. Fendler, J. H. Chem. Eng. News, 1984, Jan. 2., 62, 25 - 38.
7. Duonhong, D.; Ramdsen, J.; Gratzel, M. J. Am. Chem. Soc., 1982, 104, 2977 - 2985.
8. Meissner, D.; Memming, R.; Kastening, B. Chem. Phys. Lett., 1983, 96, 34 - 37.
9. Krishnan, M.; White, J. R.; Fox, M. A.; Bard, A. J. J. Am. Chem. Soc., 1983, 105, 7002 - 7003.
10. Meyer, M.; Wallberg, C.; Kurihara, K.; Fendler, J. H. J. Chem. Soc. Chem. Commun., 1984, 90 - 91.
11. (a) Tricot, Y. M.; Fendler, J. H. J. Am. Chem. Soc., 1984, in press; (b) Rafaeloff, R.; Tricot, Y. M.; Nome, F.; Fendler, J. H. J. Phys. Chem., 1984, in press.
12. Rafaeloff, R.; Tricot, Y. M.; Emeren, Å; Nome, F.; Fendler, J. H., unpublished results, 1984.
13. Rossetti, R.; Ellison, J. L.; Gibson, J. M.; Brus, L. E. J. Chem. Phys., 1984, 80, 4464 - 4469.
14. Brus, L. E. J. Chem. Phys., 1984, 80, 4403 - 4409.
15. Rossetti, R.; Nakahara, S.; Brus, L. E. J. Chem. Phys., 1983, 79, 1086 - 1088.
16. Henglein, A. Ber. Bunsenges. Phys. Chem., 1982, 86, 301 - 305.
17. Ramsden, J. J.; Gratzel, M. J. Chem. Soc. Faraday Trans. I, 1984, 80, 919 - 933.
18. Rosetti, R.; Brus, L. J. Phys. Chem., 1982, 86, 4470 - 4472.
19. Kuczyuski, J. P.; Milosavljeciv, B. H.; Thomas, J. K. J. Phys. Chem., 1983, 87, 3368 - 3370.
20. Nome, F.; Reed, W.; Politi, M.; Tundo, P.; Fendler, J. H. J. Am. Chem. Soc., 1984, 106, 0000 - 0000.

RECEIVED January 10, 1985

8

Organic Photochemical Reactions in Monolayers and Monolayer Systems

DIETMAR MÖBIUS

Max-Planck-Institut für Biophysikalische Chemie (Karl-Friedrich-Bonhoeffer-Institut), Abt. Molekularer Systemaufbau, Am Faßberg, D-3400 Göttingen, Federal Republic of Germany

Organic photochemical reactions in monolayer organizates are strongly influenced by the restricted molecular mobility in these systems. Reactions at the air-water interface where molecular relaxation is possible, can be followed by measuring the enhanced light reflection in the spectral range of the absorption band of the involved species. In monolayer systems, photoinduced electron transfer processes have been studied by fluorescence techniques.

The organization of molecules at interfaces provides an excellent possibility of control of the chemical reactivity. In particular at the air-water interface, the most important parameters like intermolecular separation, geometrical coordination of the reacting molecules and the molecular mobility can be controlled or at least modified by variation of the external conditions like surface pressure, temperature, subphase and monolayer composition.

The complex monolayer organizates formed by stepwise transfer of monolayers from the air-water interface onto solid substrates (Langmuir-Blodgett technique) are characterized by a strongly reduced molecular mobility. However, by gaining the third dimension as compared to the monolayer at the air-water interface, more complex systems can be built where the interacting components are appropriately arranged to meet the spatial and energetic requirements of the intended function (1-3).

One prominent aspect of the investigation of photochemical reactions in organized media is the environmental control of the reaction kinetics and product distribution (4). The increased knowledge may be used to construct more efficient catalysts since the amount of reacting monolayer-bound material is very small on the absolute scale. Therefore, a high turn-over including control of substrate binding and product release could be a way of using the potential of the monolayer structure. Another important aspect is the use of photochemical reactions in producing rapid changes in surface pressure (π-jumps) as a tool of studying the dynamics of various phenomena involving structural changes in monolayers (5, 6). This is a 2-dimensional ana-

0097-6156/85/0278-0113$06.00/0

log of the various jump techniques used in studying equilibria or fast reactions in solution.

Monolayersat the Air-Water Interface

Many photochemical reactions like photoisomerization or photopolymerization lead to changes in the area requirements of the molecules at the air-water interface. Even the topochemical polymerization of diacetylene compounds in the organic solid state where a well defined coordination of the molecules in necessary (7), when carried out with adequately substituted analogs in monolayers and monolayer assemblies, may lead to domain formation (8, 9) due to small strain imposed on the monolayer by minute changes of the molecular structure. Different to the situation in rigid, densely packed monolayers on solids, relaxation may occur at the air-water interface if such reactions are carried out under constant surface pressure and particular ambient conditions (10). This additional possibility of molecular reorganization will reduce stress in the systems and should lead to larger homogeneous arrays as compared to systems on solids.

Photoisomerization Reactions. Reversible photoisomerization reactions have been studied in monolayers by following the surface pressure change at constant area or the change in area per molecule at constant surface pressure. The photoisomerization may also lead to a change of the molecular charge distribution and/or reorientation of the molecular dipoles that may be attributed to the molecules at the air-water interface thereby causing a change of the interfacial potential. This parameter has been used to study the kinetics of the dark reaction following the exposure of a monolayer of the amphiphilic anthocyanidine AC at the air-water interface (11). In Figure 1, the simultaneous changes in surface pressure π (top) and surface potential Δv (bottom) are shown which are observed on repeated 1s exposures of a monolayer of the anthocyanidine AC with blue light at a surface pressure of 10 mN/m.

The action spectrum of the photochemically induced surface potential change has been evaluated. In combination with the observed dependence of the rate of the dark reaction on the bulk pH of the aqueous subphase a reaction scheme was established in analogy to the processes observed with the anthocyanidine analog without the octadecyl chain in solution.

The reactions observed with the anthocyanidine system were reversible. The relaxation processes in monolayers following a chemical change, however, may lead to a rearrangement of the environment which prevents the back reaction. An example for this behavior is the photochemical cis→trans isomerization of the thioindigo derivative TI at the air-water interface where the trans→cis back reaction is not observed, although the cis isomer is photochemically formed from the trans in chloroform solution before spreading (4, 12). The cis→trans isomerization has been studied in mixed monolayers with arachidic acid (AA), molar ratio TI:AA = 1:3, at constant surface pressure (12) by measuring the enhanced light reflection from the air-water interface (13). The change of the light reflection measured at 550 nm at an angle of incidence of 45° with linearly polarized light (s-polarization) upon illumination of the monolayer in the spectral range of the cis TI absorption (450-500 nm) is shown in Figure 2.

Although the photoreaction at constant surface pressure has been carried out in a small part of the total monolayer area, there is no mixing of the formed trans TI with the surrounding cis TI due to convection or diffusion during the experiment as evidenced by the constant signal after the end of exposure.

The rate constant for the cis→trans photoisomerization can be calculated from the observed time dependence. The rate constants obtained at various monolayer surface pressures under identical conditions of illumination are plotted vs surface pressure in Figure 3.

The rate constant for the photoisomerization drops with increasing surface pressure, levelling off between 20 and 30 mN/m. This dependence is attributed to the decrease in molecular mobility due to tighter packing of the monolayer head groups and stronger interactions between the long hydrocarbon chains when the surface pressure is increased. The spectroscopic method for the evaluation of the rate constant used here is more direct than using the change in surface pressure at constant area or the change in area at constant surface pressure which require a different surface behavior of the isomers. The method of monolayer enhanced light reflection from the air-water interface provides a convenient tool to study the environmental effects that modify interfacial reactions.

Photochemical Generation of an Interfacial Shock Wave. Both the anthocyanidine and the thioindigo monolayers showed a decrease in surface pressure at constant area during the photoisomerization reaction. A different behavior is observed with mixed monolayers of the surface active spiropyran SP and octadecanol (OD), molar ratio SP:OD = 1:5, on illumination with UV radiation. The isomerization of the spiropyran to the merocyanine MC causes an increase in surface pressure at constant area (5, 14). This is shown in Figure 4, where the sudden rise in surface pressure π upon repeated 0.5 s exposures (as indicated by the arrows) can be seen to occur in a wide surface pressure range (15). The kinetics of the relaxation process following the surface pressure increase depends on the surface pressure.

The fast isomerization of the spiropyran to the merocyanine provides a possibility of generating an interfacial shock wave. The methods used so far in studying the transmission of waves in monolayers and the adjacent bulk phases require mechanical (16) or electrocapillary (17) excitation of the interface which involves the displacement of the aqueous bulk phase. In addition, the range of frequencies accessible to the investigation of interfacial waves by the conventional techniques is very limited. The fast photochemical generation of an interfacial shock wave is strictly occurring in the monolayer and provides a larger spectrum of frequencies which can be fully explored only after the development of appropriate detection methods.

One way of detecting a surface shock wave is the observation of a phase transition in a detection monolayer in contact with the generation monolayer. In the case of a cyanine dye monolayer which undergoes a phase transition from monomeric to aggregate dye characterized by a strong and narrow absorption band shifted to longer waves with respect to the monomeric absorption band (13), the shock wave can be followed by monitoring the light reflection from the air-water

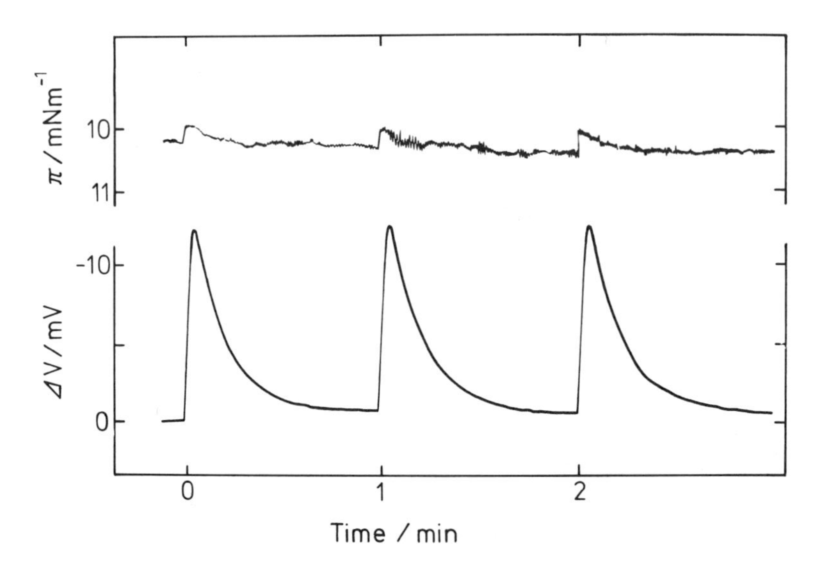

Figure 1. Simultaneous change at constant monolayer area in surface pressure π (top) and surface potential Δv (bottom) following 1s exposures of a monolayer of the anthocyanidine AC as indicated by the arrows. Subphase: 10^{-2} N H_2SO_4.

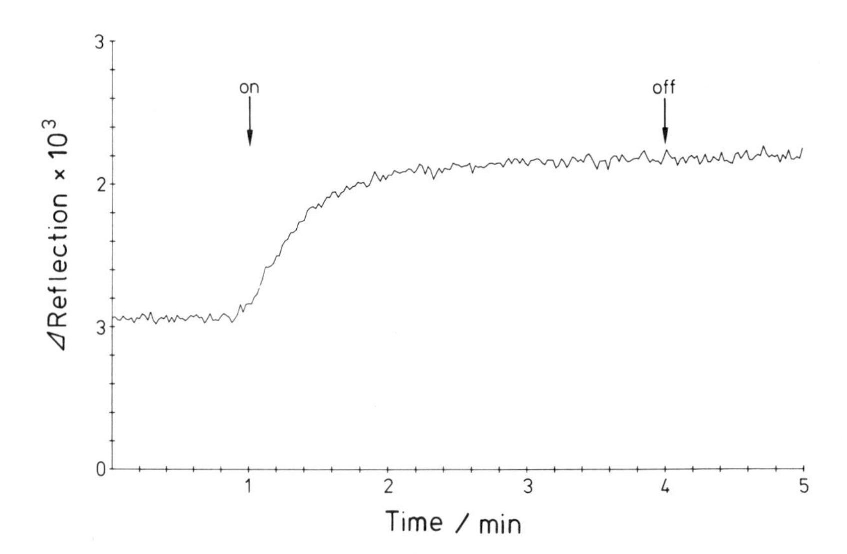

Figure 2. Time dependence of enhanced light reflection at 550 nm of a mixed monolayer of the thioindigo cis-TI and arachidic acid (AA), molar ratio TI:AA = 1:3, on illumination with blue light as indicated by the arrows. Subphase: bidistilled water; surface pressure: 10 mN/m.

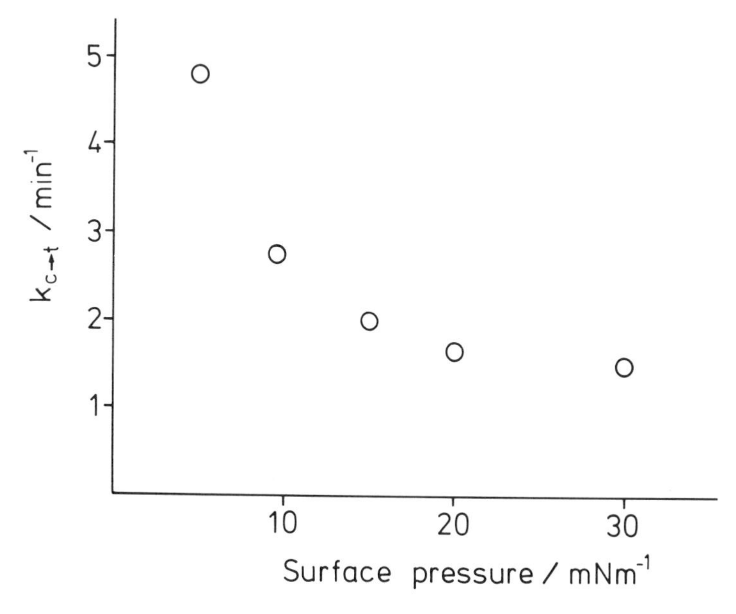

Figure 3. Rate constants of cis→trans photoisomerization obtained from enhanced reflection measurements vs. surface pressure of mixed monolayers of the thioindigo TI and arachidic acid (AA), molar ratio TI:AA = 1:3. Subphase: bidistilled water, 22° C.

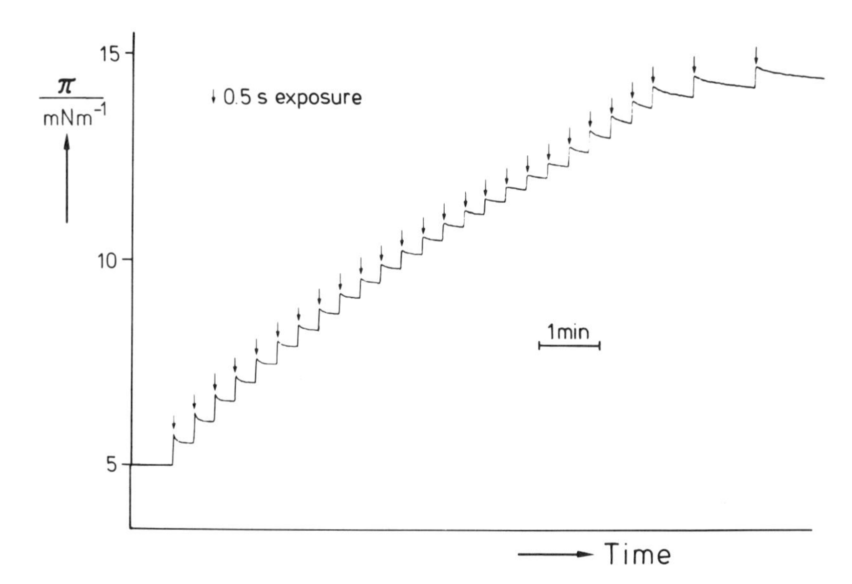

Figure 4. Fast surface pressure jumps after 0.5 s exposures with UV radiation of a mixed monolayer of the spiropyran SP and octadecanol (OD), molar ratio SP:OD = 1:5, at constant monolayer area. Subphase: bidistilled water, pH = 6.2.

interface in the range of the aggregate band (5). When the phase transition is fast compared to the rise time of the shock wave the time dependence of the reflection signal characterizes the shape of the shock wave and yields information on the damping of the various frequencies forming the shock wave initially. The photochemical system therefore provides a unique tool to investigate the interfacial propagation of shock waves and the dynamics of structural changes in the interfacial region.

Photopolymerization. The photopolymerization of appropriate diacetylene compounds with UV radiation at the air-water interface should be possible without inducing large defects since this type of reactive compounds is known to undergo topochemical transformation from monomer to polymer in the organic solid state. The behavior of various long chain carboxylic acids containing the diacetylene group has been studied at the air-water interface (8, 10) and in assemblies on glass or quartz plates built up from spread monolayers (9). Even in these systems, the formation of domains has been observed, and one mayor task has been to find the appropriate conditions for formation of particularly large homogeneous areas.

The photopolymerization has been followed by measuring the light transmission of the monolayer at the air-water interface in a double pass instrument involving light reflection at a mirror underneath (18). The particular spectroscopic feature of the polymerizing monolayers are a "red band" with maximum at 530 nm and a band at 650 nm ("blue band") which disappears when the reaction is completed. The alternative technique of measuring the enhanced light reflection has been used to study the photopolymerization of monolayers of the diacetylenecarboxylic acid DA (19). The reflection spectra obtained from the air-water interface after increasing exposure to UV radiation are shown in Figure 5. The DA monolayer has no absorption band in the visible range before exposure (curve 1), and the band with maximum 530 nm ("red band") grows in during the reaction (curves 2 to 6).

The measurement of enhanced light reflection provides direct information concerning intermediates, products and the kinetics of the photopolymerization at the air-water interface.

Monolayer Assemblies

Monolayer assemblies differ from the simple Langmuir-Blodgett multilayers in the details of the structure. Instead of depositing monolayers of one type only, the monolayers at the air-water interface are changed between the single steps, immersion and withdrawal of the solid substrate, respectively. Various interacting molecular species can be arranged in the desired geometry by using the well-known monolayers of fatty acids, esters, alcohols or amines as matrices for the interesting molecules. In order to ensure the geometry, the molecular mobility must be strongly reduced. This rigid environment also prevents large structural changes in the course of chemical reactions. Therefore, those reactions are particularly suited for monolayer systems that do not require extended motion of molecular parts.

Photoisomerization. Mixed monolayers of the thioindigo derivative TI discussed before and arachidic acid (AA), molar ratio TI:AA = 1:3, can be transferred to glass plates. When the cis TI is transferred, the thioindigo can be isomerized to the trans form by irradiation with light of the wavelength 480 nm. A series of absorption spectra of such a monolayer illustrating the spectral changes during the photoreaction is shown in Figure 6 (12). The absorption decreases between 460 nm and 495 nm and increases in the ranges from 370 nm to 460 nm and 495 nm to 570 nm. The increase in the integrated absorption is partly due to different orientations of the transition moments for the cis as compared to the trans TI. The observation of isosbestic points is evidence for a simple photochemical transformation of the cis to the trans TI. The reverse photoisomerization has not been observed in transferred monolayers. As can be seen from molecular models, the cis → trans isomerization is possible even when the long chains are rigidly fixed, since the rotating chromophore moiety rotates simultaneously around the benzene-oxygen bond and the central C-C bond in the excited state. From the point of view of steric conditions, it is difficult to understand why the trans → cis isomerization is not observed. A clarification of this question requires a more detailed investigation of the structure of the mixed monolayers.

Photoinduced Electron Transfer. Monolayer organizates are particularly suited for the investigation of photoinduced electron transfer, since the molecules are fixed and the distance between the planes at which the donor and the acceptor molecules, respectively, are located can be well defined. Therefore, complex monolayers have been arranged in order to study the distance dependence of electron transfer in these systems (2, 20). This strategy has also been used to elucidate the relative contributions of electron injection and energy transfer mechanisms in the spectral sensitization of silver bromide (21).

When the molecules of the electron donor D and the acceptor A are fixed at the same interface, the distance between D and A may vary around an average value due to the statistical distribution of the molecules in the mixed monolayers. Therefore, particular pairs of D and A are in a more favourable situation with respect to electron transfer than the average. If the excited state of the donor can reach a particular pair by exciton hopping, the primary excitation of an unfavourable donor molecule may end in the electron transfer act. This energy delocalization enhances the quantum yield of photoinduced electron transfer as evidenced by the increase of fluorescence quenching when the density of the donor molecules is enhanced at constant acceptor density (22). This effect has been found in systems of mixed monolayers of the cyanine dye CY as donor and mixed monolayers of the viologen derivative SV as acceptor, using a mixture of arachidic acid (AA) and methylarachidate (MA), molar ratio AA:MA = 9:1, as matrix. A typical result is shown in Figure 7 (bars), where the relative fluorescence intensity of the donor is plotted vs. the donor density at constant acceptor density.

The situation is different when donor and acceptor molecules are located at different interfaces, that are separated by a fatty acid monolayer of well defined thickness. There are no longer close pairs of donor and acceptor with a high probability of electron transfer as in the "contact" case. Consequently, no change in relative fluorescence intensity with increasing donor density is expected, contrary to the former case. Indeed, in systems with a spacer monolayer of

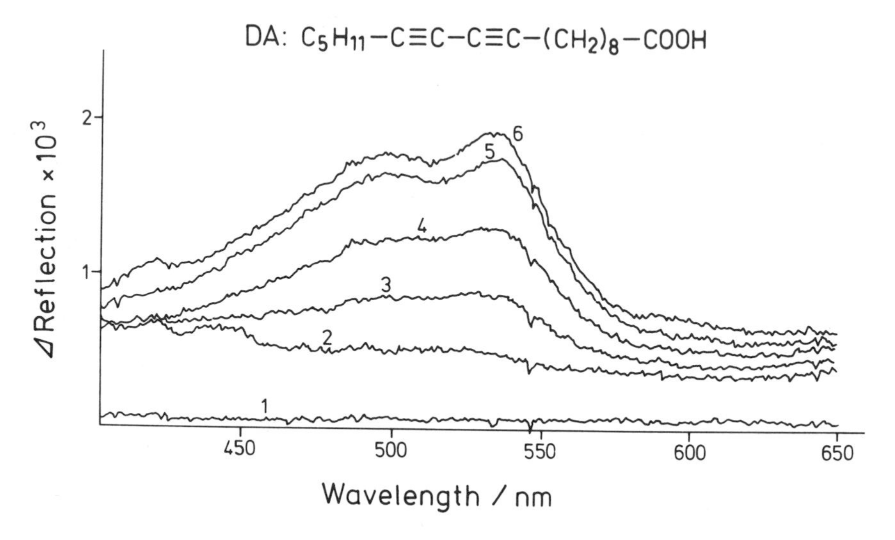

Figure 5. Enhanced reflection spectra of a monolayer of the diacetylenecarboxylic acid DA before (curve 1) and after increasing exposure to UV radiation: 5 min (2); 15 min (3); 30 min (4); 60 min (5); 90 min (6). Subphase: $3x10^{-4}$ M $CdCl_2$ and $5x10^{-5}$ M $NaHCO_3$; surface pressure: 10 mN/m.

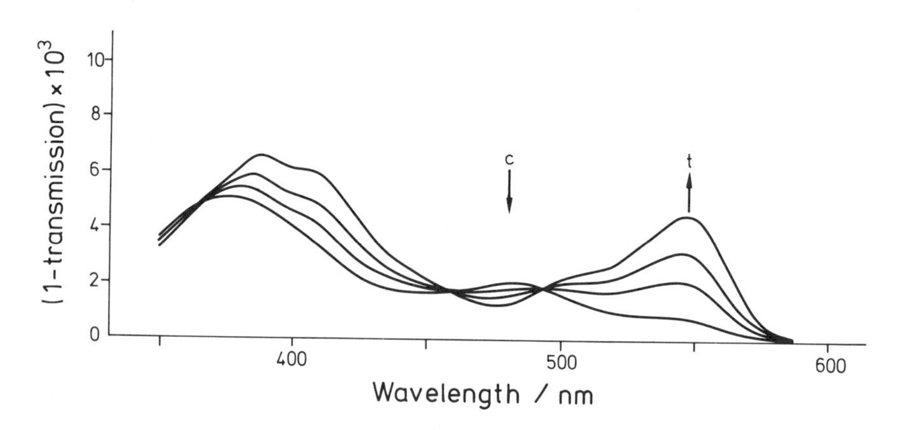

Figure 6. Absorption spectra of a mixed monolayer of the thioindigo cis TI and arachidic acid (AA), molar ratio TI:AA = 1:3, on a glass plate after various times of irradiation with 480 nm light. The absorption decreases in the range between 460 nm and 495 nm and increases due to formation of trans TI at 550 nm. The spectra were taken before and after 10 s, 19 s and 30 min of irradiation.

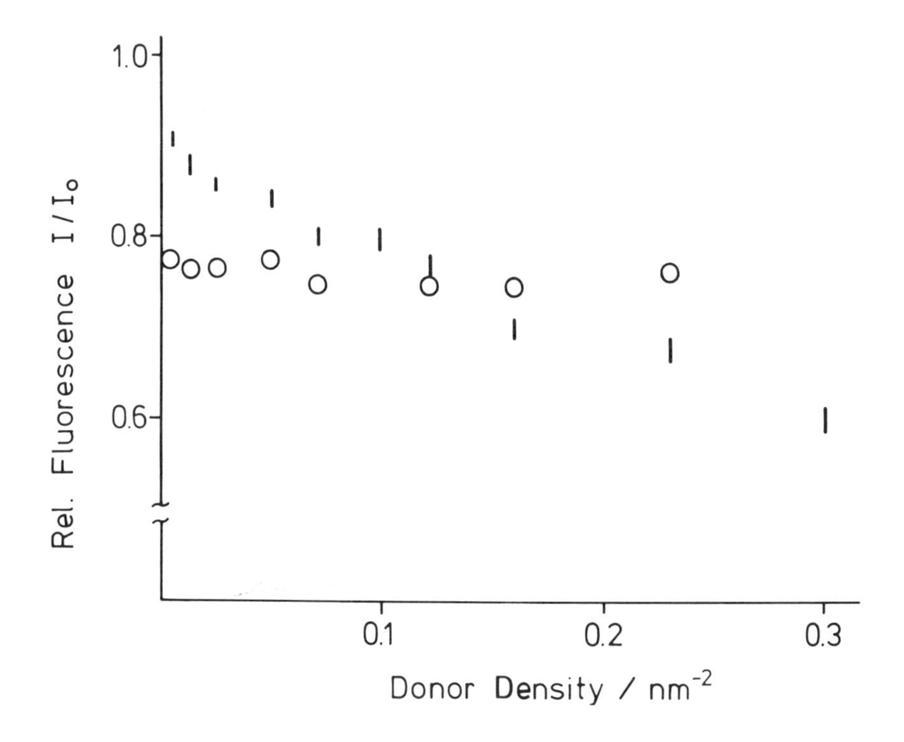

Figure 7. Photoinduced electron transfer in monolayer systems with the cyanine dye CY as donor and the viologen derivative SV as acceptor. Relative fluorescence intensity of the donor monolayer vs. donor density at constant acceptor density. Bars: Donor and acceptor at the same interface, density of A, $\sigma(A) = 0.01\ nm^{-2}$. Circles: donor and acceptor at different interfaces, distance 2.3 nm, $\sigma(A) = 0.43\ nm^{-2}$.

thickness 2.3 nm, this is found as shown in Figure 7, circles (23). This result, on the other hand, can be interpreted as evidence for the absence of close pairs, i. e. the absence of contacts between the donor and the acceptor monolayers. The fluorescence quenching, therefore, must be attributed to long distance electron transfer, presumably via electron tunneling across the insulating fatty acid monolayer.

Conclusions

Photochemical processes in monolayers at the air-water interface can be controlled externally by variation of the various parameters like matrix composition, subphase composition, temperature and surface pressure. When the product of the reactions has a different area per molecule, the surface pressure may change at constant monolayer area. An interfacial shock wave has been generated in this way. This technique permits the investigation of the kinetics of reorganization processes and the transmission of mechanical signals in monolayers.

Monolayer systems are characterized by a very limited molecular mobility and high degree of order. Photoinduced electron transfer processes have been investigated in these systems in order to evaluate the influence of energy delocalization on the quantum efficiency of the electron transfer step and the range of long distance electron transfer.

Structures :

AC : ClO_4^-

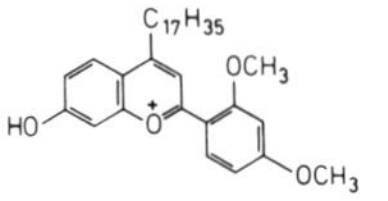

TI :

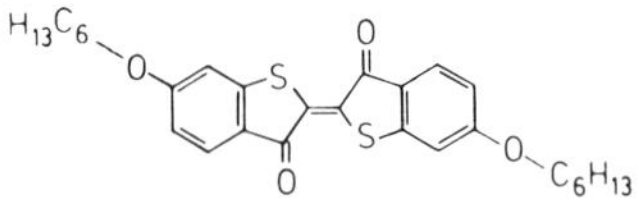

SP :

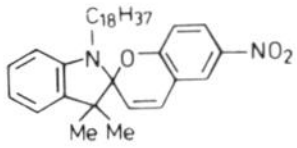

MC :

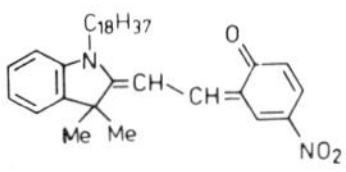

DA : $C_5H_{11}-C\equiv C-C\equiv C-(CH_2)_8-COOH$

CY : ClO_4^-

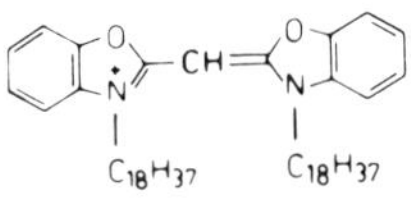

SV : $2\,ClO_4^-$

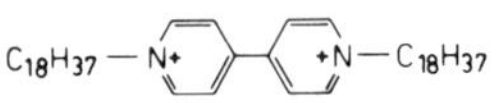

Literature Cited

1. Bücher, H.; Drexhage, K. H.; Fleck, M.; Kuhn, H.; Möbius, D.; Schäfer, F. P.; Sondermann, J.; Sperling, W.; Tillmann, P.; Wiegend, J. Mol. Cryst. 1967, 2, 199.
2. Möbius, D. Acc. Chem. Res. 1981, 14, 63.
3. Kuhn, H. Thin Solid Films 1983, 99, 1.
4. Whitten, D. G. Angew. Chem. Intern. Ed. Engl. 1979, 18, 440.
5. Möbius, D.; Grüniger, H. In "Charge and Field Effects in Biosystems"; Allen, M. J.; Usherwood, P. N. R., Eds.; Abacus Press, Tunbridge Wells 1984, p. 265.
6. Suzuki, M.; Möbius, D.; Ahuja, R. C. Thin Solid Films submitted.
7. Wegner, G. Z. Naturforsch. B 1969, 24, 824.
8. Day, D. R.; Ringsdorf, H. J. Polymer Sci., Polym. Lett. Ed. 1978, 16, 205.
9. Lieser, G.; Tieke, B.; Wegner, G. Thin Solid Films 1980, 68, 77.
10. Day, D. R.; Lando, J. B. Macromolecules 1980, 13, 1478.
11. Möbius, D.; Bücher, H.; Kuhn, H.; Sondermann, J. Ber. Bunsenges. Phys. Chem. 1969, 73, 845.
12. Whitten, D. G.; Möbius, D., unpublished results.
13. Grüniger, H.; Möbius, D.; Meyer, H. J. Chem. Phys. 1983, 79, 3701.
14. Polymeropoulos, E. E.; Möbius, D. Ber. Bunsenges. Phys. Chem. 1979, 83, 1215.
15. Möbius, D., unpublished results.
16. Lucassen, J.; van den Tempel, M. J. Colloid Interface Sci. 1972, 41, 491.
17. Sohl, C. H.; Miyano, K.; Ketterson, J. B. Rev. Sci. Instrum. 1978, 49, 1464.
18. Day, D. R.; Ringsdorf, H. Makromol. Chem. 1979, 180, 1059.
19. Möbius, D.; Milverton, D. R. J.; Veale, G., unpublished results.
20. Kuhn, H. Pure & Appl. Chem. 1981, 53, 2105.
21. Steiger, R.; Hediger, H.; Junod, P.; Kuhn, H.; Möbius, D. Photogr. Sci. Eng. 1980, 24, 185.
22. Möbius, D. Mol. Cryst. Liq. Cryst. 1983, 96, 319.
23. Möbius, D.; Debuch, G., unpublished results.

RECEIVED January 10, 1985

9

Photolabeling of Neurotransmitter Receptor Sites in the Brain

D. I. SCHUSTER[1], R. B. MURPHY[1], R. A. ASHTON[1,2], K. THERMOS[1], L. P. WENNOGLE[1], and L. R. MEYERSON[2]

[1]Department of Chemistry, New York University, New York, NY 10003
[2]Department of CNS Research, Medical Research Division of American Cyanamid Company, Lederle Laboratories, Pearl River, NY 10965

The principles of photoaffinity labeling of receptor binding sites of biologically active macromolecules are reviewed, in terms of the general problem of molecular characterization of neurotransmitter and drug receptor sites in mammalian brain. Selected examples from the literature are used to illustrate the design of compounds for the purpose of labeling of specific receptor sites located in cell membranes, and the pharmacological and biochemical methods of analysis utilized in such studies. Two examples of such studies from the authors' laboratories are described, one involving photolabeling of serotonin uptake sites in human platelets and rat brain by 2-nitroimipramine, and the other utilizing chlorpromazine to label dopaminergic sites in bovine and canine striatal tissue preparations. In both instances, specific protein fractions are labeled, according to gel electrophoretic analysis.

The study of biologically active macromolecules using photoexcited reagents was introduced into biochemistry by Westheimer and his co-workers in the early 1960's (1,2), and has since developed into one of the most important techniques for studying the interaction of ligands with active sites on the surface of macromolecules such as receptors and enzymes (3-6). The basic idea is to incorporate into the ligand a moiety that can be activated on exposure of the ligand to light while the ligand is complexed at the binding site, to produce a highly reactive chemical intermediate that will bind covalently to functional groups on the macromolecule in the immediate vicinity of the binding site, causing irreversible attachment ("labeling") of the ligand to the enzyme or receptor molecule. This is schematically illustrated in Figure 1. This method has been shown to have considerable advantages over affinity labeling using highly reactive ligand-based reagents, in that photoaffinity labels are chemically inert until activated by light, and therefore can be used in critical preliminary experiments (see below) without danger of irreversible attachment to the binding

0097-6156/85/0278-0125$06.50/0

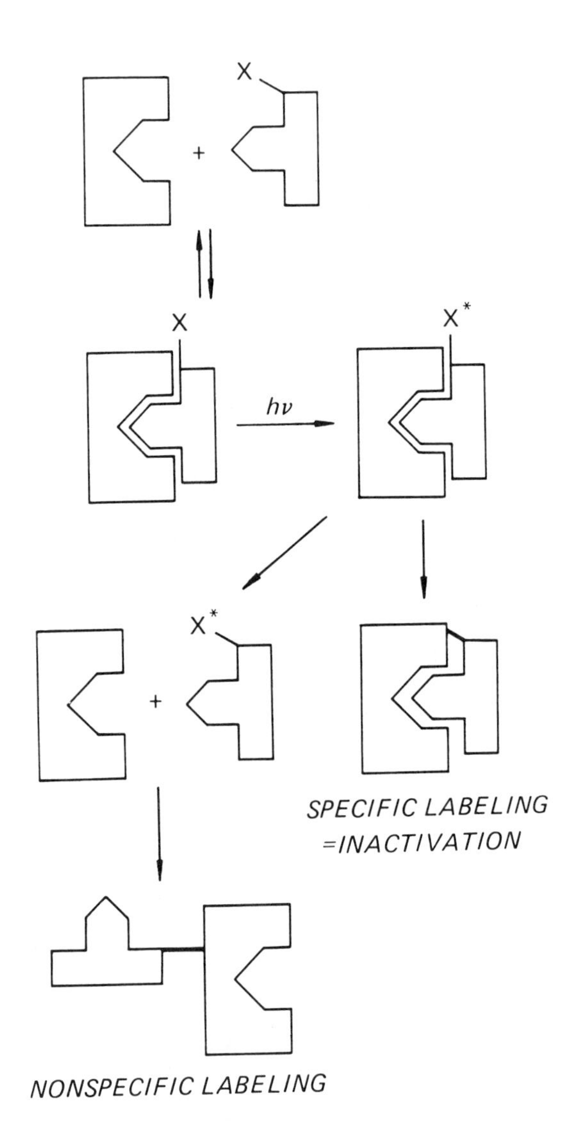

Figure 1. Schematic depiction of photoaffinity labeling of biologically active macromolecules.

sites, as well as other proteins and lipids present in the biological preparation. In general, photoaffinity labels have much greater specificity than affinity labels, and have been utilized much more frequently in studies of biologically active sites.

The initial studies using the so-called "photoaffinity labeling" technique involved purified enzyme systems, where the photolabile group was directly attached to the macromolecule prior to photolysis, as in diazoacetylchymotrypsin where the diazoacetyl group was attached to a serine residue at the active site (1,2). In most later applications the photolabile moiety is attached to a substrate or ligand of the biologically active system, and the method has become a powerful method for investigating the normally reversible interaction between a variety of pharmacological agents and their binding sites (3-6).

Ideally, the photogenerated intermediate should be highly reactive and relatively indiscriminate in its behavior, such that attachment can occur to hydrophobic as well as hydrophilic sites, i.e., insertion can occur into C-H bonds as well as nucleophilic centers. The technique has the potential of identifying such binding sites and giving information about their molecular structure, as will be illustrated below. We will be particularly interested here in characterization of biological receptor sites which are activated on attachment of an appropriate ligand to give a particular physiological or biochemical response, e.g., opening of a channel for ion transport, activation of an enzyme such as adenylate cyclase, etc., resulting in propagation of a nerve impulse, muscle contraction, release of a hormone, or some other physiological response.

A full description of the use of photogenerated reagents in biochemistry and molecular biology, including detailed experimental procedures and discussions of the pitfalls associated with this technique, is presented in a recent monograph by Bayley (6).

Criteria for Site-specific Photoaffinity Labeling

In order to achieve photolabeling of specific sites, certain requirements have to be met (4,5). If a ligand is chemically modified to allow photoactivation, it must be established that the interaction of the modified ligand with its recognition site on an enzyme or a receptor mimics that of the unmodified ligand. This can be determined in enzymatic systems by noting whether the photoprobe is a substrate for the enzyme in question, and in some receptor systems by seeing whether a physiological response induced by the unmodified ligand is observed with the photoprobe. For example, in studies of the insulin receptor, azidobenzoyl groups were specifically introduced into the B-29 lysine and B-1 phenylalanine residues, which have been shown to be part of the region of insulin which induces binding and hormonal activity; these modified insulins retained 65% and 75%, respectively, of the activity of insulin (7). In many systems, the binding of the potential photolabel to the receptor system under investigation is studied by determining the extent to which the photolabel competes for the binding site with known radioligands for that site. Thus, enkephalins in which arylazido moieties were incorporated were shown to inhibit the binding of [^{3}H]-enkephalinamide and [^{125}I]-labeled enkephalin to cell membranes (8,9). In studies of muscarinic cholinergic binding sites, azido groups were introduced into the benzilic acid moieties of 3-quinuclidinyl and N-methyl-4-piperidyl benzilates, and these photoprobes were found to be potent competitors for the binding of the tritiated unmodified compounds to muscarinic receptors in rat cortex homogenates (10). In cases where the photoprobe is also radiolabeled, which is an advantage for analysis of the results of the photolabeling reaction, as discussed below, it is possible to study directly the interaction of the probe with the receptor system of interest. Thus, [^{3}H]-p-azidobenzylcarazolol was synthesized as a labeling reagent for beta-adrenergic receptors, and it was possible to directly determine by standard procedures the rates of association and dissociation of this ligand to beta-adrenergic receptor sites in frog erythrocyte membranes, as well as the specificity of displacement of this ligand from such sites using a series of known ligands for this and other receptor systems (11,12). Thus, the rank order of potency of drugs in displacing the tritiated ligand, including enantiomeric specificity, was characteristic of the known affinities of the drugs for beta-adrenergic receptors (11).

If one can directly use a previously characterized high affinity receptor ligand as a photoaffinity label, a number of the above problems can be avoided. Notable examples of this approach to study of receptor sites in mammalian brain include labeling of benzodiazepine (e.g., valium) receptors by [^{3}H]-flunitrazepam (13) and of dopamine receptors by [^{3}H]-dopamine (14). Other examples of this type include the photolabeling of ribosomes by puromycin (15), inactivation of steroid isomerisases using photoexcited α,β-unsaturated ketosteroids (16), photoincorporation of cyclic nucleotides into receptors in a variety of tissues (17,18), and photocross-linking of proteins to nucleic acids (19-21). Often, the photochemical mechanisms involved in such labeling studies are obscure. In these studies, where the ends (i.e., irreversible photolabeling) justify the means, the absence of a rational mechanistic explanation of the observed photochemical transformation does not seem to create a problem. Indeed, the exact site of attachment of the photolabel in these complex macromolecules is rarely determined, so that the chemistry involved in the process remains unknown. It is clear, however, that the reactions occurring on photoexcitation of ligands at their binding sites often have no relationship to reactions which occur with the same ligands freely diffusing in solution, and this must be attributed to the proximity of the photoexcited state and derived intermediates to specific sites on the macromolecular surface.

In order to achieve specific labeling of a binding site, the chemically reactive intermediate generated upon photoexcitation should have a very short lifetime, in order to assure covalent association at a rate comparable to diffusion away from the binding site. The latter could lead to nonspecific labeling at sites remote from the binding site, which in heterogeneous systems (such as homogenates derived from brain dissections and most cellular preparations) could well involve other macromolecules present in the system. Such nonspecific incorporation has been termed pseudophotoaffinity labeling (22), and is well documented, particularly in examples using aryl azides as photolabels (3). One indication that pseudophotoaffinity labeling has taken place is when the extent of incorporation of radiolabeled ligand substantially exceeds the extent of inactivation of the binding site in the presence as well as the absence of a competitive inhibitor (22). Inclusion of a scavenger can sometimes reduce such effects (6,72).

Control experiments must be carried out in which the extent of incorporation of photolabel and of inactivation of the binding site is determined (a) when the putative label is irradiated in the absence of the cellular preparation, which is then added and subjected to the usual incubation and workup procedures, and (b) on irradiation of the biological preparation in the absence of the label, followed by addition of the label, incubation, and workup. Control (a) would indicate whether on irradiation of the label a long-lived intermediate is generated which can react nonspecifically with the biological preparation. For example, ketenes produced by Wolff rearrangement of photoexcited diazoacyl moieties or long-lived stabilized free radicals might show such effects. The second control provides a measure of the extent of inactivation of the biological preparation under the irradiation conditions used in photolabeling, and obviously should be minimized as much as possible. In a recent study of photolabeling of opiate receptors involving the use of diazoacetyl and azido derivatives of fentanyl which showed high affinity for such receptors, reactions were carried out using short term (less than 3 min) irradiation at 254 nm (23). However, it was shown that receptor preparations exposed to such light for only 5 min lost 50% of their binding activity. Thus, if one is forced to use short wavelength light in such studies, the incorporation of the photolabel (which should be able to absorb most of the incident light) must be particularly rapid and efficient to compete with photodegradation of the biological preparation. Indeed, one must interpret results in such cases with extreme caution. Ideally, irradiation conditions should be utilized which do not lead to such photodegradation, and this means using a label which absorbs at long wavelengths, preferably > 350 nm. Carbene precursors have been introduced which absorb at long wavelengths and which also minimize competitive reactions such as Wolff rearrangements. These include p-toluenesulfonyldiazoacetates (24), trifluorodiazopropionates (25) and 3-aryl-3-trifluoromethyldiazirines (26). Similarly, nitroarylazides which absorb at long wavelengths are generally preferred as nitrene precursors to unsubstituted aryl azides (4).

It should be evident from the material presented above that when known ligands for binding sites have been chemically modified to produce potential photoaffinity labels, in most cases these labels are precursors to carbenes and nitrenes. Because of the side reactions of carbenes which compete with insertion, aryl azides are generally thought to be preferable to diazo compounds as photolabels (4), although as indicated above these problems can be circumvented using substituted diazo compounds. However, rearrangement of aryl nitrenes to azepines is a known process, and this can also result in attachment to nucleophiles remote from the binding site (4,27,28). Nonetheless, the literature indicates that the majority of applications of photoaffinity labeling carried out to date involve nitrene precursors, and that these labels are remarkably effective (3-6,27,28). Because of the difficulty in performing adequate mechanistic studies in these heterogeneous systems, the question of the involvement of triplet or singlet states of carbenes and nitrenes in these reactions remains completely open, as does the nature of the sites (e.g., amino acid residues) which are captured (4,27,28). Until the actual sites of attachment of the labels are characterized, this question will remain unresolved.

Methods of Analysis of Photolabeled Receptor Sites

Success in photolabeling the active site on a biologically active macromolecule can be assessed in some cases by loss or reduction of that activity. In the case of enzymes, this can be readily determined by comparing the activity of the modified vs. the native enzyme towards typical substrates. In the case of receptors, the ability to bind typical radiolabeled ligands should be reduced or eliminated following successful modification of the binding site by irreversible attachment of a photolabel. This has been seen, for example, in photolabeling of enkephalin and muscarinic acetylcholine receptor sites (8,10). In such studies, the membranes after irradiation in the presence of a photolabel must be washed extensively to remove any non-attached photolabel that would compete for available receptor sites with the radioligand used in the binding assay (which can be a problem because of the high lipophilicity of many receptor ligands). The washed membranes are then subjected to the usual centrifugation and homogenization procedures to produce the preparation (usually as a homogenate) used in the radioligand binding assay under previously established conditions. As we have found in our studies of dopamine receptors, there can be problems associated with this approach because of difficulties in washing the membranes free of the unbound radiolabeled reagent. Indeed, when membrane-associated receptors are exposed to photolabels in the dark, a significant decrease in the extent of binding of the radioligand used in the analysis can often be observed (11,29,30). This is most likely due not to dark reactions of the photolabel with the receptor, but rather to the failure to completely wash the receptors free of the reversibly bound hydrophobic labeling reagent. When reduction in binding activity of the receptor system is observed, it should be demonstrated (as was done in the study of the enkephalin receptor system) (8) that known specific receptor ligands compete for the binding site with the photolabel, thus reducing the extent of inactivation due to photolabeling. This effect depends, however, on the respective rates of association and dissociation of the ligands and the receptor, since a slow off-rate of the photolabel may not allow effective competition for the binding site by known receptor ligands. We believe this is a problem with some potential labeling reagents for dopamine receptors synthesized in our laboratory (30). If the photolabel is radiolabeled, the association and dissociation kinetics of the photolabel and the receptor should be determined (10,11,13,29). However, if only non-radiolabeled material is available, these kinetics are not easily determined. Thus, there is a distinct advantage to having radiolabeled photoaffinity labels, which in the case of tritium-labeled compounds can be obtained either by derivitization of commercially available receptor ligands, by non-specific exchange of tritium for protons (usually on aromatic rings) or by organic synthesis. For ^{125}I-labeled compounds, the latter is usually required, although ^{125}I can be conveniently introduced under oxidative conditions into suitably activated aromatic systems.

The most common methodology used for analysis and identification of the sites on the receptor to which the photolabel has been irreversibly (i.e., covalently) attached as a result of exposure to UV irradiation involves gel electrophoresis. In this procedure, the receptors must first be separated from the lipid environment of the membranes by solubilization in a buffer containing a detergent, most commonly sodium dodecyl sulfate (SDS). This buffer often contains various protease inhibitors to prevent protease-induced degradation of labeled proteins and 2-mercaptoethanol to reduce disulfide bridges. Electrophoresis is usually done on polyacrylamide slab gels (PAGE) prepared by standard procedures (31). Samples are placed in the various wells of the plate, with one lane reserved for molecular weight standards, and a running buffer is used to separate the components of the mixture according to molecular weight on passage of current through the gel. The various bands on the gel can be visualized by standard staining procedures. When radiolabeled photolabeling reagents are utilized, the extent of incorporation of radiolabel into various bands on the gel can be determined by slicing the gel and measuring the radioactivity in each slice by liquid scintillation techniques, or by fluorographic analysis in which the gel after addition of a fluor is exposed to x-ray film. Depending on the specific activity of the photolabel (Ci/mmol), the exposure time may vary from a few hours to several months, so it is advantageous to use label with as high specific activity as can be conveniently obtained and handled. Thus, ^{125}I (typical activity on the order of 1,000 Ci/mmol) is preferable to ^{3}H (typical activity of tens of Ci/mmol) as the radiolabel.

A particularly nice example of the application of this methodology is shown by the work of Lefkowitz and his group (12), in which beta-adrenergic receptors in frog erythrocyte membranes partially purified by affinity chromatography were incubated with ^{125}I-p-aminobenzylcarazolol (1) in the presence and absence of competing adrenergic ligands, followed by cross-linking of the label to the receptor through the free amino moiety using the bifunctional reagent SANAH (2) upon exposure to UV light. As can be seen in Figure 2, the label is specifically incorporated into a 58 kDalton band, which is also the major band labeled when an unpurified receptor preparation is utilized (11). Furthermore, the specificity of labeling is characteristically beta-adrenergic in that labeling is stereospecifically inhibited by the (-) enantiomers of alprenolol and isoproterenol, which have much higher beta-receptor affinities than the corresponding (+) isomers. Phentolamine and haloperidol, which are characterizing agents for alpha-adrenergic and dopaminergic systems, respectively, are ineffective in protecting the 58 kDalton receptor component from labeling by [^{125}I]-PAMBC. There is now a considerable body of work which indicates that this protein is indeed a principal component of the beta-adrenergic receptor, and work is underway to obtain its molecular structure using the standard techniques of biochemistry and molecular biology (32) It should perhaps be explicitly noted that these membrane-bound receptors are not crystalline, so that x-ray crystallography, which has proved to be of great value in assigning structures to enzymes and other proteins, is of no use in receptor biochemistry, at least thus far.

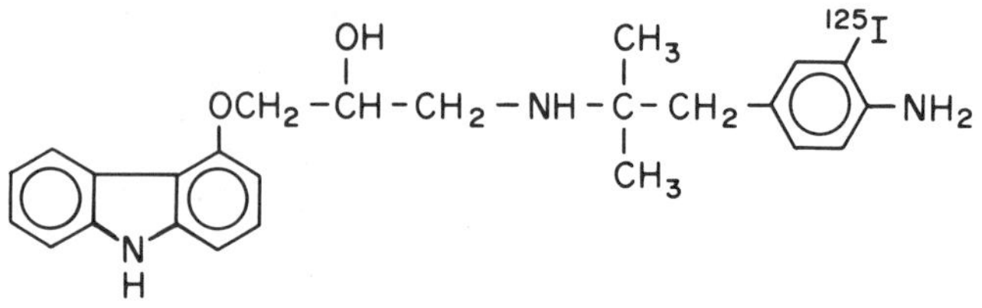

$[^{125}I]$ p-AMINOBENZYLCARAZOLOL

($[^{125}I]$ PAMBC)

1

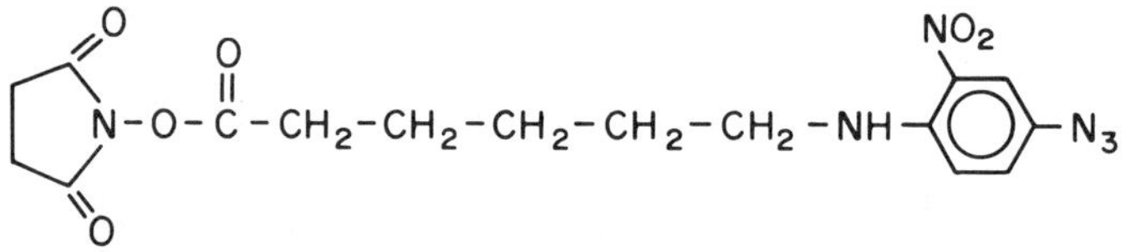

N-SUCCINIMIDYL-6 (4'AZIDO-2'-NITROPHENYLAMINO) HEXANOATE (SANAH)

2

Results of Current Studies Involving Receptor Characterization

We have been involved for several years in a program to understand the nature of the interactions at the molecular level between neurotransmitters and their receptor sites in the brain, and the mechanism of action of clinically active drugs which inhibit this interaction in neuronal systems. Such studies have significance with respect to understanding the basic pathology involved in mental disease states such as depression and schizophrenia, and in aiding in the design of specific drugs to treat these illnesses. The results of two such studies employing photoaffinity labeling will be briefly summarized.

Photoaffinity Labeling of Serotonin Uptake Sites Using 2-Nitroimipramine

Serotonin (5-hydroxytryptamine, 5-HT) (3) is a neurotransmitter found in the periphery as well as the central nervous system. Following release from storage vesicles in serotonergic neurons, 5-HT diffuses across the synaptic cleft where it is recognized at appropriate binding sites, ultimately generating a response in the receptor-containing cell by a complex process which need not concern us here. Subsequently, the neurotranmitter can be inactivated enzymatically (e.g., oxidized to 5-hydroxyindoleacetic acid by the action of monoamine oxidase and aldehyde reductase) or can be recaptured at prejunctional sites. This route, known as reuptake, is an important way in which the level of neuronal activity is controlled (33). In the periphery, 5-HT is known to be taken up by platelets in the circulatory system, where it is stored, transported and eventually released (34). Platelets do not have the machinery for producing 5-HT directly. Previous studies have established that the pharmacological properties of the 5-HT reuptake site in the CNS and on platelets are virtually identical, qualitatively as well as quantitatively (35,36). The ease of isolating and purifying platelets makes them an excellent model system for investigating 5-HT uptake sites in normal as well as diseased subjects (36,37).

Platelets also possess a high density of sites for recognition of tricyclic antidepressive agents, such as imipramine (IMI, 4a), which inhibit the uptake of biogenic amines such as 5-HT into platelets as well as synaptosomes (38,39). The characteristics of [^{3}H]-IMI binding sites in human platelets and human brain are nearly identical (40,41). It has been demonstrated that 2-nitroimipramine (2-NI, 4b) is a potent inhibitor of 5-HT uptake and IMI binding in human platelets and rat brain, and in general 2-NI and IMI have similar pharmacological properties (42). The components of the 5-HT uptake site have not been identified, and a study was undertaken to label such sites using 2-NI as a photoaffinity label, using human platelets as well as neuronal preparations (43).

Scatchard analyses of equilibrium binding of [^{3}H]-2-NI to human platelet membranes were markedly curvilinear (43). These curves could be dissected into a high affinity component (K_D = 1.34 ± 0.69 nM), in agreement with an earlier report (44), and one or more sites of lower affinity. The UV absorption spectrum of 2-NI reveals a maximum at 392 nm (ε 7,800), so that long wavelength light which does not damage the membrane preparations could be utilized in these studies. The output of a 550-watt Ace-Hanovia lamp could be filtered to exclude wavelengths below 340 nm, and the various preparations in Pyrex test tubes were irradiated for 20 min at 0°C in the presence of [^{3}H]-2-NI, following which the preparations were concentrated, briefly sonicated and then solubilized in an SDS buffer prior to polyacrylamide gel electrophoresis (31). The fluorograms in Figure 3 show that there was no measurable irreversible incorporation of radioactivity in samples which were not exposed to UV light, while irradiation resulted in efficient and selective incorporation of radioactivity into a 30 kDalton band. After subtracting background, 20-25% of the radioactivity in the gel was located in this band, although Coomassie staining revealed that this band represents only a very small fraction of the total membrane protein. The incorpora-

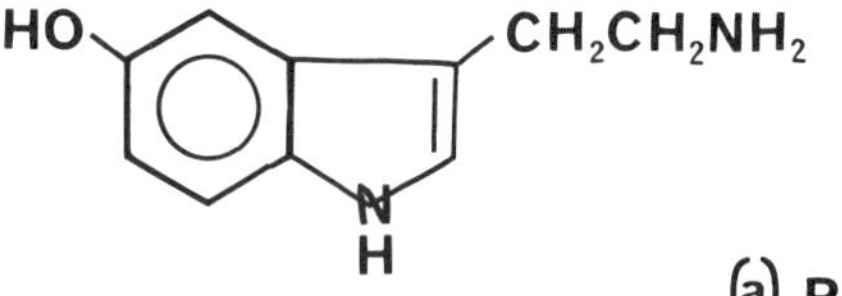

HO, $CH_2CH_2NH_2$, N, H

3

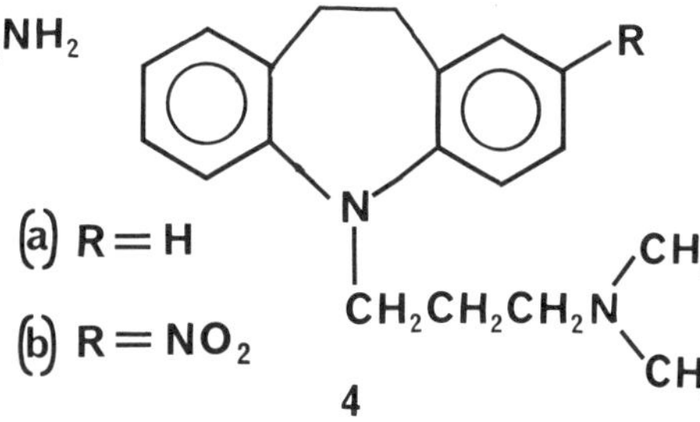

(a) R = H

(b) R = NO_2

R, N, $CH_2CH_2CH_2N$, CH_3, CH_3

4

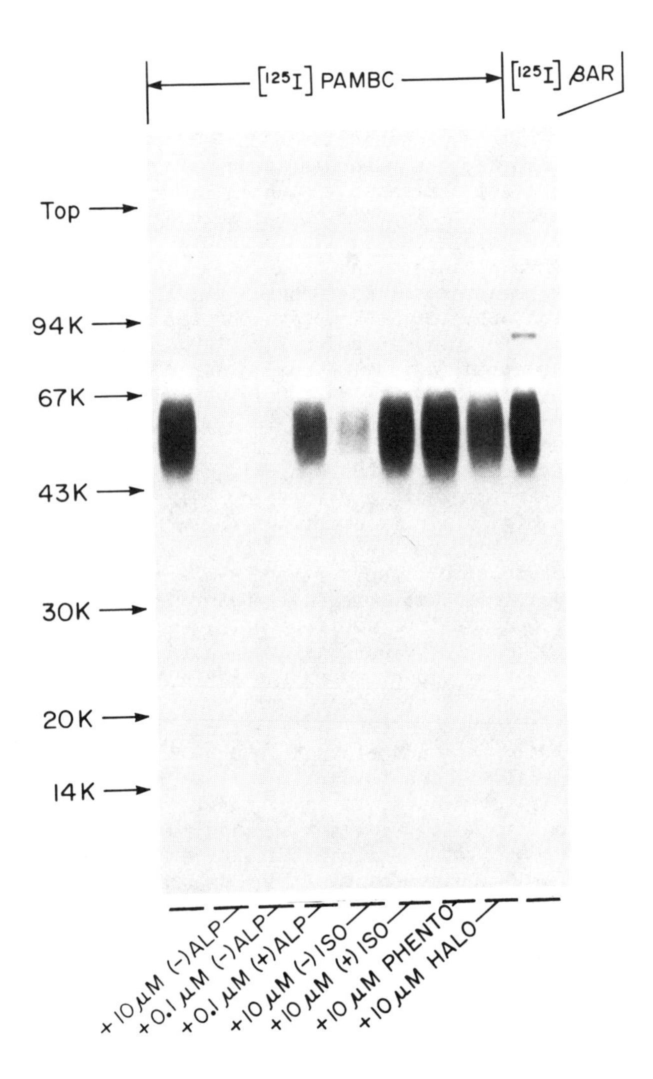

Figure 2. Specificity of photoaffinity crosslinking of ^{125}-I-p-aminobenzylcarazolol (1) to a purified beta-adrenergic receptor preparation. Arrows to the left indicate positions of iodinated molecular weight standards. Taken with permission of the copyright owners from ref. 12.

tion of [^{3}H]-2-NI into the 30 kDalton band of human platelet membranes was blocked by 10 μM IMI, as shown in Figure 4, but IMI enhanced incorporation of radioactivity in other bands, confirmed by trichloroacetic acid precipitation studies, perhaps as a result of changes in membrane fluidity (45,46). The dose-response effects of three 5-HT uptake inhibitors (IMI, fluoxetine and citalopram) on equilibrium binding of 2-NI and incorporation of [^{3}H]-2-NI into the 30 kDalton band were compared in human platelets and rat liver preparations. This relationship in the case of IMI is clearly illustrated in Figure 5. However, significant reductions in drug potency in inhibiting irreversible incorporation as compared to equilibrium binding of [^{3}H]-2-NI were observed (43).

Significant photoincorporation of [^{3}H]-2-NI into membrane homogenates derived from human cortex, rat cortex, rat hypothalamus, and rat hippocampus was also observed. In all cases a 30 kDalton band was irreversibly labeled in yields of 1.5-5.0% of total specifically and reversibly bound ligand in these regions. Finally, the pharmacological profile of reversible binding and photoincorporation of [^{3}H]-2-NI were determined using thirteen selective 5-HT uptake inhibitors and fourteen compounds selective for other neurotransmitter systems, using preparations of human platelets, rat liver and human cortex. Every 5-HT uptake inhibitor blocked labeling of the 30 kDalton band; in most cases to upwards of 90%, while only small inhibitory effects (usually less than 30%) were seen for most of the other drugs. In general, there is a good correlation between the ability of drugs to inhibit reversible labeling and photolabeling of the 30 kDalton band, with coefficients of 0.979 in rat liver and 0.866 in platelets. However, 5-HT not only did not block photolabeling of the 30 kDalton band by [^{3}H]-2-NI but actually enhanced photolabeling in a dose-dependent manner, even though 5-HT is an potent inhibitor of reversible binding of this ligand to platelet and rat liver homogenates.

There seems little doubt from these observations that [^{3}H]-2-NI is labeling a 30 kDalton subunit of the 5-HT uptake and transport complex, presumably a polypeptide. The regional distribution of binding and labeling parallels the distribution of known high affinity IMI binding sites (47), and the pharmacological profile of inhibitors of binding and labeling is as expected for the 5-HT uptake system. However, the quantitative discrepancies between the ability of three selective 5-HT uptake inhibitors (IMI, fluoxetine and citalopram) to inhibit binding and to protect the tissue against photolabeling suggests that the 30 kDalton protein is not itself the IMI binding site but is a second heretofore unidentified unit which may be coupled to, but is distinct from, the IMI binding site. Thus, although the labeled protein has a high affinity for 5-HT uptake inhibitors, that affinity is not as high as that of the IMI binding protein itself (40,42,44). Thus, the effect of 5-HT, which actually promotes rather than inhibits photolabeling of the 30 kDalton peptide, can be explained if this component is allosterically coupled to the serotonin binding site in such a manner that binding of 5-HT to its site increases the affinity of the 30 kDalton protein for [^{3}H]-2-NI or leads to an increase in the number of available sites for labeling. Further studies should clarify the relationship between the two sites.

Some additional bands at 16, 17, 35, and 37 kDaltons are labeled

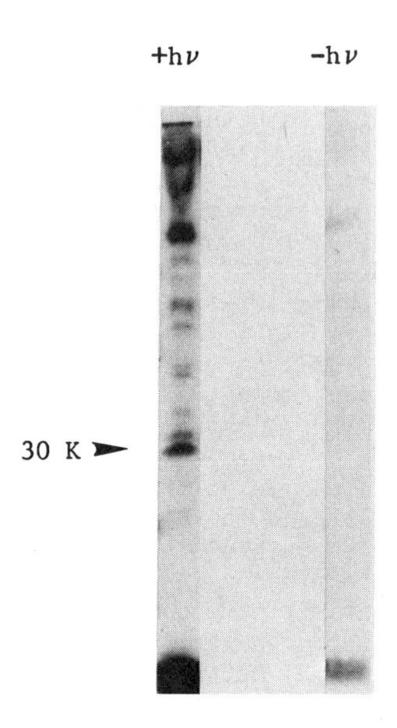

Figure 3. Effect of UV light (340 nm) on incorporation of [^{3}H]-2-nitroimipramine into human platelet membranes, as shown by SDS-PAGE and fluorography.

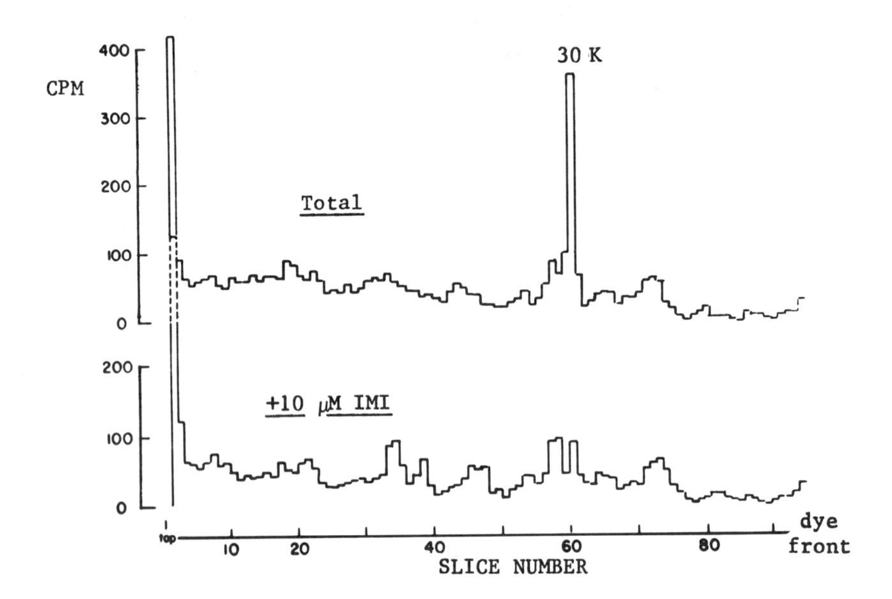

Figure 4. Photolabeling of human platelet membranes with [^{3}H]-2-NI in the presence or absence of 10 M imipramine. Proteins were separated by SDS-PAGE, slices were solubilized in NCS/toluene and counted by scintillation spectrometry.

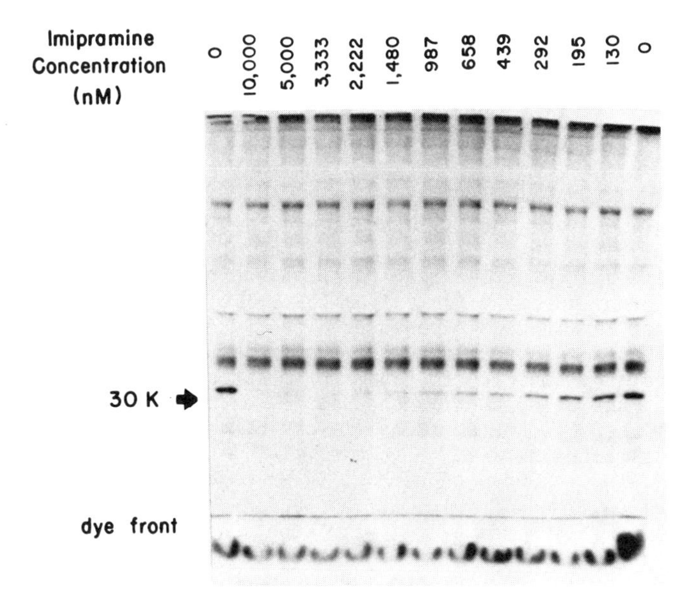

Figure 5. Dose response inhibition by imipramine of photolabeling of the 30 KDalton fraction in human platelet membranes in the presence or absence of varying concentrations of imipramine as indicated, followed by solubilization and SDS-PAGE/fluorography.

in human platelet but not in rat brain homogenates, and the pattern of inhibition of labeling of these bands is complex. This raises the possibility that the 5-HT uptake systems in the platelet and the CNS are not in fact identical in all respects. Finally, it should be noted that the photochemistry involved in photolableling of proteins by 2-NI remains obscure. Photoexcited aromatic nitro compounds are known to undergo hydrogen abstraction reactions and reduction (48,49), so that interaction of photoexcited 2-NI with C-H bonds on amino acid moities at or near the binding site, followed by radical coupling, is a plausible explanation for the covalent attachment of 2-NI to the binding protein (50). However, one can not exclude other mechanisms which may have little or no analogy to solution photochemistry of nitroaromatics.

Photoaffinity Labeling of Striatal Dopaminergic Receptor Sites by Chlorpromazine

A principal interest in our laboratory is the molecular characterization of CNS receptor sites of the neurotransmitter dopamine (DA, 5). These sites are strongly implicated in the biochemical etiology of schizophrenia and Parkinson's Disease, as well as other diseases of the CNS (50,51). Thus, the rank order of clinical potency of antipsychotic drugs (neuroleptics) correlates with the affinity of these drugs for dopaminergic sites (52,53). It is also well established that Parkinson's disease is directly related to deterioration in dopaminergic neurotransmission in the corpus striatum, which is a brain region rich in dopamine receptor sites (54). The use of L-DOPA, the biosynthetic precursor of dopamine, in treatment of patients with Parkinson's disease is one of the best examples of biochemically directed medical treatment.

We are using a number of methodologies to characterize these dopaminergic sites, one of which is photoaffinity labeling. There are a number of problems associated with such an endeavor. First of all, pharmacological studies indicate that there is more than one type of DA receptor in the CNS, the exact number being a question of considerable debate (55,56). The density of DA receptors as determined by radioligand binding assays is only on the order of a few hundred femtomoles per milligram of protein in the optimal region (corpus striatum) and even less in other macroscopic areas (frontal cortex, mesolimbic system, etc.) (55). As there is no direct physiological response associated with these CNS receptors (e.g., muscle contraction, vasodilation) aside from single unit electrophysiology, the only practical way of studying them conveniently is through

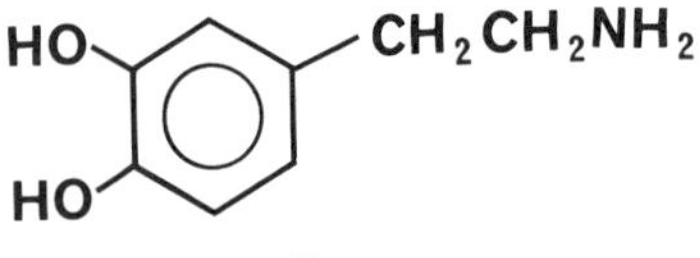

HO, HO, $CH_2CH_2NH_2$

5

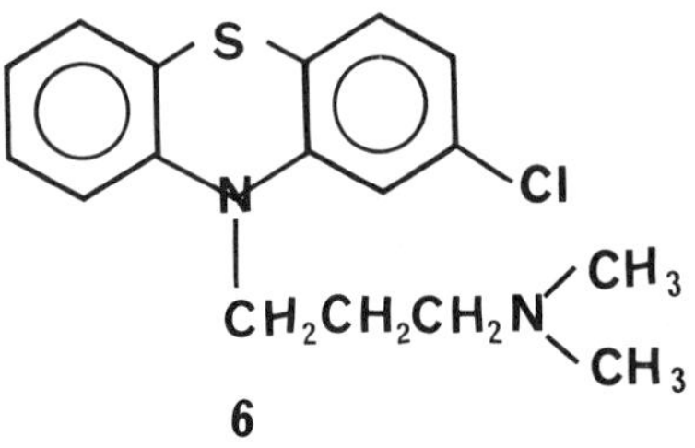

S, N, Cl, $CH_2CH_2CH_2N(CH_3)_2$

6

radioligand binding assays. Finally, most dopaminergic ligands including DA itself bind to other types of receptors sites, such as 5-HT, adrenergic and nor-adrenergic sites, so that there is a problem of specificity, although some progress has been made recently in developing highly specific dopaminergic receptor ligands (57,58). Thus, it is critical to establish that the sites in a given neuronal preparation that are irreversibly photolabeled using a compound derived from a dopaminergic ligand are indeed dopaminergic in nature. It should perhaps be mentioned that the considerable success achieved in recent years in characterizing cholinergic and beta-adrenergic receptor sites is in no small part due to the extraordinary abundance of such sites in easily obtainable non-neuronal tissues, namely the electric organ of *Torpedo californica* and related electric fish (cholinergic receptors) and frog and turkey erythrocytes (beta-adrenergic receptors). The practical problems associated with the direct characterization of analogous receptor sites in the brain remain formidable.

We have synthesized several carbene and nitrene precursors based on a variety of dopaminergic receptor ligands, and have carried out some preliminary pharmacological and photolabeling studies with these potential photoaffinity labels (59). Although the results are moderately encouraging, the lack of specificity of these compounds toward neuronal dopaminergic sites remains a serious problem. The studies which will be summarized below (29) utilize chlorpromazine (CPZ, 6), which was the first drug that was found to be effective in treatment of schizophrenia and whose use revolutionized the practice of psychiatry (60,61). This compound is still the most commonly used antipsychotic drug. We chose CPZ for out initial attempts at DA receptor labeling since it had been shown that photoexcitation of CPZ leads to a variety of reactions attributable to cleavage of the C-Cl bond. Thus, CPZ can be irreversibly attached to a variety of proteins and nucleic acids on photoexcitation,(62,63), including calmodulin, a small protein (17 kDaltons) found in neuronal tissues (64).

Although the affinity of CPZ for DA receptors is well known (55), complete pharmacological characterization of CPZ in the CNS has been lacking. We have recently carried out such a study using homogenates of canine striatum and [^{3}H]-CPZ with specific activity 39 Ci/mmol (New England Nuclear), and demonstrated that there is a very high affinity site with a dissociation constant of 0.98 ± 0.24 nM and a maximum binding capacity of 197 ± 15 fmol/mg protein (29). The ability of a large number of dopaminergic and non-dopaminergic drugs to displace [^{3}H]-CPZ revealed a pharmacology not at all typical of dopaminergic sites, suggesting that the high affinity site may in fact represent a specific phenothiazine binding site. At concentrations above 10 nM, CPZ binds to yet additional sites and its behavior becomes quite complex. Ideally, then, one should use CPZ as a photoaffinity label at concentrations less than 10 nM, but that turns out to be impractical because of the very low extent of incorporation of radioactivity into the neuronal preparation as measured by SDS-PAGE and fluorography, even after exposure of the gel for several months. We were therefore forced to use CPZ at relatively high concentrations in order to get measurable incorporation into the membrane preparations, leading not unexpectedly to complex results.

Initially (65), we utilized a digitonin-solubilized bovine striatal preparation whose pharamacology has been characterized (66) and [^{3}H]-CPZ at a concentration of 42 nM under conditions in which CPZ would not associate with calmodulin (67). Gel filtration of the photolysate (Figure 6), which separates fractions in order of decreasing molecular weight, revealed a high molecular mass fraction (> 200 kDaltons) which was not observed in controls when the two components (CPZ and the striatal preparation) were irradiated separately before addition of the other component. A low molecular mass fraction was seen in the photolysate as well as the controls. Thermal denaturation of the photolyzed mixture leads to no loss of the high mass component and significant loss of the low mass fraction, indicating the high mass fraction represents irreversible labeling while in the low mass fraction the labeling is largely reversible. The incorporation of radioactivity into the high mass components was significantly reduced when irradiations were carried out in the presence of (+)-butaclamol, which is the pharmacologically active enantiomer of this potent dopaminergic ligand (68), while the (-)-isomer offered much less protection. PAGE analysis of the high mass fraction revealed a 61 kDalton component which was not present in the low mass fraction (Figure 7). Attempts to further characterize photolabeling of the 61 kDalton component were frustrated by experimental problems associated with the low activity of the [^{3}H]-CPZ and the failure to obtain satisfactory resolution on gels in the presence of test ligands. Thus, we can only surmise that the 61 kDalton band _may_ represent a component of the solubilized DA receptor site, although this remains to be established with certainty.

Better results were obtained more recently in studies using higher activity [^{3}H]-CPZ and canine striatal homogenates (29). Using cold CPZ (10 nM), it was established that the ability of a standard dopaminergic assay ligand, [^{3}H]-spiroperidol, to bind to the irradiated (> 300 nm) preparation was reduced by up to 40% compared to nonirradiated controls. Irradiation of the membrane preparation under the same conditions in the absence of CPZ resulted in no reduction in specific binding of [^{3}H]-spiroperidol. Using hot CPZ (100 nM), the pattern of irreversible incorporation was determined using SDS-PAGE, as shown in Figure 8, which compares irradiated with non-irradiated preparations. The densitometric scan of the fluorogram reveals major peaks at 120, 57, 34, and 32 kDaltons which were not observed on SDS-PAGE analysis of the unphotolyzed samples. A number of pharmacological agents were tested for their ability to protect the preparation against photolabeling by CPZ at concentrations in the micromolar range which would be sufficient for saturation of their respective receptor sites. None of the classic DA antagonists used, with the exception of cold CPZ, were particularly effective in blocking photolabeling of the 57 and 34 kDalton bands, while incorporation of hot CPZ into the 32 kDalton band was blocked non-selectively by all the ligands. The latter effect was concluded, from analysis of UV absorption spectra, to be a light screening effect not related to receptor occupancy. At lower concentrations light screening should not be a problem, but then the ligands would not be expected to saturate their respective binding sites. Thus, it is not clear whether the canine striatal membrane components labeled by [^{3}H]-CPZ at a concentration of 100 nM represent portions of the

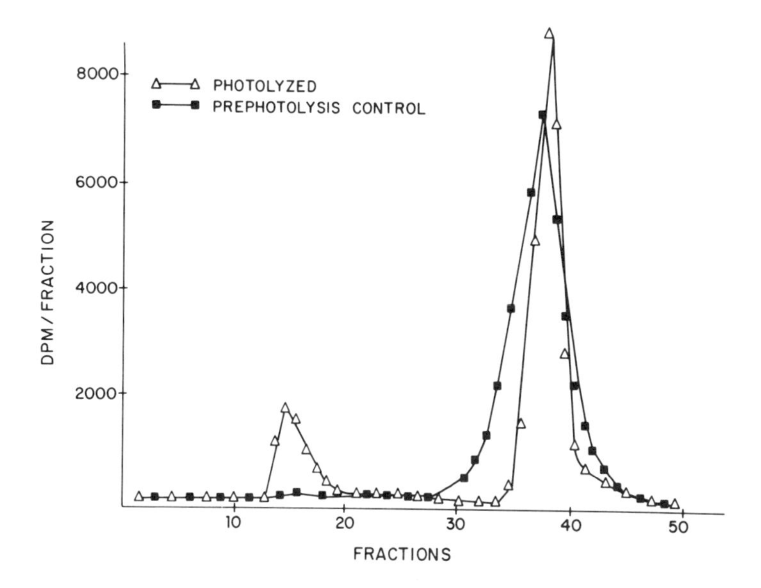

Figure 6. Gel filtration on Sephadex G-200 of [^{3}H]-CPZ photolyzed in the presence of a digitonin-solubilized bovine striatal homogenate. Open triangles represent specific activity of fractions obtained from photolyzed samples and solid squares are data from nonirradiated controls. Taken with permission of the publisher and authors from ref. 65.

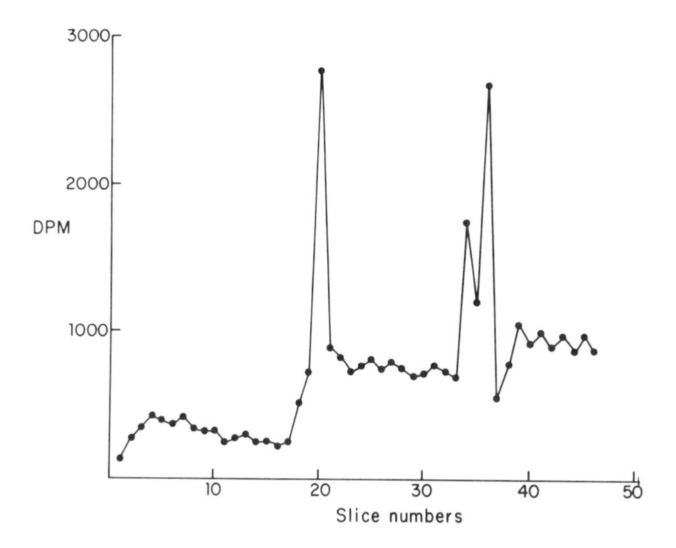

Figure 7a. Gel electrophoresis of (A) high molecular mass fraction from Figure 3. Reproduced with permission from Ref. 65. Copyright 1982.

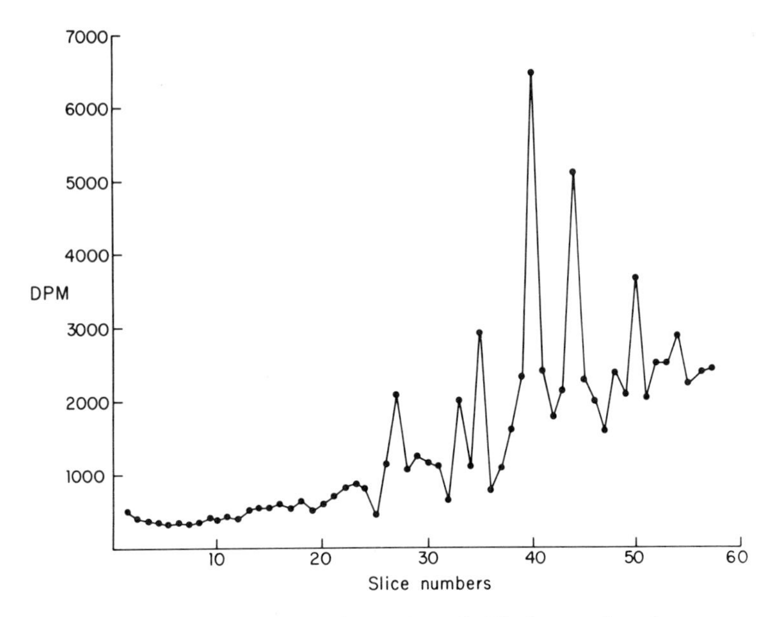

Figure 7b. Gel electrophoresis of (B) low molecular mass fraction from Figure 3. Reproduced with permission from Ref. 65. Copyright 1982.

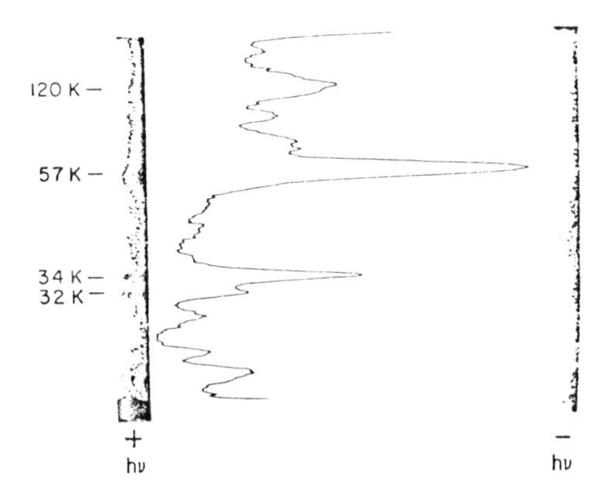

Figure 8. Effect of UV light on binding of [^{3}H]-CPZ to canine striatal homogenates. Fluorogram of irradiated (left) and non-irradiated samples (right) after SDS-PAGE analysis. Taken from the Ph. D. Thesis of K. Thermos.

dopamine receptor or an alternative pharmacological entity. In this connection, Lilly et al. (69) have reported the molecular size of the canine striatal D-2 receptor subtype to be 123 kDaltons using target size analysis, while Kuno et al. (70,71) observed that irradiation of 5 μM [^{3}H]-dopamine in the presence of membrane preparations of rat striatum and frontal cortex at 0°C for 120 min led to predominant incorporation (SDS-PAGE analysis) into a band at 57 kDaltons, which they concluded from protection studies to be a subunit of the D-1 class of dopamine receptors. It is noteworthy that the two most prominent bands labled by CPZ in our experiments correspond in mass to these suggested dopaminergic receptor components.

It is clear at this point that successful photolabeling of dopaminergic receptors will require ligands which show greater selectivity and higher affinity for such receptors compared to other sites in striatal tissue and other brain tissues containing dopaminergic sites. The low density of such sites in the brain under the best of circumstances also necessitates the development of labels with high specific activity, to allow for eventual separation of receptor components, which implies labels containing ^{125}I. The synthesis and pharmacological characterization of such materials is now underway in our laboratory.

Conclusion

Although considerable progress has been made in identification of receptor components using the photoaffinity labeling technique, the only system in which the detailed molecular structure has been elucidated to date is the nicotinic cholinergic receptor (73). It seem to be only a matter of time before such information is also available for beta-adrenergic receptors (74). In both cases, these represent receptors in special tissues (see above) outside the central nervous system. The full molecular characterization of neurotransmitter receptor sites in mammalian brain, which will be a great aid in understanding the biochemical basis of mental illness and the mechanisms of action of psychoactive drugs, remains a major research goal.

Literature Cited

1. Singh, A.; Thornton, E.R.; Westheimer, F.H. J. Biol. Chem. 1962, 237, PC3006.
2. Shafer, J.; Baronowsky, P.; Laursen, R.; Finn, F.; Westheimer, F.H. J. Biol. Chem. 1966, 241, 421.
3. Chowdhry, V.; Westheimer, F.H. Ann. Rev. Biochem. 1979, 48, 293.
4. Knowles, J.R. Acc. Chem. Res. 1972, 5, 155.
5. Bayley, H.; Knowles, J.R. Methods Enzymol. 1977, 46, 69.
6. Bayley, H. "Photogenerated Reagents in Biochemistry and Molecular Biology"; Elsevier: New York, 1983.
7. Yip, C.C.; Moule, M.L. Federation Proc. 1983, 42, 2842.
8. Lee, T.T.; Williams, R.E.; Fox, C.F. J. Biol. Chem. 1979, 254, 11787.
9. Hazum, E.; Chang, K.-J.; Shechter, Y.; Wilkinson, S.; Cuatrecases, P. Biochem. Biophys. Res. Comm. 1979, 88, 841.

10. Amitai, G.; Avissar, S.; Balderman, D.; Sokolovsky, M. Proc. Natl. Acad. Sci. USA 1982, 79, 243.
11. Lavin, T.N.; Heald, S.L.; Jeffs, P.W.; Shorr, R.G.L.; Lefkowitz, R.J.; Caron, M.C. J. Biol. Chem. 1981, 256, 11944.
12. Shorr, R.G.L.; Heald, S.L.; Jeffs, P.W.; Lavin, T.N.; Strohsacker, M.W.; Lefkowitz, R.J.; Caron, M.G. Proc. Natl. Acad. Sci. USA 1982, 79, 2778.
13. Mohler, H.; Battersby, M.K.; Richards, J.G. Proc. Natl. Acad. Sci. USA 1980, 77, 1666.
14. Nishikori, K.; Noshiro, O.; Sano, K.; Maeno, H. J. Biol. Chem. 1980, 255, 10909.
15. Cooperman, B.S.; Grant, P.G.; Goldman, R.A.; Luddy, M.A.; Minnella, A.; Nicholson, A.W.; Strycharz, W.A. In "Methods in Enzymology", Vol. LIX, 1978, 796.
16. Smith, S.B.; Benisek, W.F. J. Biol. Chem. 1980, 255, 2690.
17. Guthrow, C.E.; Rasmussen, H.; Brunswick, D.J.; Cooperman, B.S. Proc. Natl. Acad. Sci. USA 1973, 70, 3344.
18. Kallos, J. Nature (London) 1977, 265, 705.
19. Markovitz, A. Biochim. Biophys. Acta 1972, 281, 522.
20. Strniste, G.F.; Smith, D.A. Biochem. 1974, 13, 485.
21. Schimmel, P.R.; Budzik, G.P.; Lam, S.S.M.; Schoemaker, H.J.P. In "Aging, Carcinogenesis and Radiation Biology", Smith, K.C., Ed; Plenum: New York, p. 123.
22. Ruoho, A.E.; Kiefer, H.; Roeder, P.E.; Singer, S.J. Proc. Natl. Acad. Sci. USA 1973, 70, 2567.
23. Maryanoff, B.E.; Simon, E.J.; Gioannini, T.; Gorissen, H. J. Med. Chem. 1982, 25, 913.
24. Chowdhry, V.; Westheimer, F.H. J. Am. Chem. Soc. 1978, 100, 309.
25. Chowdhry, V.; Vaughan, R.; Westheimer, F.H. Proc. Natl. Acad. Sci. USA 1976, 73, 1406.
26. Brunner, J.; Senn, H.; Richards, F.M. J. Biol. Chem. 1980, 255, 3313.
27. Nielsen, P.E.; Buchardt, O. Photochem. Photobiol. 1982, 35, 317.
28. Staros, J.V. Trends. Biochem. Sci 1980, 320.
29. Thermos, K. Ph.D. Thesis, New York University, New York, 1984; Thermos, K.; Schuster, D. I.; Murphy, R. B.; Wennogle, L.P.; Meyerson, L. R. Neurosci. Abs. 1983, 9, 1116.
30. Libes, R.B. Ph. D. Thesis, New York University, New York, 1983.
31. Laemmli, U.K. Nature (London) 1970, 227, 680.
32. Stiles, G.L.; Caron, M.G.; Lefkowitz, R.J. Physiol. Rev. 1984, 64, 661.
33. Iversen, L. L. Biochem. Pharmacol. 1974, 23, 1927.
34. Pletscher, A.; Da Prada, M.; Berneis, K.H. Experientia 1971, 27, 993.
35. Sneddon, J.M. In "Progress in Neurobiology"; Kerkut, A. and Phillis, J.W., Ed.; Pergamon: Oxford, 1973; Vol. 1, Part 2, p. 151.
36. Stahl, S.M.; Meltzer, H.Y. J. Pharmacol. Exp. Ther. 1978, 205, 118.
37. Pletscher, A. Brit. J. Pharmacol. Chemother. 1968, 32, 1.
38. Drummond, A.H. In "Platelets in Biology and Pathology";

Gordon, J.L., Ed.; Elsevier: Amsterdam, 1976; p. 203.
39. Wennogle, L.P.; Meyerson, L.R. Eur. J. Pharm. 1982, 86, 303.
40. Rehavi, M.; Ittah, Y.; Skolnick, P.; Rice, K.C.; Paul, S.M. Biochem. Biophys. Res. Comm. 1981, 99, 954.
41. Paul, S.M.; Rehavi, M.; Rice, K.C.; Ittah, Y.; Skolnick, P. Life Sci. 1981, 28, 2753.
42. Rehavi, M.; Tracer, H.; Rice, K.C.; Skolnick, P.; Paul, S.M. Life Sci. 1983, 32, 645.
43. Wennogle, L. P.; Ashton, R. A.; Schuster, D. I.; Murphy, R.B.; Meyerson, L. R. EMBO Journal 1984, in press.
44. Rehavi, M.; Ittah, Y.; Skolnick, P.; Rice, K.C.; Paul, S.M. Naunyn-Schmiedeberg's Arch. Pharmacol. 1982, 320, 45.
45. Cater, B.R.; Chapman, D.; Hawes, S.M.; Saville, J. Biochem. Biophys. Acta 1974, 363, 54.
46. Delmelle, M.; Wattiaux-De Coninck, S.; Dubois, F.; Wattiaux, R. Biochem. Biophys. Acta 1980, 600, 791.
47. Biegon, A.; Rainbow, T.C. Neurosci. Lett. 1983, 37, 209.
48. Cu, A.; Testa, A.C. J. Am. Chem. Soc. 1974, 96, 1963.
49. Cahnmann, H.J.; Matsuura, T. Photochem. Photobiol. 1982, 35, 23.
50. Carlsson, A. Biol. Psychiatry 1978, 13, 3.
51. Iversen, L.L. In "Chemical Communication within the Nervous System and its Disturbance in Disease"; Taylor, A. and Jones, M.T., Ed.; Pergamon: Oxford, 1978; p. 17.
52. Seeman, P.; Lee, T. Science 1975, 188, 1217.
53. Burt, D.R.; Enna, S.J.; Creese, I.; Snyder, S.J. Proc. Natl. Acad. Sci. USA 1975, 72, 4655.
54. Hornykiewicz, O. Pharmacol. Rev. 1966, 18, 925.
55. Seeman, P. Pharmacol. Rev. 1980, 32, 229.
56. Sibley, D.; DeLean, A.; Creese, I. J. Biol. Chem. 1982, 257, 6351.
57. Iorio, L.C.; Barnett, A.; Leitz, F.H.; Houser, V.P.; Korduba, C.A. J. Pharmacol. Exp. Ther. 1983, 226, 462.
58. Hjorth, S.; Carlsson, A.; Wikstrom, H.; Lindberg, P.; Sanchez, D.; Hacksell, U.; Arvidsson, L.-E.; Svensson, U.; Nilsson, J.L.G. Life Sci. 1981, 28, 1225.
59. Narula, A.P.S., unpublished results from this laboratory. See also ref. 30.
60. Delay, J.; Deniker, P. In "Congres de Medecines Alienistes de Neurologistes de France"; Cossa, P., Ed., 1952; p. 497.
61. Davis, J.M. Arch. Gen. Psychiatry 1965, 13, 552.
62. Rosenthal, I.; Ben-Hur, E.; Prager, A.; Riklis, E. Photochem. Photobiol. 1978, 28, 591.
63. Fujita, H.; Hayashi, H.; Suzuki, K. Photochem. Photobiol. 1981, 34, 101.
64. Prozialeck, W.C.; Cimino, M.; Weiss, B. Molec. Pharmacol. 1981, 19, 264.
65. Thermos, K.; Murphy, R.B.; Schuster, D.I. Biochem. Biophys. Res. Comm. 1982, 106, 1469.
66. Davis, A.; Madras, B.; Seeman, P. Eur. J. Pharm. 1981, 70, 321.
67. Weiss, B.; Prozialeck, W.; Cimino, M., Barnette, M.S.; Wallace, T. L. Ann. N. Y. Acad. Sci. 1980, 356, 319.
68. Humber, L.G.; Bruderlein, F.T.; Voith, K. Mol. Pharmacol. 1975, 11, 833.
69. Lilly, L.B.; Fraser, C.M.; Jung, C.Y.; Seeman, P.; Venter, J.C. Mol. Pharmacol. 1983, 24, 10.

70. Kuno, T.; Tanaka, C. Brain Research 1981, 230, 417.
71. Tanaka, C.; Kuno, T.; Mita, T.; Ishibe, T. In "Molecular Pharmacology of Neurotransmitter Receptors"; Segawa, T. et al., Ed.; Raven: New York, 1983; p. 135.
72. Ruoho, A. E.; Kiefer, H.; Roeder, P. E.; Singer, S. J. Proc. Natl. Acad. Sci. USA 1973, 70, 2657.
73. Changeux, J.-P.; Devillers-Thiery, A.; Chemouilli, P. Science 1984, 225, 1335.
74. Stiles, G. L.; Caron, M. G.; Lefkowitz, R. J. Physiol. Rev. 1984, 64, 661.

RECEIVED January 10, 1985

10

Liquid-Crystalline Solvents as Mechanistic Probes

The Properties of Ordered Chiral Media That Influence Thermal and Photochemical Atropisomeric Interconversions of 1,1′-Binaphthyl

SRINIVASAN GANAPATHY and RICHARD G. WEISS

Department of Chemistry, Georgetown University, Washington, DC 20057

We have investigated various factors which contribute to solvent-induced partial resolution or racemization of 1,1'-binaphthyl (BN). Only photochemical interconversions of BN conducted in cholesteric mesophases influenced the steady state concentration of atropisomers. Thermal equilibrium in cholesteric media or photochemical interconversions in chiral isotropic solvents did not alter appreciably the atropisomeric ratio of initially racemic BN. Solvent order accelerates the rate of BN thermal racemization. A discussion of the physical properties of the solvents and BN responsible for the observations is presented.

The role of liquid-crystalline solvents in affecting the rates and specificities of solute reactions is not clear. In some cases, no detectable influence of solvent order has been reported (2-6) while in others, seemingly quite similar, large effects are found (1,7-16). In the extreme, different laboratories have published conflicting claims for the same reaction performed in the same solvent (2,17-19). The need for care in performing these experiments and in analyzing results from them cannot be emphasized too strongly.

In an attempt to discern the factor(s) most responsible for ordered solvent induced alterations of reaction rates and specificities, we have investigated the influence of cholesteric liquid-crystalline and other optically active media upon the induction or loss of optical activity in the atropisomers of 1,1'-binaphthyl (BN, equation 1). We find that optical induction is negligible from thermal (ground-state) isomerizations (usually <0.1%) but is larger for excited-state isomerizations conducted in cholesteric mesophases (up to 1.1%). The factors responsible appear to be the geometry and polarizability of the BN triplet state and rather specific solvent-solute interactions in ordered

NOTE: This chapter is Part 15 in a series.

0097-6156/85/0278-0147$07.00/0

optically-active phases. Other leading candidates such as the presence of chiral centers within unordered solvent molecules, circularly polarized light, high viscosities and low solvent polarities cannot explain the data.

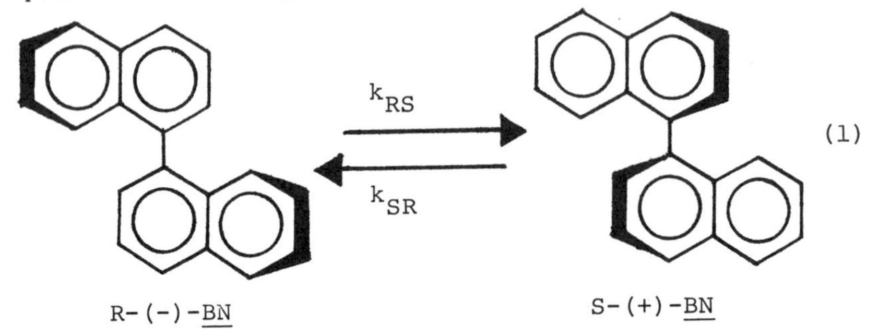

Experimental

Instrumentation. Absorption spectra (uv-vis) were recorded on a Cary 14 or a Perkin-Elmer 552 spectrophotometer using 1 cm quartz cuvettes for isotropic solutions or quartz plates separated by varied widths of Teflon spacers for liquid-crystalline samples. Chromatographic analyses were performed on a Perkin-Elmer 3920 B dual flame-ionization gas chromatograph equipped with an M-2 electronic integrator and a 7' x 1/8" 10% SE-30 on Anakrom column (glpc) or a Waters high performance liquid chromatograph using a 10μ Rad-Pak B silica column (10 cm x 0.8 cm) and both constant wavelength (254 nm) and refractive index detectors (hplc). For preparative hplc separations, two 1' x 7.8 mm 75-125 μ porasil columns were used in series. Optical rotations, accurate within ±0.003°, were measured on Perkin-Elmer 241 and 141 polarimeters using a 1 dm cell. All reported rotations (from which residual rotations from solvent impurities have been subtracted) are the difference between solution and pure solvent measurements. Uncorrected steady-state emission spectra were obtained from room temperature samples on a Perkin-Elmer MPF-2A or Spex Fluorolog spectrofluorometer. Melting points and transition temperatures are corrected. Elemental analyses were performed by Guelph Laboratories, Guelph, Ontario, Canada.

Materials. Benzene was spectrograde (Baker) or reagent grade (Baker, distilled shortly before use). Hplc grade solvents were used as received. 5α-Cholestan-3β-yl nonanoate (CHN), mp 80.5-81.0°C (lit: (9) mp 81.2°C), was available from previous work (20). 5α-Cholestan-3β-yl acetate (CHA) was synthesized by the method of Nerbonne (9), mp 110-111°C (lit: (21) mp 109°C). Cholesteryl nonanoate (CN) from Aldrich was purified by column chromatography (silica; benzene as eluant) and recrystallization from 2-butanone/95% ethanol to yield an enantiotropic cholesteric phase (K = crystal, c = cholesteric, i = isotropic): $T_{K \to c}$ 79-80°C; $T_{c \to i}$ 92.50°C (lit: $t_{K \to c}$ 77.5°C (22), 76.3°C (23); $T_{c \to i}$ 92.0°C (22), 92.1°C (23)).

Cholesteryl oleate (CO; Aldrich, 97%) contained impurities which absorb strongly below 320 nm. Even after purification (see below) it developed extraneous absorptions unless stored and handled in the absence of air and below room temperature. The purification procedure has been described (10). The isolated material exhibited an enantiotropic phase ($T_{K \to c}$ 41-42°C and $T_{c \to i}$ 55.0°C [lit:(24) mp 50.5°C]) and no discernible absorption above 300 nm. Cholesteryl chloride (CCl) from Aldrich was recrystallized from 95% ethanol, mp 96.5-99°C (lit: (9) 95-96°C).

5α-Cholestan-3β-yl dimethylamine (CA) was prepared from reduction of the Schiff's base of dimethyl amine and 5α-cholestan-3-one (25). After 4 recrystallizations from ether, an 8% yield of CA, mp 107-108°C (lit:(26) mp 106°C) and $[\alpha]_D^{35}$ +24±3° ($CHCl_3$) (lit:(26) $[\alpha]_D^{20}$ +23±2° ($CHCl_3$)), was obtained. Calculated for $C_{29}H_{53}N$: C, 83.79; H, 12.85. Found : C, 84.11; H, 12.10.

Racemic BN was synthesized by the method of Sakellarios and Kyrimis (27) or purchased (ICN). After sublimation and two recrystallizations from benzene/abs. ethanol, both samples were >99.5% pure by hplc analyses and displayed mp 145.5-147.5°C (lit:(28) 144.5-145.0°C). Racemic BN was resolved in the solid state by the heating-cooling-recrystallization cycles suggested by Wilson and Pincock (29). The ultimate material exhibited $[\alpha]_D^{29}$ +220°(benzene) (lit:(30) $[\alpha]_{578}^{23}$ 268°(benzene)) corresponding to ca. 90% optical purity of the S(+) atropisomer.

Narrow molecular weight range polybutadiene oligomers from Pressure Chemicals Co., PBD-500 (mol. wt. avg. 420) and PBD-2500 (mol. wt. avg. 2350), were passed through a silica column and stored at -30°C under N_2 in the dark until being used. The fractions were transparent above 355 nm.

Irradiation procedures. Mesophase solutions and neat solid samples of BN were prepared and sealed under N_2 or vacuum in Kimax capillary tubes. Isotropic samples were either degassed (freeze-pump-thaw techniques) and sealed in pyrex tubes or saturated with N_2 in pyrex tubes. Nitrogen was bubbled through the latter solutions during irradiation periods. When thermostatted, samples were placed in a temperature controlled (±1°) water bath. All samples were irradiated with a 450 W Hanovia medium pressure Hg arc and were stored at -30°C until their futher use. Usually, a "dark" sample was prepared and treated in an identical fashion to the irradiated samples except that it was shielded from the light. BN from each tube was recovered by either column chromatography (silica or alumina and pentane eluant) at 4°C followed by solvent removal at 0°C and reduced pressure or by hplc (n-hexane) at room temperature followed by solvent removal at 0°C and reduced pressure. Neat solid samples were dissolved in one of either benzene, tetrahydrofuran or toluene and were frozen until analyzed.

Thermal racemizations of S(+)-BN in mixture C. Solutions of S(+)-BN and the mesophase components (Table I) were prepared in ether. The ether was removed slowly at 20-25°C by reducing the pressure gradually to 0.25 torr over 90 min. After being stirred at 0.25 torr for 90 min. more, the mesophase was saturated with N_2 and transferred to a tightly stoppered 1 cm cuvette. The cuvette was thermostated for 1-2 h prior to recording kinetic data. Aliquots (200-300 mg) were withdrawn periodically over several hours and were frozen immediately. The BN from each aliquot and from the equilibrated mixture (32 h at 22.5°C) was isolated by preparative hplc (hexane). The eluate peaks corresponding to BN were collected in ice-cold flasks (The total elution time of each sample was less than 10 min.) and the solvent was removed at 0°C under reduced pressure. The BN residues were dissolved in benzene just prior to their analysis. Analyses were conducted sequentially for optical rotations, ultraviolet absorptions (from which concentrations were calculated), and rotations after BN racemizations (which yielded residual cholesteric ester contributions).

Thermal racemizations of S(+)-BN in isotropic phases. Samples of optically active BN in PBD oligomers were prepared in dim light as described above. Benzene solutions of S(+)-BN were prepared directly. Each isotropic solution was transferred to a thermostated (22.8°C) polarimeter tube and rotations at several wavelengths were recorded as a function of time after a ca. 1 h equilibration period. After several hours, the temperature was raised rapidly to 43.5°C and rotations were measured as a function of time. Finally, after no further changes in rotation were discernible, the temperature was lowered to 22.8°C and an infinity rotation was taken.

Results

Several liquid-crystalline mixtures have been employed in this work. They and several of their physical characteristics are collected in Table I. As expected, the bulky BN molecules disturb mesophase order causing transition temperatures to be lowered. Monotropic c→K transitions are approximate since they depend upon the rate of cooling and other factors.

Another manifestation of BN disruption of solvent order can be obtained from changes in the pitch band whose wavelength is directly proportional to the distance between cholesteric "layers" with parallel constituent molecules (31). At 18°C, the pitch band of either neat mixture C or containing 1% BN is in the infrared region, beyond the limit of our spectrophotometer. Increased BN loading moves the pitch band progressively through the visible and into the ultraviolet region: with 1.5% BN, the pitch band maximum occurs at 880±50 nm; with 3% BN, the maximum shifts to 430±50 nm.

Significantly, BN excitation and emission spectra in mixture B (monotropic phase) at room temperature are red-shifted with respect to those obtained in hexane: for λ_{excit} 290 nm, the emission maxima were 362 nm (mixture B) and 359 nm (hexane); for $\lambda_{emission}$ 365 nm, the excitation maxima were 302 nm (mixture B) and 292 nm (hexane). A very small solvent shift was observed in the wavelength maxima for the BN absorption spectra (293 nm in mixture B and 291 nm in hexane). However, as in the excitation spectra, the absorption spectrum in mixture B tailed significantly farther into the red than did the spectrum in hexane.

The properties of the PBD oligomers have been summarized by us previously (11). The most important of these for our work are the bulk viscosity (13.1 cp for PBD-500 and 506 cp for PBD-2500 at 25°C) and the dielectric constant (estimated to be near that of benzene (11)).

Table I. Cholesteric mesophases and their transition temperatures.

Mixture Designation	Cholesteric Mixture (by wt %)	phase type[a]	BN (by wt %)	K→c[b,c] (°C)	c→i[b] (°C)	i→c[b] (°C)
A	50/50 CHN/CHA	m	0	<35		60-59
		m	2	<35		53
B	65/35 CN/CHA	e	0	50-51	78	
		m	0	~21		
		e	2	~50	70-71	
		m	2	~21		
C	60/26/14 CO/CN/CHA	e	0	<17	61-62	
		e	0.5	<17	59	
		e	2.0	<17	52	
D	70/30 CN/CCl	e	0		80	
		e	0.5	~49	76	
E	60/25/10/4 CO/CN/CA/CHA	e	2	<20	51.5	

a) e = enantiotropic, m = monotropic; b) K = crystal, c = cholesteric, i = isotropic; c) transitions are c→K for monotropic phases.

Sample preparation and equilibration for following the thermal racemization of S(+)-BN in mixture C or the PBD oligomers entails significant (>40%) loss of optical activity. Fortunately, the initial rotation of our S(+)-BN was high ($[\alpha]_D$ +220°) and allowed us to follow the kinetics over at least two half-lives. Rate constants, k, could be extracted from raw data using a standard first-order treatment (equation 2). Since the equilibrium constants between R- and S-BN are very near one in optically-active media and are exactly one in achiral solvents, we can take $k_{RS} \simeq k \simeq k_{SR}$ (see equation 1). The α_o, α_t, and α_∞ represent, respectively, rotations at an arbitrarily defined initial time, at t minutes thereafter, and when no further rotational changes can be detected. The analytical procedure with mixture C as solvent involved isolation of partially racemized BN from aliquots taken at various intervals. Since the amount of recovered BN varied from aliquot to aliquot, equation 2 for mixture C was modified to replace observed rotations with specific rotations. Upper limits to solvent component impurities (CO, CN, CHA) in these samples were determined from residual rotations after allowing the polarimeter solutions to remain at room temperature for periods which ensure >99% racemization of BN. This method indicated that no more than ca. 10% of the initial rotation could have been due to residual solvent.

$$\ln \frac{\alpha_t - \alpha_\infty}{\alpha_0 - \alpha_\infty} = -2kt \quad (2)$$

Representative rate plots for data obtained at 23°C in PBD and benzene using equation 2 are shown in Figure 1. The same type of plot for data from mixture C (Figure 2) shows much more scatter. Intrinsically, each of the points in this Figure is very difficult to obtain. In fact, data from three separate runs conducted on three different days have been used in order to decrease somewhat the experimental uncertainty. Rate contants obtained from the slopes of these and other plots are collected in Table II. The agreement between our rate constant in benzene and that of Colter and Clemens (32) is excellent.

Table II. Rate constants for thermal racemization of 0.5% (by weight) S(+)-BN in various solvents.

Solvent	Temperature (°C)	10^4k^a (min^{-1})	
benzene	23.0	3.67±0.01	
	43.8	53.8±0.8	(50.7^b)
PBD-500	22.8	4.75±0.02	
	43.5	67±2	
PDB-2500	22.6	4.68±0.08	
	43.3	59±3	
mixture C	22.5-23.0	15±2	

a) Average rate contant from determinations of rotations at 589, 546, 436, and 365 nm. Errors are one standard deviation.

b) From Colter and Clemens (32) at 43.9°C. The rate constant reported by these authors is 101.4 min^{-1} for racemization (i.e., $2k = k_{RS} + k_{SR}$).

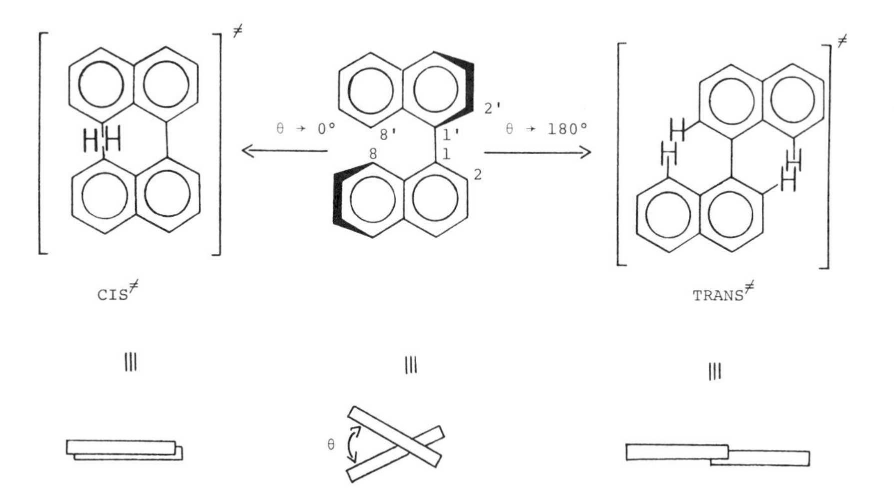

Figure 1. "Cis" and "trans" transition state geometries for interconversion of atropisomers of BN with representation along 1-1' bond.

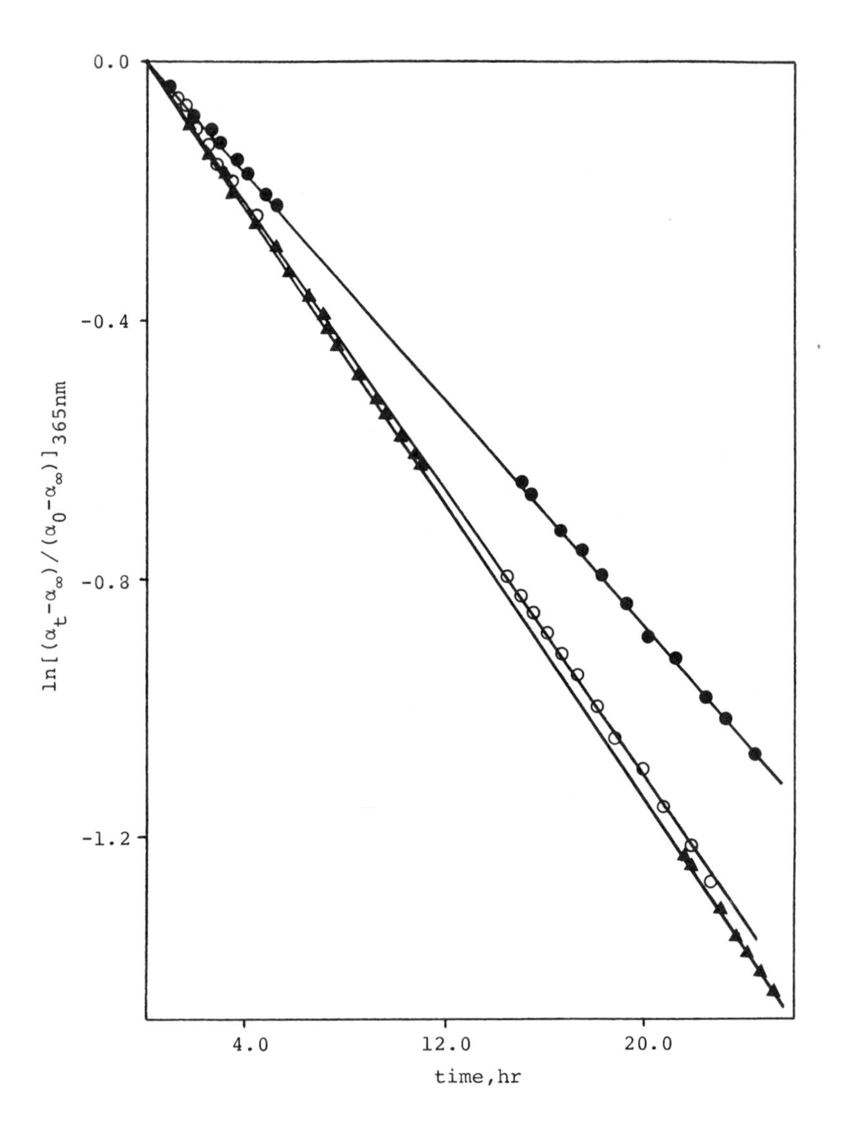

Figure 2. Plots of $\ln[(\alpha_t-\alpha_\infty)/(\alpha_0-\alpha_\infty)]$ from data at 365 nm versus time for the thermal racemization of S(+)-BN in benzene (•), PBD-500(▲), and PBD-2500(o) at 23.0°C.

The monotropic nature of mixture A created foreseen difficulties in its use as a medium for thermal or photochemical BN resolutions (9): sporadically, some mesophases of mixture A solidified during experiments. Mixture B at monotropic phase temperatures was well-behaved and its BN solutions could be kept for hours without noticeable change.

In each mesophase mixture, the dominant atropisomer after photo or thermal resolution experiments was S(+). The sole exception occurred during irradation of a 3% BN solution in mixture C. Since the sample solidified partially during the experiment, the mechanism by which the R(-) atropisomer arose is unclear. The thermal lability of the atropisomers toward interconversion and the possible contribution of cholesteric contaminants in recovered BN samples make an accurate assessment of atropisomeric excess a formidable task. Extreme care was taken to handle all solutions containing BN during work-up at temperatures which preclude significant thermal racemization: at 25°C, the half-life for racemization is ca. 10 h in normal isotropic solvents; all manipulations were conducted at 4°C or below. Since some of the cholesteric molecules are dextrorotatory and others are levorotatory, only minimum limits of their contamination (assuming that only one species is present in each BN solution) could be calculated from residual rotations after allowing BN to racemize for several days. Results from an independent analysis, a ferric chloride color test (33), confirmed the conclusions from the prior analyses that rotations of solutions of recovered BN contained only a small percentage contribution from residual cholesteric impurities. Rotations in Tables III and IV are corrected for the cholesteric contributions.

Although the atropisomeric excesses in the mesophases are difficult to reproduce exactly, a clear pattern has emerged: photoresolution produces much greater optical activity in BN than thermal resolution.

Addition of a ca. 10% concentration of CA to a mesophase solution (mixture E) of BN quenches completely its photoresolution. CA quenches singlet states of BN in n-hexane at a nearly diffusion-controlled rate ($k_q \simeq 10^{10}$ $M^{-1}s^{-1}$ assuming τ_{BN} = 3 ns (34-36)). Thus, even in a very viscous medium like cholesteric mixture E, static and dynamic quenching should preclude formation of BN triplets.

Irradiation at room temperature of BN isotropic solutions containing largecon centrations of optically active molecules, CN or CA, led to no measurable resolution. When heated in an optically active solvent, (-)-2-methyl-1-butanol, recovered BN remained racemic. On the other extreme, irradiation of crystals of optically active BN at room temperature produced no measureable racemization. When recrystallized very slowly at room temperature from (-)-2-methyl-1-butanol, BN displayed no measureable optical activity.

Table III. Attempted photoresolutions of BN in various media.

Solvent	BN (% by wt or M)	T (°C)	irradiation time (h)	$[\alpha]^{18}_{578}$(solvent)[a]	atropisomeric[b] excess (%)
mixture A	2	40	1	+2.4°(ether)	1.1
	2	40	1	+2.1°(ether)	1.0
	2	40	2	+1.0°(ether)	0.5
mixture B	2	40	0.5	+1.7°(THF)	0.8
	2	40	0.75	+0.5°(THF)	0.2
mixture C	1	18	10	+0.6°(benzene)[c]	0.2
	1	18	10	+0.4°(benzene)[c]	0.2
	1.5	18	10	+0.9°(benzene)[c]	0.4
	1.5	18	10	+0.4°(benzene)[c]	0.2
	2	20	2.5	+0.8°(THF)	0.4
	3[d]	18	10	+1.4°(benzene)[c]	0.6
	3[d]	18	10	−0.6°(benzene)[c]	0.2
	3[d]	18	15	0.4°(benzene)[c]	0.2
	3[d]	18	15	−0.6°(benzene)[c]	0.2

mixture E	2	20	5	0.1°(benzene)[c]	<0.1
neat crystal	-	19	20	+182°(benzene)[c]	e
	-	20-21	90	+184°(benzene)[c]	e
n-hexane + 10^{-2} M CA	2.1×10^{-3} M	20	5	0.1°(benzene)[c]	0.1
4/6(w/w) benzene/CN	2	17.5	10	0° (benzene)[c]	0

a) $[\alpha]_{578}$ 256° (benzene), 217° (ether), and 213° (THF) are taken for optically pure BN. They are correlated to Wilson and Pincock's reported $[\alpha]_D$ 245° (benzene) (29) via a common sample of partially resolved BN. The values in ether and THF are significantly different from those reported by Browne et al. (30); b) calculated using [α] values at 578, 546, 436, and 365 nm; error estimated to be ±0.2%; c) rotations observed at 37±1°C; d) solid formed during irradiation; e) no change within experimental error from initial rotation.

Table IV. Attempted thermal resolution of 2% BN in cholesteric mesophases.

Reaction Solvent	Temperature (°C)	heating time (h)	$[\alpha]^{18}_{578}$(solvent)	atropisomeric excess (%)[b]
mixture A	40	1	0° (ether)	0
	40	1	0° (ether)	0
	40	2	0.8° (ether)	0.4
mixture B	40	0.75	0.2° (ether)	0.1
mixture C	20	2.5	0.2° (THF)	0.1
	30	24	0.2° (benzene)[c]	0.1
	50	26	0.2° (benzene)[c]	0.1
	18[a]	10	0.2° (benzene)[c]	0.1
mixture E	20	5	0.1° (benzene)[c]	<0.1
(-)-2-methyl-1-butanol[d]	50	4.25	0° (benzene)[c]	0

a) 1% BN; b) calculated using [α] values at 578, 546, 436 and 365 nm; error estimated to be ±0.2%; c) rotations observed at 36°C; d) 3×10^{-2} M BN.

Discussion

The most favored conformation for BN is calculated to have a twist angle θ for the naphthyl rings of 75-105° (37). From x-ray data, θ in the optically active and racemic crystalline forms is 102-103° (38, 39) and 68° (39, 40), respectively.

The ability to separate atropisomers of BN results from severe steric interactions which inhibit the two naphthyl rings from becoming coplanar (θ = 0° or 180°). In the less favored "cis" transition state, strong interactions between hydrogens on C_8 and C_8' create a calculated activation energy for interconversion which exceeds 35 kcal mol^{-1} (41). This value is probably near the correct one since the calculated activation energy for the "trans" transition state which brings the hydrogens on C_2 and C_8' and on C_2' and C_8 to less than the sum of their van der Waals radii (20.2 kcal mol^{-1} (41)) and the observaved activation energy for racemization (32) are within 2 kcal of each other.

CNDO/S calculations (42) predict that the first excited singlet state of BN will remain non-planar. Although some conformational relaxation does occur in ^{1}BN (35, 43) on a picosecond time scale (44), it does not appear to lead to atropisomeric equilibration at room temperatures.

Our spectroscopic studies of BN in mixture B and in hexane support our contention that ground state conformers are forced by cholesteric mesophases toward extremes of θ (i.e., closer to 0° or 180° than in hexane solvent). As the two naphthyl groups become more coplanar, their π-overlap increases. Consequently, the o-o transitions in absorption (and excitation) occur at longer wavelengths (lower energies) (43). For the same reasons, the cholesteric solvent compresses excited singlets of BN, causing their fluorescence spectra to be red-shifted with respect to those in hexane.

Compelling evidence for equilibration from ^{3}BN, with an attendant small activation barrier has been provided by Irie and coworkers (45). They claim that $E_a \simeq 1.9$ kcal mol^{-1}. However, at lower temperatures, a conformational relaxation process with a 9.3 kcal mol^{-1} barrier is identified. It is ascribed to a motion which brings twisted ^{3}BN to a nearly planar (achiral) structure (45). The latter activation barrier is much more compatible with the measured rate constant for photoracemization at room temperature ($k_r = 2.4 \times 10^6$ s^{-1} (46)) than the former which has been selected by Irie et al. (45). To complicate matters further, a triplet chain process dominates at higher concentrations of BN (46).

It is interesting to speculate upon the activation barrier for racemization from the singlet state. In order for racemization to occur, it must compete kinetically with unimolecular deactivation pathways like fluorescence and intersystem crossing. The singlet lifetime of BN is ca. 3 ns (34-36) so that if 10% of the singlets were to racemize, the rate constant for this process would be ca. 3×10^7 s^{-1}. From the Arrhenius equation and taking the preexponential factor $A = 10^{13}$ s^{-1}, E_a is calculated to be no more than 6 kcal mol^{-1}. Given the absence of measurable singlet derived racemization at room temperature, E_a must exceed this value.

It is unlikely that a triplet chain mechanism is important in our photoresolution experiments conducted in liquid-crystalline media. In scrupulously oxygen-free tetrahydrofuran, the ^{3}BN lifetime, $^3\tau$, is ca. 10^{-5} s and the rate constant for energy transfer of a triplet of one atropisomer to the ground state of another, k_{et}, is ca. 10^8 $M^{-1}s^{-1}$ (46) (or only 10^{-2} of k_{diff}, the rate constant for self-diffusion). We can estimate an upper limit to k_{diff} = $3x10^7$ M^{-1} s^{-1} at room temperature in liquid-crystalline media from the rate constant for fluorescence quenching of plate-like pyrene singlets by pyrene in a 59.5/24.9/15.6 mixture of CO/CCl/CN (47). Using a 2% (≃$8x10^{-2}$ M) loading of BN, a reasonable estimate of the rate of chain propagation in the liquid-crystalline phases is $10^{-2}x3x10^7x8x10^{-2}$ = $2.4x10^4$ s^{-1}. Since the lifetime of ^{3}BN in our liquid-crystalline media is probably signficantly shorter than that observed in tetrahydrofuran (i.e., $1/^3\tau > 10^6$ s^{-1} due to pseudo zero-order quenching by residual oxygen and trace solvent impurities), the ability of ^{3}BN to energy transfer to BN before decaying unimolecularly is slim. However, we cannot eliminate completely the presence of the chain transfer mechanism.

Resolution attempts in cholesteric phases. The body of data collected to date clearly indicates that unless specific solute-solvent interactions occur, the stereochemistry of reactions will be little affected by chiral solvents, whether they be macroscopically ordered or isotropic (48-50). In fact, the low optical activity in products from irradiations in cholesteric solvents may arise from the ability of a chiral mesophase to produce circularly polarized light from normal incident radiation (51).

In spite of these problems, induction or disappearance of optical activity in products affords a sensitive measure of the degree to which cholestric liquid-crystalline or other chiral solvents interact with reacting solutes. When absolute rotations are large, as they are for BN (30) even small interactions can be detected. On a molecular level, these interactions can be considered diastereomeric and arising via differential solvation (52) of the R or S form of BN by enantiomerically pure neighbors. The macro order imposed by a cholesteric phase may result in more complex interactions. It is these which we seek to probe.

Thermal experiments with BN in cholesteric phases. Solvent-solute interactions and, specifically, whether the solvent molecules exert aggregate effects upon a reacting solute can be probed from the kinetics of racemization of optically active BN. The results in Table II show a clear rate acceleration for racemization conducted in the cholesteric phase of mixture C. One possible explanation, that changes in the dielectric constant of our solvents induce changes in the rate, can be dispatched easily: the dielectric constant of mixture C ($\varepsilon \sim 3-6$ (53, 54)) is much lower than that of dimethylformamide (ε 36.7) although at 23°C the rate constant for racemization in mixture C ($1.5x10^{-3}$

min^{-1}) is higher than in dimethylformamide ($1.1x10^{-3}$ min^{-1} (32)). Viscosity, another experimental parameter, changes with reaction rate oppositely to the expected. It is known that high viscosities slow molecular rotations (55-57); our results indicate that a 40-fold increase in viscosity (between PBD-500 and PBD-2500) has no measurable effect upon the racemization rate. Furthermore, in cholesteric mixture C, which is much more viscous than PBD-2500 (58-60), racemization proceeds more rapidly.

Aggregate order of a cholesteric mesophase is described crudely as a layer-like arrangement in which the long axes of the constituent molecules within a "layer" are (on average) parallel to one another. The distance between "layers" and the twist angle between them determines the position of the pitch band (31, 61). Changes in both the pitch bands and transition temperatures demonstrate that introduction of BN guest molecules into a cholesteric matrix has a large disturbing influence. We believe that this can be attributed to the globular shape of BN which cannot be accomodated easily by the cholesteric matrix. Previous investigations by others (62-64) and us (11, 65) indicate that plate-like or rod-like guests have a much smaller disturbing influence upon local memosphase order than do globular guests. In its attempt to force BN into a shape which is more amenable to its macrostructure, the mixture C may compress the twist angle θ. Although this would increase the internal energy of BN, it would decrease the solvent free energy. Thus, the energy required to attain the "trans" transition state (θ = 180°) would be lower in mixture C than in isotropic solvents of comparable polarity and viscosity (Figure 3). The bathochromic shifts observed in the excitation and emission spectra of BN in mixture B are completely consistent with this hypothesis. We have observed a completely analogous deceleration effect upon isomerization rates when a plate-like reactant adopts a globular transition state (11). Pincock et al. (68) find that thermal racemization of BN can be catalyzed (probably for the same reasons as those given above) by carbon-black and other graphite-like particles into which BN can intercalate.

Figure 4 shows the ground-state energies of R- and S-BN to be almost identical even though the mesophase solvents are chiral. We arrive at this conclusion from the lack of more than a slight atropisomeric excess (ca. 0.1% in all but one anomalous experiment) after equilibration of racemic BN in the cholesteric phases at several temperatures (Table IV). The lack of change in the ratio of atropisomers in the cholesteric phases is consistent with our observation that liquid-crystal induced circular dichroism spectra (67) of BN in cholesteric mixture D are due to a macroscopic property of the solvent: the LCICD spectra disappear when mixture D is heated to an isotropic temperature.

Photochemical experiments with BN in cholesteric and optically active isotropic phases. Photoinduced interconversions of BN atropisomers are in competition with thermally induced racemization at the temperatures of our experiments. Thus, the observed rotations reflect lower limits to the actual atropisomeric pho-

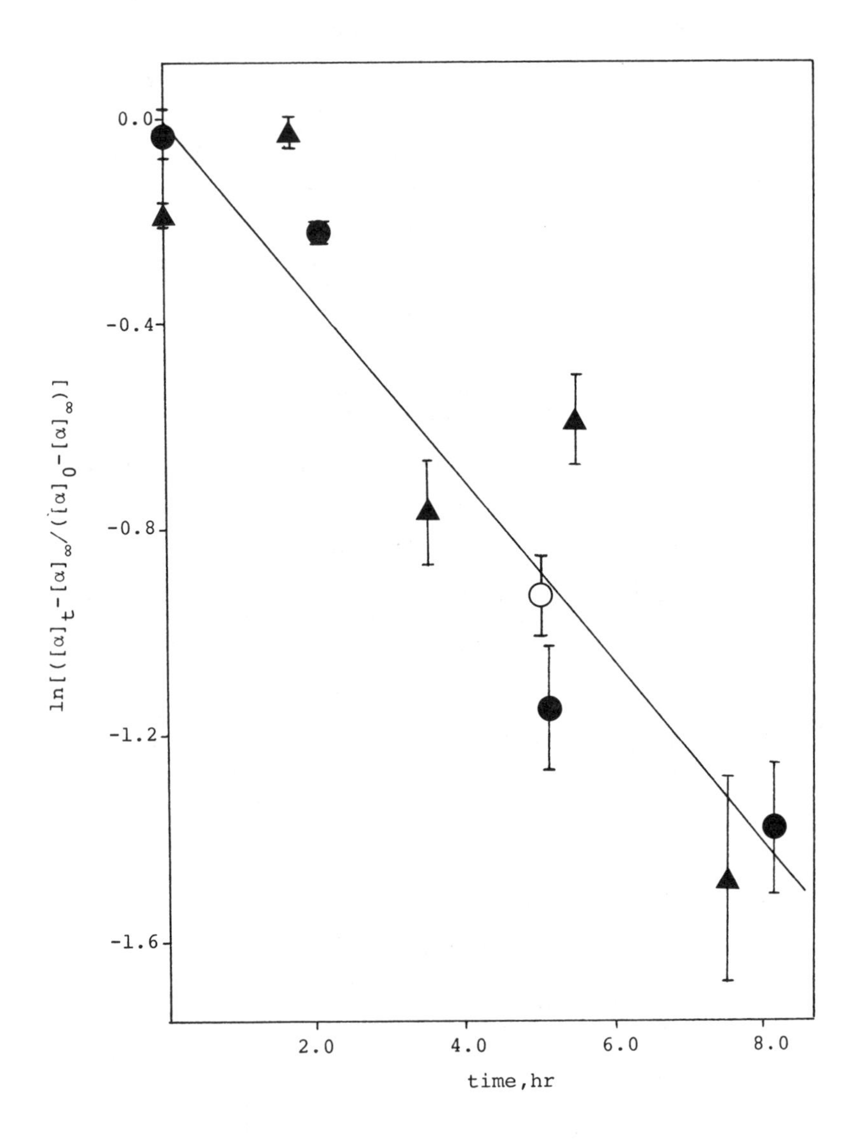

Figure 3. Plot of $\ln[([\alpha]_t-[\alpha]_\infty)/[\alpha]_0-[\alpha]_\infty)]$ versus time for the thermal racemization of S(+)-BN in cholesteric mixture C at 23.0°C. Data from three independent kinetic runs represented by O, ●, and ▲ are fitted to a single line. Each data point represents the mean of a set of data points obtained from at least three different wavelengths (589, 436, and 365 nm). Error bars indicate one standard deviations.

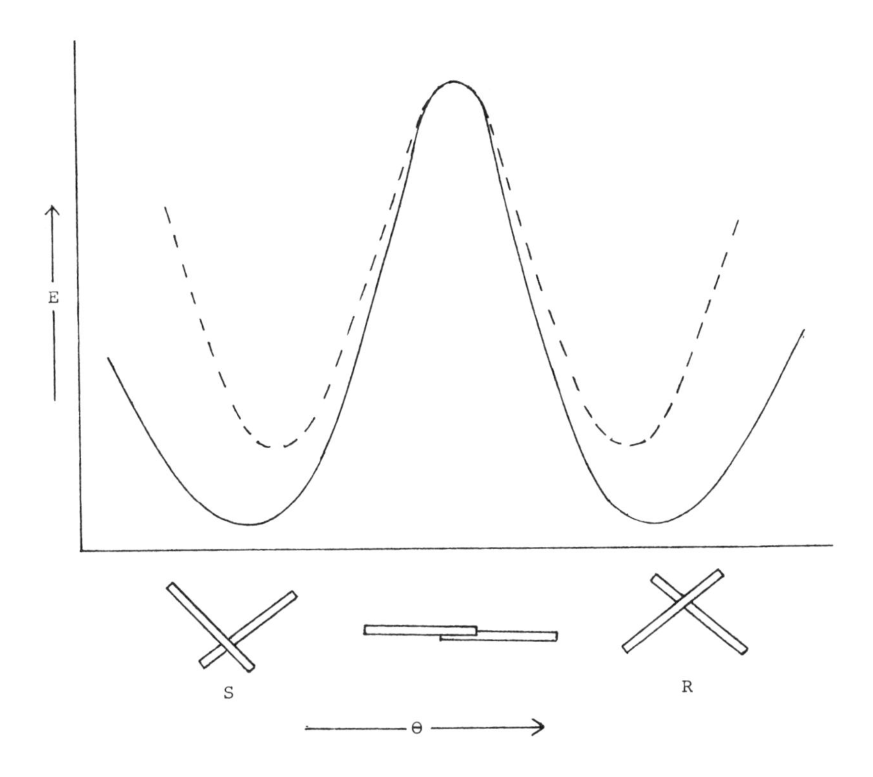

Figure 4. Hypothetical potential energy curves of BN along the coordinate for atropisomeric interconversion in a cholesteric liquid-crystalline phase (---) and an isotropic phase (——).

tostationary states. A part of the scatter associated with replicated experiments may be due to differences in the light flux bathing the samples.

The absence of optical activity in BN isolated after its irradiation in a benzene/CN solution (Table III) is completely consistent with the need for macroscopic order as well as a chiral environment. However, in another phototransformation, the synthesis of hexahelicene in chiral isotropic solvents, 0.2-2.0% enantiomeric excesses were found (68).

Dynamic constraints imposed by the crystalline lattice of racemic BN can be overcome by heat. However, the activation energy for solid state atropisomerism is estimated by Wilson and Pincock to be ca. 60 kcal mol^{-1} (69). In spite of the fact that the reactive triplet state of BN lies ca. 60 kcal mol^{-1} above its ground state (70), our attempts to photoracemize or photoresolve further crystals of 68% optically pure BN at 20°C were unsuccessful. Although it may be premature to derive a great deal from this result, it is tempting to speculate that much of the excitation energy is dissipated throughout the lattice before it can be localized in the torsional motion specific to isomer interconversion. Since our lamp is capable of exciting only a small percentage of molecules at any one time, the total lattice never finds itself with sufficient energy to allow a solid-solid transformation. If this conjecture is correct, solid state atropisomerism is a cooperative phenomenon.

The lack of photoresolution from irradiations of BN in an isotropic hexane/CA solution or in cholesteric mixture E is somewhat surprising. Irie *et al.* (71, 72) have shown that optically active tertiary amines quench the excited singlet states of BN atropisomers at two different rates. The disparity between them is greatest in non-polar solvents (hexane) and diminishes to zero in polar ones (acetonitrile). The quenching mechanism involves formation of a singlet exciplex which, presumably, is sterically most specific and holds the quencher and quenchee most closely in non-polar solvents. Although it is well-established that ^{3}BN is the photoreactive state when it is uncomplexed, the fate of BN in chiral exciplexes has not been determined. Neither our rough lower limit to adiabatic interconversion on the excited singlet potential surface (*vide ante*) nor the singlet excitation energy (a lower limit (35) estimated from the overlap point of normalized absorption and fluorescence spectra in hexane (34)), 88 kcal mol^{-1}, precludes interconversion. Yet, we find that racemic BN remains so when its excited singlets are quenched by CA in both isotropic and anisotropic media of low polarity. We have not investigated the fate of ^{3}BN with CA. Presumably, it is quenched also and in a fashion which either respects the integrity of the original atropisomer or is insensitive to the chirality of CA. The conditions of our experiments allow for quenching of the majority of the singlet states of BN by CA.

The photoresolution of BN proceeds most efficiently in chiral media without singlet quenchers and with macroscopic order. Several factors may contribute to the necessity of long-range order. One of these is the ability of cholesteric media to reflect selectively one component of circularly unpolarized light and transmit the other (73). Manifestations of macroscopic order include observation of LCICD spectra (67) and the induction of optical activity in photoreactive, achiral solutes (74).

That both phenomena arise as a consequence of macroscopic solvent order and not intimate solvent-solute interactions is clear: Saeva and Olin (75) have shown that solute LCICD spectra can be observed in twisted nematic phases only; Nakazaki *et al.* (76) find an excess of one enantiomer of hexahelicene is produced photochemically from achiral precursors in twisted nematic phases; no LCICD spectra or optical induction occurs in untwisted nematic phases and the handedness of the twist can be correlated with the sign of the LCICD and the preferred product enantiomer. Furthermore, isotropic phases of cholesteric mixtures display no discernible LCICD spectra (12, 67) and the enantiomeric excesses in products of photolabile reactants in isotropic phases are near zero (51).

The enantiomeric excesses in chiral products reported from circularly polarized irradiation of achiral reactants in achiral solvents are usually less than 0.2-0.3% (77, 79). The largest enantiomeric excess reported is 1.6% for an unrelated rearrangement (80). Furthermore, it is known that circularly polarized irradiation of BN in dichloromethane yields very low (<0.2%) atropisomeric excesses (81). Therefore, the 1.1% atropisomeric excess of BN observed upon irradiation of solutions in mixture A cannot be ascribed to circularly polarized excitations.

Possible explanation for increased chiral recognition from experiments with solvent order and irradiation. The necessity of both macro-solvent order and interconversion via an excited state of BN in order to observe the largest atropisomeric excesses indicates that a combination of factors, working in concert, may be responsible. In toto, these must result in a greater interaction energy (and, therefore specificity of interaction) between solvent molecules and excited BN.

Dispersion energies, E_D, between two molecules may be approximated by London's expression (82) (equation 3) in which I and α are the ionization potential and polarizability, respectively, of BN and a solvent molecule (S), and r is the distance between them. Several variables contribute to changes in E_D. Any one of them may dominate. A lack of knowledge of experimental properties for S, BN, and their solutions requires that our arguments concerning these variables remain qualitative. The ionization potentials of the cholesteric solvent components are higher than that of BN. In this equation, both α and r are treated as scalar quantities. The shapes of BN and the cholesteric molecules and their electronic distributions clearly indicate that some orientations between them will be favored over others. This means that no single r value can represent the true nature of the distance dependence between an S and a BN. The r^{-6} dependence requires that S and BN be very close if dispersion forces are non-negligible. Additionally, more than one S molecule may interact separately with the naphthyl rings of BN (52, 83).

$$E_D = -\frac{3}{2}\,\frac{\alpha_S\alpha_{BN}}{r^6}\,\frac{I_S I_{BN}}{I_S + I_{BN}} \qquad (3)$$

To compare dispersion energies between S and either ground state or triplet BN, α, I, and r must be discussed. In all likelihood r should be treated as a sum of vectors or their expectation value $\langle r \rangle$. Either should differ slightly when S interacts with ground-state and triplet BN: $\langle r \rangle$ is probably not very different since the ordering of the cholesteric matrix should be a primary influence upon the location of BN. On the other hand, the distribution of locations at which a BN is in contact with an S may be narrowed when the solute is in an excited state. If true, a change in E_D (due to a factor related to r but different from it) would occur. In any case, the sixth power dependence upon r means that even small differences in $\langle r \rangle$ between ground and excited state BN may change E_D significantly. Since $I_{^3BN} \simeq I_{BN} - E_T$ (where $E_T \simeq 2.5$ eV is the triplet excitation energy (70)) and $I_S > I_{BN}$, the term $I_S I_{BN}/(I_S + I_{BN})$ must be smaller when BN is a triplet than when it is a ground-state singlet. However, the decrease caused in E_D may be more than compensated for by an increase in $\alpha_{^3BN}$ over α_{BN} or a decrease in r.

The value of α is dependent upon the distortion of electron clouds from their "normal" shape. In quantum mechanical terms, α is a function of the second power of d, the change in the average distance between a nucleus and an electron under the influence of an external electric field (84). Since the triplet state places an electron in a more loosely bound anti-bonding orbital, it should be much more easily distorted than electrons in bonding orbitals of the ground state which do not protrude effectively as far into space. The first excited singlet (^1L_a) of naphthalene has a measured polarizability (ca. 27Å^3) which is much larger than the ground state (ca. 17Å^3)(85). A very large increase in α is calculated to occur along the long molecular axis upon excitation (86). Unfortunately, corresponding data do not exist for BN triplets and the extent to which naphthalene can be taken as a model of BN is unclear. Even with this uncertainty, the data suggest that $\alpha_{^3BN}$ will be significantly larger than α_{BN}. Furthermore, if the long axes of the BN naphthyl groups lie parallel to the long axis of the cholesteric solvent molecules (along which the largest polarizability vector obtains (87)), the effect of BN excitation on E_D will be much larger than predicted by the average (scalar) α.

Similarly, the reaction field, R (88-90), associated with a group of solvent molecules with cholesteric phase order is much larger when operating on a triplet of BN: R increases with increasing α. The limitations of the Onsager model to the very anisotropic environment experienced by ^{3}BN preclude a reasonable quantitative discussion. The solute cavity is not spherical; ^{3}BN may be described better for the purposes of elucidating its interactions with neighboring solvent molecules as a quadrupole (i.e., a sum of two dipoles (91)); the solvent dielectric is anisotropic instead of homogeneous (53). In spite of this, it is obvious that ^{3}BN will be more sensitive to the reaction field than will ground-state BN. The effect of solvent molecule alignment as exhibited by cholesteric phases should increase the reaction field over that experienced by a solute in an isotropic medium. Thus, the greater chiral recognition between solvent and ^{3}BN may be related to the larger reaction field and dispersion energy.

In conclusion, we believe that our ability to observe higher atropisomeric excesses from irradiations of BN in cholesteric mesophases than from thermal isomerizations can be traced to the larger interaction energies associated with the excited state species and its environment. The cumulative effect of these interactions is manifested more specifically on a reactive solute when the solvent molecules are uniquely ordered than when they are isotropically dispersed.

Acknowledgments. We wish to thank Drs. Niel Glaudemans and Kent Rice of National Institutes of Health for their gracious help and for allowing us to use their electronic polarimeters. Dr. Miklos Kertesz is thanked for several enlightening discussions. Mrs. Kay Bayne is thanked for her expert help in preparation of the manuscript. The National Science Foundation (Grant No. CHE 83-01776) is acknowledged for its support of this work.

Literature Cited

1. Part 14: Hrovat, D.A.; Liu, J.H.; Turro, N.J.; Weiss, R.G. J. Am. Chem. Soc., in press 1984.
2. Eskenazi, C.; Nicoud, J.F.; Kagan, H.B. J. Org. Chem. 1979, 44, 995.
3. Dondoni, A.; Medici, A.; Colonna, S.; Gottarelli, G.; Samori, B. Mol. Cryst. Liq. Cryst. 1979, 55, 47.
4. Bacon, W.E.; Brown, G.H. Mol. Cryst. Liq. Cryst. 1971, 12, 229.
5. Dewar, M.J.S.; Mahlovsky, B.D. J. Am. Chem. Soc. 1974, 96, 460.
6. Cassis, E.G., Jr.; Weiss, R.G. Photochem. Photobiol. 1982, 35, 439.
7. Seuron, P.; Solladie, G. J. Org. Chem. 1980, 45, 715.
8. De Maria, P.; Lodi, A.; Samori, B.; Rusticelli, F.; Torquati, G. J. Am. Chem. Soc. 1984, 106, 653.
9. Nerbonne, J.M.; Weiss, R.G. J. Am. Chem. Soc. 1979, 101, 402.
10. Anderson, V.C.; Craig, B.B.; Weiss, R.G. J. Am. Chem. Soc. 1981, 103, 7169.
11. Otruba, J.P., III; Weiss, R.G. Mol. Cryst. Liq. Cryst. 1982, 80, 165.
12. Anderson, V.C.; Weiss, R.G. J. Am. Chem. Soc., in press 1984.
13. Hrovat, D.A.: Liu, J.H.; Turro, N.J.; Weiss, R.G. J. Am. Chem. Soc., 1984, 106, 5291.
14. Aviv, G.; Sagiv, J.; Yogev, A. Mol. Cryst. Liq. Cryst. 1976, 36, 349.

15. Liebert, L.; Strzelecki, L.; Vacogne, D. Bull. Soc. Chim. Fr. 1975, 2073.
16. Kunieda, T.; Takahashi, T.; Hirobe, M. Tetrahedron Lett. 1983, 5107.
17. Saeva, F.D.; Sharpe, P.E.; Olin, G.R. J. Am. Chem. Soc. 1975, 97, 204.
18. Verbit, L.; Halbert, T.R.; Patterson, R.B. J. Org. Chem. 1975, 40, 1649.
19. Pirkle, W.H.; Rinaldi, P.L. J. Am. Chem. Soc. 1977, 99, 3510.
20. Nerbonne, J.M. Ph.D. Thesis, Georgetown University, Washington, DC, 1978.
21. North, B.E.; Shipley, G.G.; Small, D.M. Biochem. Biophys. Acta 1976, 424, 376.
22. Gray, G.W. J. Chem. Soc. 1956, 3733.
23. Ennulat, R.D. Mol. Cryst. Liq. Cryst. 1969, 8, 247.
24. Davis, G.J.; Porter, R.S.; Steina, J.W.; Small, D.M. Mol. Cryst. Liq. Cryst. 1970, 10, 331.
25. Dodgson, D.P.; Haworth, R.D. J. Chem. Soc. 1952, 67.
26. Borch, R.F.; Beinstein, M.D.; Durst, H.D. J. Am. Chem. Soc. 1971, 93, 2897.
27. Sakellarios, E.; Kyrimis, T. Chem. Ber. 1964, 57, 324.
28. Badar, Y.; Ling, C.C.K.; Cooke, A.S.; Harris, M.M. J. Chem. Soc. 1965, 1543.
29. Wilson, K.R.; Pincock, R.E. J. Am. Chem. Soc. 1975, 97, 1474.
30. Browne, P.A.; Harris, M.H.; Manzengo, R.Z.; Singh, S. J. Chem. Soc. C 1971, 3990.
31. Baessler, H.; Labes, M.M. Mol. Cryst. Liq. Cryst. 1970, 6, 419.
32. Colter, A.K.; Clemens, L.M. J. Phys. Chem. 1964, 68, 651.
33. Kates, M., In "Laboratory Techniques in Biochemistry and Molecular Biology"; Work, T.S.; Work, E., Eds.; Elsevier: Amsterdam, 1972; Vol. 3, p. 360.
34. Berlman, I.B. "Handbook of Fluorescence Spectra of Aromatic Molecules," 2nd Ed.; Academic Press: New York, 1971; p. 352.
35. Post, M.F.; Langelaar, J.; Van Voorst, J.P.W. Chem. Phys. Lett. 1975, 32, 59.
36. Luo, X.-J.; Beddard, G.S.; Porter, G.; Davidson, R.S. JCS, Faraday Trans. I 1982, 78, 3477.
37. Gamba, A.; Rusconi, E.; Simonetta, M. Tetrahedron 1970, 26, 871.
38. Pauptit, R.A.; Trotter, J. Can. J. Chem. 1983, 61, 69.
39. Kress, R.B.; Duesler, E.N.; Etter, M.C.; Paul, I.C.; Curtin, D.Y. J. Am. Chem. Soc. 1980, 102, 7709.
40. Kerr, K.A.; Robertson, J.M. J. Chem. Soc. B 1969, 1146.
41. Carter, R.E.; Liljefors, T. Tetrahedron 1976, 32, 2915.
42. Bigelow, R.W.; Anderson, R.W. Chem. Phys. Lett. 1978, 58, 114.
43. Post, M.F.M.; Eweg, J.K.; Langelaar, J.; Van Voorst, J.D.W.; Ter Maten, G. Chem. Phys. 1976, 14, 165.
44. Teschke, O.; Eisenthal, K.B.; Shank, C.V.; Ippen, E.P. J. Chem. Phys. 1977, 67, 5547.
45. Irie, M.; Yoshida, K.; Hayashi, K. J. Phys. Chem. 1977, 81, 969.

46. Yorozu, T.; Yoshida, K.; Hayashi, K.; Irie, M. J. Phys. Chem. 1981, 85, 459.
47. Anderson, V.C.; Craig, B.B.; Weiss, R.G. J. Am. Chem. Soc. 1982, 104, 2972.
48. Faljoni, A.; Zinner, K.; Weiss, R.G. Tetrahedron Lett. 1974, 1127.
49. Morrison, J.D.; Mosher, H.S. "Asymmetric Organic Reactions"; Prentice-Hall: Englewood Cliffs, N.J., 1971; pp. 411ff.
50. Mason, S.F. "Molecular Optical Activity and the Chiral Discrimination"; Cambridge University Press: Cambridge, 1982; Secton 10.3.
51. Hibert, M.; Solladie, G. J. Org. Chem. 1980, 45, 5393.
52. Gottarelli, G.; Hibert, M.; Samori, B.; Solladie, G.; Spada, G.P.; Zimmermann, R. J. Am. Chem. Soc. 1983, 105, 7318.
53. Kelker, H.; Hatz, R. "Handbook of Liquid Crystals"; Verlag Chemie: Weinheim, 1980; Section 4.5.1.
54. Baessler, H.; Labes, M.M. J. Chem. Phys. 1969, 51, 1846.
55. Brey, L.A.; Schuster, G.B.; Drickamer, H.G. J. Am. Chem. Soc. 1979, 101, 129.
56. Wilhelmi, B. Chem. Phys. 1982, 66, 351.
57. Rothenberger, G.; Negus, D.K.; Hochstrasser, R.M. J. Chem. Phys. 1983, 79, 5360.
58. Benicewicz, B.C.; Johnson, J.F.; Shaw, M.T. Mol. Cryst. Liq. Cryst. 1981, 65, 111.
59. Sakamoto, K.; Porter, R.S.; Johnson, J.F. Mol. Cryst. Liq. Cryst. 1969, 8, 443.
60. Porter, R.S.; Griffen, C.; Johnson, J.F. Mol. Cryst. Liq. Cryst. 1974, 25, 131.
61. Adams, J.E.; Haas, W.E. Mol. Cryst. Liq. Cryst. 1971, 15, 27.
62. Schnur, J.M.; Martire, D.E. Mol. Cryst. Liq. Cryst. 1974, 26, 213.
63. Martire, D.E. In "The Molecular Physics of Liquid Crystals"; Luckhurst, G.R.; Gray, G.W., Eds.; Academic Press: New York, 1979; Chapter 11.
64. Oweimreen, G.A.; Martire, D.E. J. Chem. Phys. 1980, 72, 2500.
65. Otruba, J.P., III; Weiss, R.G. J. Org. Chem. 1983, 48, 3448.
66. Pincock, R.E.; Johnson, W.M.; Haywood-Farmer, J. Can. J. Chem. 1976, 54, 548.
67. Saeva, F.D. In "Liquid Crystals. The Fourth Sate of Matter"; Saeva, F.D., Ed.; Marcel Dekker: New York, 1979; Chapter 6.
68. Laarhoven, Wm. H.; Cuppen, Theo. J.H.M. JCS Perkin II 1978, 315.
69. Wilson, K.R.; Pincock, R.E. Can. J. Chem. 1977, 55, 889.
70. Kira, A.; Thomas, J.K. J. Phys. Chem. 1974, 78, 196.
71. Irie, M.; Yorozu, T.; Hayashi, K. J. Am. Chem. Soc. 1978, 100, 2236.
72. Yorozu, T.; Hayashi, K.; Irie, M. J. Am. Chem. Soc. 1981, 103, 5480.
73. de Gennes, P.G. "The Physics of Liquid Crystals"; Clarendon Press: Oxford, 1974; Chapter 6.
74. Nakazaki, M.; Yamamoto, K.; Fujiwara, K. Chemistry Lett. 1978, 863.

75. Saeva, F.D.; Olin, G.R. J. Am. Chem. Soc. 1976, 98, 2709.
76. Nakazaki, M.; Yamamoto, K.; Fujiwara, K.; Maeda, M. JCS, Chem. Commun. 1979, 1086.
77. Moradpour, A.; Nicoud, J.F.; Balavoine, G.; Kagan, H.; Tsoucaris, G. J. Am. Chem. Soc. 1971, 93, 2353.
78. Kagan, H.; Moradpour, A.; Nicoud, J.F.; Balavoine, G.; Martin, R.H.; Cosyn, P. Tetrahedron Lett. 1971, 2479.
79. Berstine, W.J.; Calvin, M.; Burkhardt, O. J. Am. Chem. Soc. 1973, 95, 524.
80. Zandomeneghi, M.; Cavazza, M.; Festa, C.; Pietra, F. J. Am. Chem. Soc. 1983, 105, 1839.
81. Hayashi, K.; Irie, M. Japan. Kokai 78 18549, 20.2.78; Chem. Abst. 1978, 89, 6151j.
82. London, F. Z. Physik. 1930, 63, 245.
83. Solladie, G.; Zimmermann, R. Angew. Chem., Int. Ed. Eng. 1984, 23, 348.
84. Eyring, H.; Walter, J.; Kimball, G.E. "Quantum Chemistry"; John Wiley: New York, 1944; pp. 118-123.
85. Mathies, R.; Albrecht, A.C. J. Chem. Phys. 1974, 60, 2500.
86. Marchese, F.T.; Jaffe, H.H. Theor. Chim. Acta 1977, 45, 241.
87. Shivaprakash, N.C.; Abdoh, M.M.M.; Srinivasa; Prasad, J.S. Mol. Cryst. Liq. Cryst. 1982, 80, 179.
88. Mataga, N.; Kubota, T. "Molecular Interactions and Electronic Spectra"; Marcel Dekker: New York, 1970; pp. 377ff.
89. Onsager, L. J. Am. Chem. Soc. 1936, 58, 1486.
90. Wilson, J.N. Chem. Rev. 1939, 25, 377.
91. Stien, M.-L.; Claessens, M.; Lopez, A.; Reisse, J. J. Am. Chem. Soc. 1982, 104, 5902.

RECEIVED January 10, 1985

11

Photochemical and Thermal Reactions of Hydrophobic and Surfactant Stilbenes in Microheterogeneous Media

P. E. BROWN, T. MIZUTANI, J. C. RUSSELL, B. R. SUDDABY, and D. G. WHITTEN

Department of Chemistry, University of Rochester, Rochester, NY 14627

A wide number of surfactant and hydrophobic *trans*-stilbene derivatives have been prepared and studied in media ranging from detergent micelles and vesicles to amylose inclusion complexes and microemulsions. Phenomena that have been investigated include fluorescence and *trans-cis* photoisomerization yields which provide considerable information concerning the microviscosity and degree of organization of the medium investigated as well as a number of ground state processes. The ground state reactions investigated include formation of charge-transfer complexes with organic cations such as methyl viologen and other processes including oxidation and bromination. These phenomena in each case involve interaction of the hydrophobic stilbene chromophore with relatively hydrophilic reagents. Observation of the rate and extent of these processes provide some indication of the degree of hydrophobic-hydrophilic compartmentalization occurring in the media and "protection" of hydrophobic species. The broad spectrum of results obtained for these phenomena in the several media investigated provide insights into the type of "solubilization" provided by different media.

The reactivity of molecules bound to surfaces, located at various kinds of interfaces, solubilized in microheterogeneous media, or incorporated as "guests" in various "hosts" as inclusion complexes has been the subject of much recent study. Indeed the structure of the medium, the nature of "solubilization sites" and reactivity in these environments have all been the focus of independent or interrelated investigations (1-12). Photochemistry has played a major role in these studies both in terms of studies of the media and also in terms of modified or controlled reactivity (1,5,8,9). In the course of these investigations numerous questions have arisen; many of these have developed from differing pictures of solute-environment interactions which are furnished by different studies using different molecules as "probes" (5,10-12). Controversies arising

0097-6156/85/0278-0171$06.00/0

from these studies have been most pronounced in investigations involving detergent micelles; however the problems or questions encountered with micelles extend to other media including, for example, lipid bilayers, microemulsions and inclusion complexes (13-15). In this paper we summarize a body of work from our laboratories involving one chromophore--*trans*-stilbene--which we have incorporated into a variety of surfactant and hydrophobic molecules and whose reactivity we have investigated in a number of different media. In these studies, as in others previously mentioned, we have used reactivity of the stilbene probe both as a tool to investigate solute-environment interactions in various media and to demonstrate the type of control or modification of reactivity which can be obtained by the various media. Our studies have revealed that the different kinds of *trans*-stilbene molecules used in this study experience a wide variety of environments with a consequent array of varying reactivity in different organized media. The results of these studies provide some interesting bases for conclusions which should be fairly general and extend to many other classes of compounds.

The "Stilbene Probe" and Its Reactivity

Of the numerous chromophores studied over the last thirty years, *trans*-stilbene is probably one of those whose photochemistry and photophysics has been most thoroughly investigated (16-25). Although interest in the photochemistry and photophysics of *trans*-stilbene (TS) was undoubtedly stimulated due to its being an ethylene analog having strong characteristic absorption in the relatively long wavelength ultraviolet, it has developed that TS has an atypical and in many ways peculiar excited state behavior. These "peculiarities" appear both in the excited singlet and triplet states; in this regard it is interesting to compare TS with the corresponding double bond-methylated analog 1,2-diphenylpropene. A large number of investigations have established that the excited state potential surfaces for rotation about the olefinic bonds have profiles for TS approximately as described in Figure 1 (22). In contrast 1,2-diphenylpropene has a potential surface for both singlet and triplet excited states which closely resembles those for ethylene. Consequently it has been found that the photochemistry and photophysics of 1,2-diphenylpropene is quite uncomplicated; on excitation to either singlet or triplet excited states of either isomer one observes photoisomerization as the only detectable photoprocess (19). It is evident that this photobehavior in both sensitized and direct excitation processes can be described by rapid decay of the initially formed excited states to a common intermediate of twisted geometry which has approximately a 1:1 decay ratio for both triplets and singlets to give ground state *cis* and *trans* isomers. In contrast TS fluoresces upon direct excitation and for both singlets and triplets there is evidence that the excited state capable of obtaining transoid geometry can be selectively quenched to produce TS in the ground state (19,21, 23). For the excited singlet state it is clear that the fluorescent state of approximately transoid geometry undergoes competitive decay by fluorescence or by crossing a small energy barrier to reach the twisted isomerization precursor state (22). *cis*-Stilbene under excitation does not produce the fluorescent transoid singlet; although its photochemistry is complicated by the cyclization process (19)

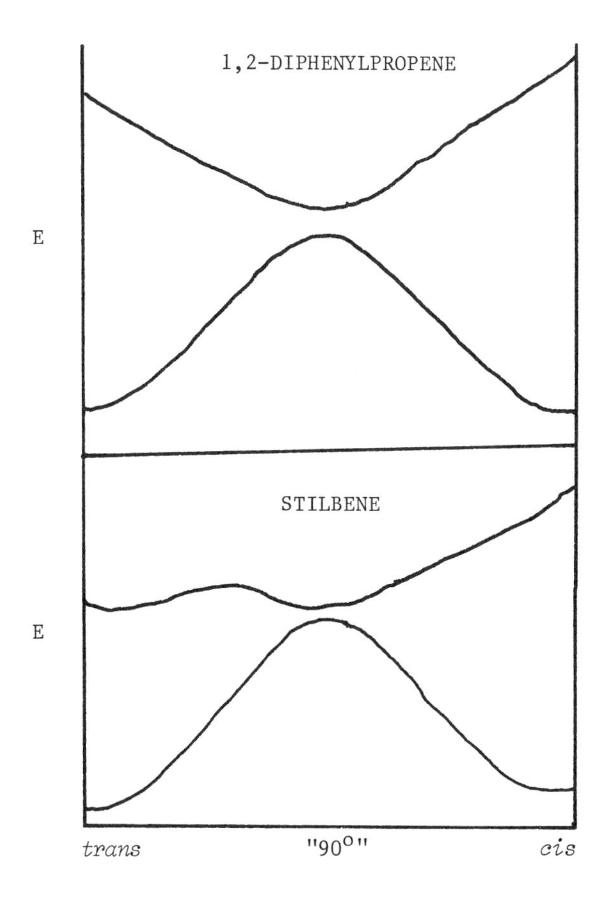

Figure 1. Profiles of the excited state potential surfaces for rotation about the olefinic bonds for 1,2-diphenylpropene and stilbenes.

its isomerization (the major process) occurs by rapid population of the same twisted form produced from the *trans*. Activation of either TS or *cis*-stilbene via triplet sensitization produces a common intermediate; however in contrast to 1,2-diphenylpropene, this intermediate has a sufficiently long lifetime to be quenched by species having low energy triplet states (azulene, oxygen, etc.) (19,22) in processes which in some cases result in selective formation of ground state TS. The lifetime of the stilbene triplet is dependent on viscosity and substituents.

Due to the rather unique features described above the stilbene chromophore is a particularly interesting one as a probe especially sensitive to effects of medium or the "microenvironment." For example, it is found that incorporation of TS into a rigid glass results in an increase in the fluorescence yield to almost unity with a concomitant decrease in the *trans* to *cis* isomerization efficiency (22). In viscous media one sees intermediate behavior (20,22). The photoisomerization of *cis* to *trans* is relatively unaffected by increases in viscosity and goes on even in semirigid glasses in contrast to that of TS (22). As will be developed in more detail later, another feature of TS which makes it especially attractive for microheterogeneous media composed of hydrocarbon components which are largely linear polymethylene units, is that the chromophore itself is a relatively rod-like molecule whose molecular dimensions and packing behavior allow it to be incorporated in such media in such a way as to presumably provide minimal disruption and yet simultaneously a chromophore absorbing at relatively long wavelengths. Finally, in addition to its photochemical and photophysical properties, TS offers both the classical reactivity of a relatively electron rich aromatic hydrocarbon and an olefin. Thus it has been found that TS forms donor-acceptor (CT) complexes which strong electronic acceptors (11,14) and undergoes characteristic reactions of alkenes such as electrophilic addition with reagents including bromine and other halogens (26-32).

The specific probes that we have used in our investigations have included TS itself and a variety of hydrophobic and surfactant *trans*-stilbenes having the structures shown below (SNA, MSNA and MSM).

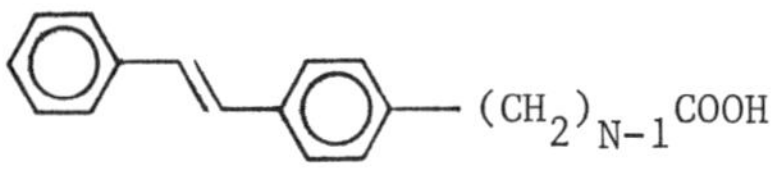

$(CH_2)_{N-1}COOH$

SNA

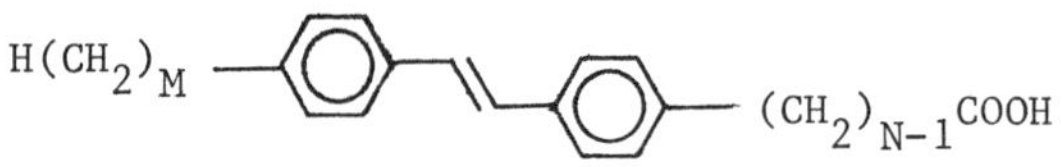

$H(CH_2)_M$ $(CH_2)_{N-1}COOH$

MSNA

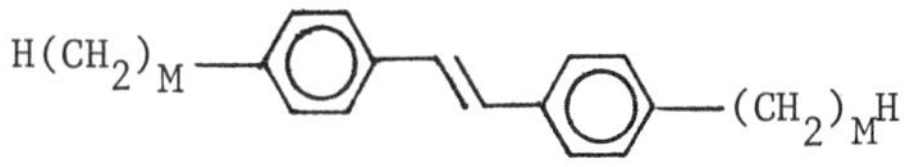

$H(CH_2)_M$ $(CH_2)_MH$

MSM

These molecules were in general synthesized as outlined in scheme I. In general, the solution photochemistry of the single chain derivatives, SNA, has been found to be remarkably similar to that of TS. In non-viscous organic solvents the fluorescence efficiency of most SNA derivatives is comparable to that of TS as are the yields of *trans* to *cis* photoisomerization (33,34). The double chain derivatives, MSNA and MSN, show substantially higher fluorescence yields (0.2 ± 0.02) in non-viscous solvents with a corresponding decrease in the *trans* to *cis* isomerization efficiency (34). In the sensitized isomerization, the MSN derivatives investigated thus far show only photoisomerization but there is an increase in the triplet lifetime by a factor of 2; however there is no change in the decay ratio of the isomerization precursor and the sum of the benzophenone sensitized isomerization efficiencies of 4S4 is approximately unity. When water insoluble SNA or MSNA molecules are spread as a film, either pure or in mixtures with insoluble fatty acids, at the air-water interface, it is found that the films show comparable behavior on compression to those of pure fatty acid with an indicated area per molecule of the surfactant stilbenes very close to that of a linear fatty acid (*ca.* 20^2 Å/molecule) (35). This suggests strongly that the TS chromophore in these molecules offers suitable molecular dimensions to pack into a crystalline-like array of linear parafin chains.

The Media Under Investigation

Our initial interest as far as the TS probes synthesized above is concerned was with the incorporation of the stilbenes into oriented supported multilayer assemblies (35,36). However, since these molecules are surfactants having hydrophobic-hydrophilic relationships similar to detergents and bilayer-forming lipids, it became clear that they might be equally interesting as probes to incorporate into micelles, vesicles and microemulsions. Much of the work described in this paper will deal with these three media and the use of various stilbene probes to determine simultaneously something about the medium under investigation as well as its "solubilization" properties. In addition, since several of the stilbene probes synthesized show limited water solubility but are obviously hydrophobic, it appeared that they might be good candidates for incorporation into reagents such as cyclodextrins or amylose derivatives which are known to entrap relatively small hydrophobic molecules to form host-guest inclusion complexes. We have found that this ability to form inclusion complexes is especially true of several of the surfactant and hydrophobic stilbenes with amylose and our results in this area will be discussed and compared with those obtained in the aqueous surfactant media mentioned above.

The Reactivity of Stilbene Probes in Detergent Micelles

trans-Stilbene, the hydrophobic 4S4 and the whole range of surfactant *trans*-stilbene derivatives can all be incorporated into a variety of aqueous detergent solutions, including cationic, anionic and neutral surfactants. In most cases the stilbene component would not be soluble in water by itself so it is clear that solubilization is occurring via association of the stilbene derivative with either detergent

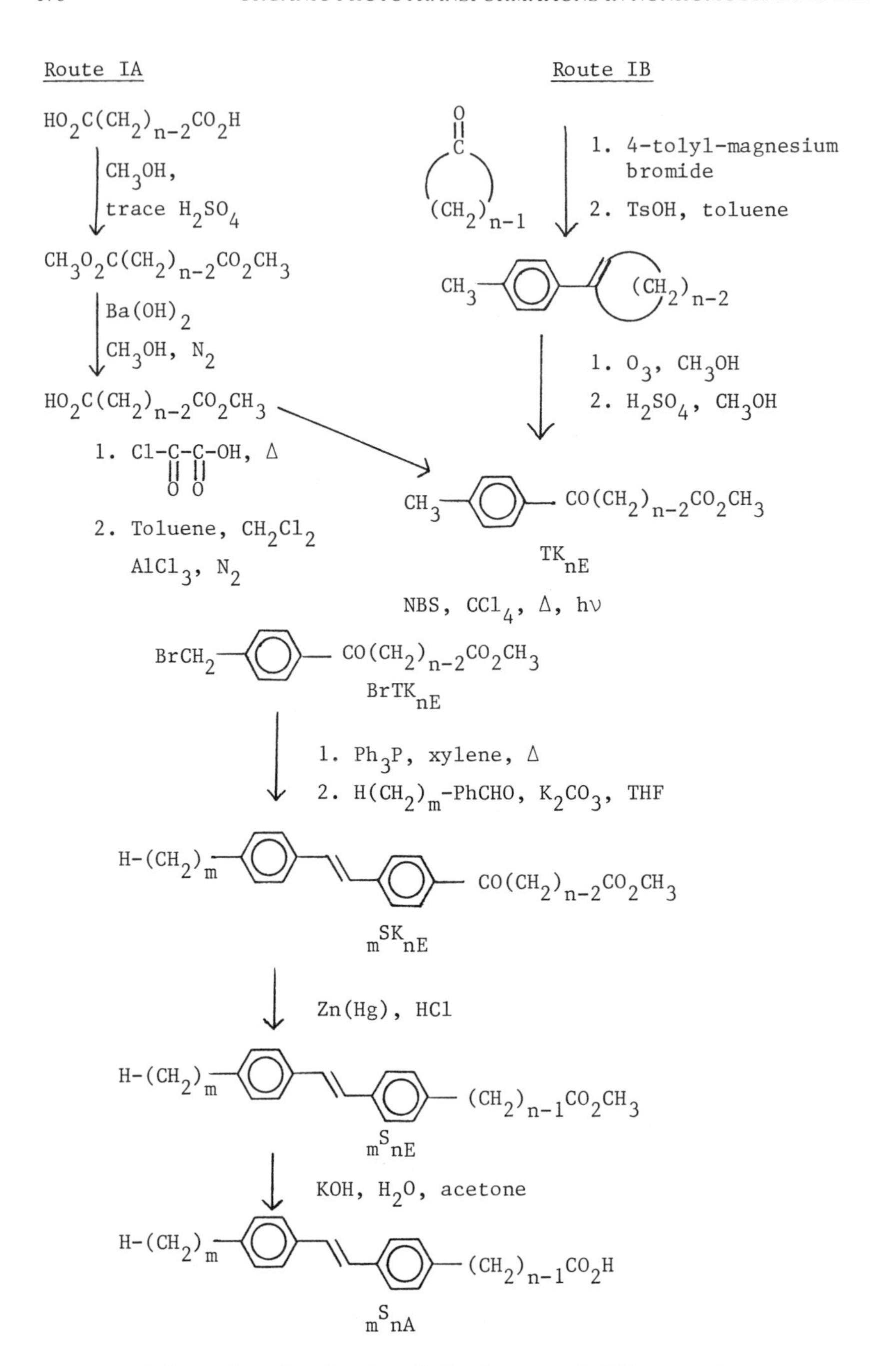

Scheme I. Synthesis of Surfactant Stilbenes ${}_mS_{nA}$

monomer or aggregates. In almost every case clear monomeric stilbene absorption and fluorescent spectra are obtained for the stilbene derivative thus stabilized in micelle media. Table I lists fluorescent quantum yields obtained for several stilbene derivatives comparing homogeneous methylcyclohexane and aqueous sodium dodecylsulfate (SDS) solutions. Similar data have been obtained in most cases when neutral or cationic surfactants have been studied; in general the major conclusion that can be drawn from the fluorescence yields is that the photophysics of the stilbene chromophore is relatively little affected by incorporation of the hydrophobic or surfactant stilbenes into micellar aggregates. Although photoisomerization has been less thoroughly studied in micellar media, it seems clear that in general there is little change upon taking the various stilbenes from homogeneous organic solvents to aqueous detergent solutions. For example with S4A the *trans* to *cis* photoisomerization in aqueous SDS is 0.44 compared to 0.48 in methylcyclohexane. The decay ratio determined (fraction *cis* isomer formed) is 0.5; this compares again very closely with a value of 0.52 obtained in methylcyclohexane. In general the results are consistent with a very slight effective increase in viscosity on going to the micellar media but otherwise with no significant changes. From these data it is clear that the microenvironment of the stilbene chromophore is generally fluid but little can be said regarding whether the environment is hydrophobic, hydrophilic or interfacial.

It has been found that addition of the organic cation, methyl viologen (MV^{2+},N,N'-dimethyl-4,4'-bipyridinium), to solutions of stilbene, pyrene and other aromatic or heteroaromatic derivatives can lead to formation of donor-acceptor complexes which are distinguished by spectral changes in the absorption of the aromatic, new bands (frequently weak) to the red of the aromatic transitions and quenching of the aromatic fluorescence (11,14,37). For stilbene itself with MV^{2+} the equilibrium constant for formation of the complex is extremely small ($K_{acetonitrile} = 15$) (11). In the case of stilbene derivatives the complex is most easily detected by fluorescence quenching and a measure of the overall equilibrium constant can be obtained by plotting the fluorescence intensity ratio (I_0/I) *vs.* the concentration $[MV^{2+}]$. Table II compares these constants for TS and several hydrophobic and surfactant stilbene derivatives in SDS micelles. Two features of the constants determined are noteworthy. First of all the overall equilibrium constants measured in the micellar media are much larger than those obtained in a simple homogeneous solution. Furthermore there seems to be very little variation between the constants measured for different molecules having the stilbene chromophore in quite different sites. The overall variation between TS (the largest) and 4S4 (the smallest) is only a factor of 2.5. The reason that the values are larger than those obtained in homogeneous solution is readily apparent; it has been previously shown that MV^{2+} is strongly associated with anionic micelles or vesicles in aqueous solution. Therefore the two reactants, stilbene molecule and MV^{2+}, are effectively "concentrated" into a very small portion of the total volume in the microheterogeneous medium. In fact, it is fairly easy to correct the fluorescence quenching constants by considering that both of the reagents are restricted only to the portion of the solution actually occupied by the surfactant micelle. When this is done the values listed as K' are obtained.

Table I. Fluorescence Quantum Yields for SNA and MSNA in Different Media at 25°C.

Compound	Homogeneous MCH	Micellar SDS	Vesicles DCP	Vesicles 1:1 DCP/DPL	Vesicles DPL
TS	0.05	0.07 ± 0.006	0.12 ± 0.003	0.19 ± 0.009	0.36 ± 0.01
S4A	0.08 ± 0.011	0.12 ± 0.004	0.20 ± 0.020	0.20 ± 0.020	0.51 ± 0.02
S6A	0.09 ± 0.009	0.12 ± 0.005	0.43 ± 0.030	0.21 ± 0.003	0.43 ± 0.030
S7A	0.09 ± 0.004	0.09 ± 0.004	0.37 ± 0.010	0.18 ± 0.009	0.41 ± 0.020
S10A	0.08 ± 0.003	0.12 ± 0.010	0.37 ± 0.003	0.24 ± 0.002	0.44 ± 0.02
6S4A	0.21 ± 0.002	0.39 ± 0.011	0.66 ± 0.04	0.76 ± 0.02	0.93 ± 0.10
4S6A	0.25 ± 0.011	0.38 ± 0.030	0.65 ± 0.03	0.78 ± 0.05	0.88 ± 0.04
2S8A	0.21 ± 0.012	0.30 ± 0.017	0.56 ± 0.03	0.62 ± 0.05	0.76 ± 0.03
4S4	0.20 ± 0.013	0.37 ± 0.030	0.58 ± 0.04	0.70 ± 0.02	0.81 ± 0.01

Interestingly for a typical surfactant stilbene S6A the value obtained is very close to that measured in pure acetonitrile. In a somewhat contrasting situation the value obtained for 4S4 in aqueous SDS is roughly one-fourth that obtained in acetonitrile. The results obtained here have considerable significance with regard to the structure and solubilization properties of the SDS micelles. Based on several widely held tenets (1-4) it might be supposed that the stilbene chromophores would be confined to an environment very closely resembling liquid hydrocarbon while the positively charged MV^{2+} would be associated either with the anionic head groups or at least the water-surfactant interfacial region. If such were the case and the detergent micelle consisted of rigidly compartmentalized regions, it would be expected that association constants should be much lower in the micelles than in homogeneous solution and that further the molecules having stilbene in a more "hydrophobic" site would show lower association constants. The fact that association constants generally are comparable to those obtained in homogeneous solutions suggests that there is very little sequestering of the stilbene from MV^{2+} in the micelles. It does appear that any sequestering observed is more or less in line with what one would intuitively expect; however the overall effect is not very large and striking result one observes is that even for the most hydrophobic stilbene, 4S4, there seems to be little sequestering of hydrophobic reagents from hydrophilic ones. The results obtained here offer an interesting contrast to those obtained in "swollen" micelles and vesicles which will be discussed below.

A characteristic "ground state" reaction of stilbene and other alkenes is electrophilic addition. This reaction has been very well studied and although there are some mechanistic controversies at present (38-40), both the classical "bromonium" and single-electron-transfer mechanisms predict that considerable separation of charge develops on proceeding from starting materials towards the transition state (38-40). In fact it has been found that the rates of bromination of TS are quite solvent dependent with a sharp increase in reaction rates observed with increase in solvent polarity proceeding from hydrocarbon solvents to alcohols or water (26-32). We have examined the bromination rates of TS and several surfactant and hydrophobic stilbenes in different homogeneous solutions and in SDS micelles. Rate data obtained in this study are listed in Table III. As far as the comparison of reactivity in micelles and homogeneous solutions is concerned, it is clear that the reactivity observed for the different stilbenes in aqueous SDS is consistent in each case with the stilbene occupying an extremely polar site. In fact the rates obtained are faster than those for 50% aqueous ethanol in each case. The differences observed in rates for different compounds studied are more attributable to differing substitution effects (alkyl groups enhance rates such that the reactivity in homogeneous solution in the micelles is 4S4 > S4A > TS) than to any kind of differences in the microenvironment. The fact that only a single reaction rate constant is obtained in each case suggests either that the different stilbenes reside in a unique environment in the micelles or, perhaps more likely, that the stilbene location is changing but on a time scale more rapid than the rate of the bromination process.

The important result of our studies of the bromination of different stilbene derivatives in the micelles is that the rapid rates

Table II. Stern-Volmer Constants for Quenching of SNA and MSNA Fluorescence by MV^{2+} in 0.028 M SDS at 25°C.

Compound	K_{sv}	K'
TS	2450	13.9
S4A	2150	12.2
S5A	2200	12.5
S6A	2440	13.8
S7A	2120	12.0
S10A	2250	12.8
S12A	1870	10.6
S16A	1020	5.8
6S4A	1840	10.4
4S6A	1780	10.0
2S8A	1490	8.3
4S4	960	5.3

Table III. Bromination of Stilbenes in SDS Micelles or $EtOH:H_2O=1:1(v/v)$

Compound	Medium	$k_2(M^{-1}s^{-1})$
TS	SDS	$(5.4 \pm 1.8) \times 10^3$
4S4	SDS	$(9.3 \pm 0.9) \times 10^4$
S4A	SDS	$(8.9 \pm 1.0) \times 10^4$
S10A	SDS	$(7.3 \pm 0.8) \times 10^4$
4S6A	SDS	$(4.9 \pm 1.0) \times 10^4$
TS	$EtOH-H_2O$	210 ± 40
S4A	$EtOH-H_2O$	490 ± 40
4S4	$EtOH-H_2O$	1140 ± 270

and lack of indication of different environments for different stilbenes provide once again a picture of the micellar solubilization site for the TS chromophore being one in which little sequestering from polar reagents and strong evidence of a high "micropolarity." The picture one obtains is certainly not consistent with the hydrocarbon TS located in an environment like liquid hydrocarbon but much more in accord with the picture in which the chromophore lies either at an interfacial site or in an environment which is both freely accessible and rich in water.

Swollen Micelles *vs.* Simple Detergent Micelles

In contrast to simple detergent micelles, so-called "swollen micelles" consisting of detergent and water-immiscible linear alcohol such as 1-pentanol, 1-hexanol or 1-heptanol in water as the principal solvent have been found to have different solubilization properties and perhaps offer quite different microenvironments. For several of the surfactant stilbenes it has been found that fluorescent yields increase very slightly as the linear alcohol is added up to an alcohol/surfactant mole ratio of 4 or 5:1. However while fluorescence is not appreciably affected, it is found that complexation with MV^{2+} is virtually eliminated for several of the surfactant stilbenes (11, 33). In fact one can observe that the fluorescence quenching, by which the process is most conveniently measured, is rapidly attenuated by addition of the linear alcohol. Very similar results have been observed with both pyrene and various surfactant pyrene derivatives with MV^{2+} (41). In this case it is found that addition of the alcohol both reduces the fluorescence quenching and changes the band shape of the fluorescence spectrum of the pyrene (41). Since the intensity ratios of the varying vibronic components of pyrene fluorescence are a good measure of solvent polarity (42), it is instructive to look at the changes which occur as linear alcohols are added. In the case of adding 1-heptanol to SDS micelles, one observes that the apparent "micropolarity" of the medium decreases sharply (41). All of this then is consistent with a picture in which the alcohol addition gives more definition or compartmentalization to the micelle and leads to the development of distinctly differing solubilization sites. Here we begin to see evidence of a true compartmentalization similar to what one might expect by the so-called "Hartley" model or traditional concept of the micelle. Additional studies of these phenomena are continuing in our laboratories using other reactions with the stilbenes and other "probe" molecules. The "swollen micelles" could be regarded as way stations en route from simple micelles to an oil in water microemulsion. We have studied extensively some of these oil/water microemulsions using the stilbene and other probes; it is beyond the scope of this paper to go into these studies in detail; however in general one finds for the oil/water systems where discrete droplets occur, the various stilbene chromophores experience moderately hydrophobic sites consistent with the stilbene residing "inside" the oil droplet portion. The question of water penetration into the droplet and surface-interior equilibration is one which needs to be further explored and clarified.

Detergent and Phospholipid Bilayer Vesicles (Liposomes)

It has been known for some time that typical phospholipids form a variety of microheterogeneous structures when dispersed in aqueous solution; these dispersions upon sonication can be induced to convert to fairly monodisperse small unilamellar vesicles consisting of a closed bilayer structure containing water in an interior compartment as well as the exterior solvent or bulk water (47,48). Relatively similar structures can be formed from a number of 2-chain detergents such as dicetylphosphate (DCP) or dioctadecyldimethylammonium bromide (DODAB) (43,44). These "liposomes" or vesicles are osmotically active and can, in common with micelles, solubilize a variety of diverse reagents. Clearly hydrophilic reagents such as inorganic ions are normally associated with the water, or if ionic and at opposite charge to the surfactant head group, with the vesicle-water interface (45,46). On the other hand, more hydrophobic reagents are believed associated with sites in the vesicle "interior." Although water can penetrate the bilayer wall of different kinds of surfactant vesicles with relative ease, it is generally accepted that at temperatures below the "phase transition temperature" the vesicles exist in a gel form in which the hydrocarbon chains are packed together in a regular, semicrystalline fashion (43). Small bilayer vesicles formed from different surfactants undergo a phase transition at component-dependent temperatures which is commonly associated with a "melting" of the hydrocarbon chains to form a more liquid or liquid crystalline array (43). The studies described herein have all been carried out at temperatures below the phase transition temperature and thus under conditions where the hydrocarbon chains should be relatively ordered. Table I compares the fluorescence quantum yields for TS and the various stilbene derivatives mentioned earlier in several different bilayer systems as well as in the reference homogeneous solution. The data tabulated are instructive in several ways: first, comparing the fluorescence yield obtained in the vesicle systems with those measured earlier in micelles, it is clear that for all of the surfactant stilbenes fluorescent yields are much higher in all three vesicle media than in aqueous SDS. The second point is that while the fluorescent yields are all higher, there are some notable differences between different stilbenes and the different media. Regarding media, it is clear in general that the highest fluorescence yields are obtained in the phospholipid DPL. DPL (dipalmitoyl lecithin) is a zwitterionic lipid which should experience minimum head group repulsion and thus perhaps the highest order of the three surfactant systems. For the "intrachain" stilbenes 6S4A, 4S6A, 2S8A and 4S4 the quantum yields in DPL approach the limiting value of unity. The picture then obtained with these probes is one in which the stilbene chromophore is effectively "locked" from the twisting process which in fluid media competes with fluorescence. Even for the other two vesicular media, these stilbenes show high quantum yields indicative of a relatively ordered and restrictive structure. The other stilbenes shown in the table show relatively high quantum yields but ones which indicate that an appreciable fraction of the molecules can undergo isomerization. Table IV lists values for isomerization of several of the surfactant and hydrophobic stilbenes and the different vesicular media at 25° together with calculated "decay ratios" which together give a good picture of the "restrictiveness" of the surrounding medium. These data show a sharp reduction in the isomerization yields in the vesicular media, especially for the "intrachain

Table IV. Photoisomerization Quantum Yields for SNA and MSNA in Different Media at 25°C.

Compound	MCH	SDS	DCP	DCP/DPL	DPL
S4A	0.48	0.44	0.38	0.38	0.15
S10A	0.50		0.33	0.34	0.15
6S4A	0.37		0.07	0.05	0.014
4S6A	0.36		0.08	0.06	
4S4	0.34		0.09		0.016

Decay Ratios Representing the Fraction of Twisted "P" Intermediates Decaying to *cis* Isomers for SNA and MSNA in Different Media at 25°C.

Compound	MCH	SDS	DCP	DCP/DPL	DPL
S4A	0.52 ± 0.06	0.50 ± 0.02	0.48 ± 0.02	0.49 ± 0.02	0.31 ± 0.02
S10A	0.54 ± 0.03		0.52 ± 0.03	0.38 ± 0.02	0.25 ± 0.02
6S4A	0.47 ± 0.02		0.21 ± 0.02	0.21 ± 0.03	0.20 ± 0.05
4S6A	0.48 ± 0.02		0.23 ± 0.02	0.27 ± 0.05	
4S4	0.43 ± 0.04		0.21 ± 0.02		0.11 ± 0.01

stilbenes;" for the latter compounds both the isomerization yields and the decay factor are reduced substantially below those obtained in fluid media or in SDS micelles. Here the difference between the stilbenes at the end of an alkyl chain and those within the chain is most pronounced. This probably reflects the likelihood that the former molecules have available more paths to permit isomerization than the latter including displacement from the plane of a single monolayer.

As mentioned above in the discussion of reactivity in micellar media, the various surfactant stilbenes form donor-acceptor complexes with MV^{2+} which can be detected by quenched fluorescence of the stilbene chromophore as well as by the appearance of a colored CT band in the visible. Somewhat analogous behavior is observed with the same stilbenes in vesicles which are composed at least partially of anionic surfactant. We have previously shown that MV^{2+} binds strongly to vesicles composed of surfactants such as dicetylphosphate (DCP). In fact, interestingly enough, we find that the driving force for binding MV^{2+} to DCP is largely an entropic one (46). This suggests that the binding of MV^{2+} need not be a coulombic or ion-exchange phenomenon but rather that it is more likely a "hydrophobic effect" (45,46). The quenching of fluorescence of the various stilbene chromophores by MV^{2+} shows a somewhat different behavior not observed for SDS micelles. The most striking effect is the leveling off of the quenching which occurs at values of $I^o/I \sim 2$ in several cases. This clearly indicates that only some of the stilbenes incorporated in the vesicles are susceptible to quenching by MV^{2+}; the most attractive explanation for these results is that only those stilbenes in the outer layer of the bilayer system are "quenchable." In general it is found (Table V) that the quenching constants (corrected as in the case of the micelles for the reduced volume in which the reactants are constrained) are slightly lower than those obtained in micelles and somewhat more dependent on the specific stilbene derivative used. Thus we find quenching constants are smaller for the intrachain stilbenes and for those in which the number of methylene groups between the carboxyl and stilbene groups is greater. That the quenching is due to formation of a complex between MV^{2+} and the stilbene chromophore is indicated by our finding that equilibrium constants measured by fluorescence quenching are, within experimental error, the same as those determined by spectrophotometric (Benesi-Hildebrand) techniques (34). It is tempting to ascribe the differences in measured equilibrium constants as due to difference in "mean" location of the stilbene chromophore in the organized assembly. Thus we would suggest that MV^{2+} should reside relatively near the water-lipid interface such that there would be an effective concentration grading it from the "surface" to the hydrophobic bilayer interior. The fall off in equilibrium constants given in Table V would then be attributed to a burying of the stilbene chromophore in progressively more hydrophobic sites. In this regard it is interesting to note that *trans*-stilbene itself has the highest equilibrium constant and that the hydrophobic 4S4 has a value considerably lower than that for TS. This is the reverse of what is observed in homogeneous solution and infers that TS and 4S4 may occupy quite different sites in the different vesicular media. In this regard it should be noted that 1,4-diphenyl-1,3-butadiene shows even larger values for complex formation constants than does TS (34). This might imply that both TS and the diene are

Table V. Corrected Stern-Volmer Constants (K'_{sv}) for Quenching of SNA and MSNA Fluorescence by MV^{2+} in 0.005 M DCP and 0.005 M DCP/DPL Vesicles at 25°C.

Compound	K'_{sv}(DCP)	K'_{sv}(DCP/DPL)
TS	19.2 ± 1.7	6.4 ± 0.6
S4A	12.8 ± 0.8	4.6 ± 0.3
S5A	4.6 ± 0.4	3.7 ± 0.3
S6A	6.2 ± 0.2	4.5 ± 0.3
S7A	4.9 ± 0.4	2.5 ± 0.2
S10A	7.3 ± 0.7	2.9 ± 0.1
S12A	11.7 ± 1.1	2.2 ± 0.2
6S4A	7.8 ± 0.5	3.1 ± 0.2
4S6A	7.2 ± 0.8	2.9 ± 0.2
2S8A	6.6 ± 0.7	3.0 ± 0.2
4S4	3.8 ± 0.2	2.4 ± 0.2

near the lipid-water interface and not aligned with the hydrocarbon chains in an ordered region. This result may be significant and suggests caution in using molecules such as TS or DPB as established "probes" which can be assumed to align themselves with the polymethylene chains. It is tempting to suggest that 4S4, which behaves very similarly to the intrachain stilbenes, is a much better type of probe molecule and that the "hydrophobic capping" present in 4S4 favors its incorporation as a "well received" guest into the bilayer structure.

Bromination of the stilbenes can also be carried out very easily in certain bilayer vesicle media and we have studied this process in both DCP and DPL. In many respects the results obtained in the bromination study correlate very nicely with those previously mentioned for formation of the complex between MV^{2+} and the stilbene chromophore. The rates observed in DCP are much faster than those observed in DPL; there are distinct differences for different members of the series investigated. The generally slower rates in DPL point once again, as suggested above, to a more "organized" structure for DPL throughout the bilayer. The different rate constants obtained in this study can be attributed to two general sites for reaction in each case. As mentioned above, the rate of the bromination reaction can be associated generally with the "micropolarity" of the medium and also of course with the effective concentration of bromine. As the stilbene chromophore is located generally farther from the lipid-water interface, reactivity is slower (rate constants decrease from 2600 M^{-1} s^{-1} to 2 M^{-1} s^{-1}) and the rate constants reflect perhaps a gradient of water concentrations across the monolayer. It should be expected that bromine is reasonably soluble in the hydrophobic portions of the bilayer due to its high solubility in hydrocarbons and other organic solvents (47). However one might argue that in a region approaching hydrocarbon crystal in structure that bromine might show a somewhat lower solubility. In either case, it is clear that in the latter situation the development of a high degree of charge separation in the transition state would be relatively unfavorable and consequently occur with a slower rate. Once again the difference between TS and the hydrophobic 4S4 is striking and suggests that these two molecules occupy quite different sites in the bilayer.

Host-Guest Inclusion Complex Formation between Surfactant Stilbenes and Amylose

A totally different kind of microheterogeneous medium is that formed by the incorporation of a small "guest" molecule into a "guest-cavity" in the formation of an inclusion complex. Among the most common types of inclusion complexes studied are those formed with various small molecules or ions as guests and zeolites, cyclodextrins, cryptates and crown ethers as hosts (11,15,48-54). Our particular interest in this area has been focused largely on complexes formed using common amylose or carboxymethylamylose as hosts (55-60). A number of studies have established that linear amylose and related molecules can form helical coils which can encapsulate small, usually relatively hydrophobic molecules. In contrast to cyclodextrins which offer a relatively rigid cavity of fixed dimensions (17,61-66), the amylose cavities are relatively flexible and apparently can accommodate a wider variety of molecular sizes and shapes. We have concentrated our studies on molecules which are reasonably hydrophobic or surfactant and thus, among the molecules we have studied are the various hydrophobic and surfactant stilbenes.

Our studies thus far have focused on TS, S4A, S6A, S12A and 6S4A. We have found for the stilbenes, as in the case where the stilbene is sequestered in other forms of microheterogeneous media, complex formation can be detected in enhanced fluorescence quantum yields and reduced isomerization efficiencies. Indeed for the stilbenes listed above, we find a very good correlation between hydrophobicity and the extent or ease of complex formation. Thus TS itself forms no complex whatsoever and addition of 1% amylose to solutions of TS in 1:1 dimethylsulfoxide/water results in no changes in its photochemistry or photophysics. The other hydrophobic stilbenes listed above all form complexes with the extent of complex formation being S4A < S6A < S10A < 6S4A. Our results with the last of the aforementioned compounds have been the most interesting. For this compound it is found that solubility in water or water/dimethylsulfoxide is very limited; in fact in the latter solvent the stilbene that is solubilized tends to exist mainly as H-aggregates. We find that the addition of amylose decomposes the aggregates with a resultant increase of monomer; however the monomer occurring is very largely in the form of the amylose complex. Within the amylose complexes the fluorescence yield is increased to nearly unity and isomerization is almost completely suppressed. It is interesting then to note that the intrachain stilbene in the amylose complex apparently experiences an environment very similar to that in a phospholipid vesicle below the phase transition temperature. The photochemical reactivity is clearly profoundly modified by incorporation into the amylose complex.

In addition to having its unimolecular photochemical reactions modified, it turns out that excited state quenching processes involving amylose-incorporated stilbenes are also effected strongly. For example iodide ion has been previously shown to quench TS in partially aqueous solutions with the rate constant, $k_q = 2.2 \times 10^{10}\ M^{-1}\ s^{-1}$ (67). Preliminary studies indicate that amylose incorporation substantially reduces, but does not eliminate, quenching of the stilbene chromophore singlet. Thus we find that Stern-Vollmer quenching constants are 2.5 and 0.5 for 6S4A in the absence and presence of amylose, respectively. Using lifetimes estimated from fluorescence efficiencies we obtain k_q values of $4.4 \times 10^{8}\ M^{-1}\ s^{-1}$ and $1.3 \times 10^{10}\ M^{-1}\ s^{-1}$ in DMSO-water in the presence and absence of amylose. Thus quenching is reduced about 30-fold for iodide by amylose incorporation of the stilbene chromophore. While it is somewhat uncertain as to what precisely the nature of the quenching of stilbene by iodide is, it is reasonable to assume that the reduced quenching constants imply a more difficult approach of the iodide ion to the complexed stilbene than to the free. We are currently exploring many aspects of reactivity of amylose-incorporated chromophores. We find for example that amylose is able to extract totally insoluble hydrophobic stilbene molecules into water and we are presently trying to obtain crystal structural data on the complex molecule. The dynamics of complex formation and dissociation are currently under investigation.

Summary

Studies presented here show that the various stilbene probes we have prepared and examined in different microheterogeneous media show a wide range of reactivity which in turn provides useful information about the "solvent properties" of the different media. Other studies using stilbenes and other molecules to probe media such as reverse micelles and microemulsions are currently under way; in general

it is safe to conclude that the stilbenes and other linear chromophores such as alpha, omega diphenylpolyenes probably are reasonably non-perturbing probes for the hydrocarbon or interfacial portions of several microheterogeneous media. While these rod-like hydrocarbon molecules are perhaps as inoffensive a series of probes as one can imagine (and still incorporate easily monitored and environment-sensitive reactivity), it is clear that small structural changes can produce dramatic effects on both binding sites and reactivity.

Acknowledgments

We are grateful to the Department of Energy (contract DE-AC02-84ER13151) and the National Science Foundation (grant CHE 8315303) for support of different phases of this research.

Literature Cited

1. Turro, N. J.; Grätzel, M.; Braun, A. M. Angew. Chem., Int. Ed. Engl. 1980, 19, 675.
2. Tanford, C. "The Hydrophobic Effect: Formation of Micelles and Biological Membranes", 2nd ed.; Wiley Interscience: New York, 1980.
3. Lianos, P.; Lang, J.; Strazielle, C.; Zana, R. J. Phys. Chem. 1982, 86, 1019. Candau, S.; Zana, R. J. Colloid Interface Sci. 1981, 84, 206.
4. Lianos, P.; Lang, J.; Zana, R. J. Colloid Interface Sci. 1983, 91,276. Zana, R.; Picot, C.; Duplessix, R. ibid. 1983, 93, 43.
5. Zachariasse, K. A.; Van Phuc, N.; Kozankiewica, B. J. Phys. Chem. 1981, 85, 2676.
6. Mukerjee, P.; Cardinal, J. R.; Desai, N. R. In "Micellization, Solubilization, and Microemulsions"; Mittal, K. L., Ed.; Plenum Press: New York, 1977; Vol. I.
7. Menger, F. M. Acc. Chem. Res. 1979, 12, 111.
8. Schanze, K. S.; Mattox, T. F.; Whitten, D. G. J. Am. Chem. Soc. 1982, 104, 1733.
9. Russell, J. C.; Whitten, D. G.; Braun, A. M. J. Am. Chem. Soc. 1981, 103, 3219.
10. Menger, F. M.; Doll, D. W. J. Am. Chem. Soc. 1984, 106, 1109.
11. Menger, F. M.; Chow, T. F. J. Am. Chem. Soc. 1983, 105, 5501.
12. Otruba, T. P.; Whitten, D. G. J. Am. Chem. Soc. 1983, 105,6503.
13. Geiger, M. W.; Turro, N. J. Photochem. Photobiol. 1977, 26,221.
14. Turro, N. J.; Cox, G. S.; Li, X. Photochem. Photobiol. 1983, 37, 149.
15. Tabushi, I. Acc. Chem. Res. 1982, 15, 66.
16. Lewis, G. N.; Magel, T. T.; Lipkin, D. J. Am. Chem. Soc. 1940, 62, 2973.
17. Förster, Th. Z. Elektrochem. 1952, 56, 716.
18. Zimmerman, G.; Chow, L.; Paik, U. J. Am. Chem. Soc. 1958, 80, 3528.
19. Hammond, G. S. et al. J. Am. Chem. Soc. 1964, 86, 3197.

20. Gegiou, D.; Muszkat, K. A.; Fischer, E. J. Am. Chem. Soc. 1968, 90, 3907.
21. Saltiel, J. et al. J. Am. Chem. Soc. 1966, 88, 2336.
22. Saltiel, J. et al. Org. Photochem. 1973, 3, 1.
23. Saltiel, J.; Megarity, E. D. J. Am. Chem. Soc. 1972, 94, 2742.
24. Dyck, R. H.; McClure, D. S. J. Chem. Phys. 1962, 36, 2326.
25. Malkin, S.; Fischer, E. J. Phys. Chem. 1962, 66, 2482.
26. Atkinson, J. R.; Bell, R. P. J. Chem. Soc. 1963, 3260.
27. Bartlett, P. D.; Tarbell, D. S. J. Am. Chem. Soc. 1936, 58, 466.
28. Buckles, R. E.; Miller, J. L.; Thurmaier, R. J. J. Org. Chem. 1967, 32, 888.
29. Ruasse, M.-F.; Dubois, J. E. J. Org. Chem. 1972, 37, 1770.
30. Heublein, G.; Schütz, E. Z. Chem. 1969, 9, 147.
31. Dubois, J. E.; Ruasse, M.-F. J. Org. Chem. 1973, 38, 493.
32. Rothman, J. E.; Davidowicz, E. A. Biochemistry 1975, 14, 2809.
33. Russell, J. C. Ph.D. Thesis, University of North Carolina, Chapel Hill, 1981.
34. Brown, P. E. Ph.D. Thesis, ibid. 1984.
35. Mooney, W. M.; Brown, P. E.; Russell, J. C.; Costa, S. B.; Pedersen, L. G.; Whitten, D. G. J. Am. Chem. Soc. 1984, 106, 5659.
36. Mooney, W. M. Ph.D. Thesis, University of North Carolina, Chapel Hill, 1983.
37. Martens, F. M.; Verhoeven, J. W. J. Phys. Chem. 1981, 85, 1773.
38. Fukuzumi, S.; Kochi, T. K. J. Org. Chem. 1981, 46, 4116.
39. de la Mare, P. B. D. "Electrophilic Halogenation"; Cambridge University Press: London, 1976.
40. Fukuzumi, S.; Kochi, T. K. J. Am. Chem. Soc. 1982, 104, 7599, and references therein.
41. Russell, J. C., unpublished data.
42. Dong, D. C.; Winnik, M. A. Photochem. Photobiol. 1982, 35, 17.
43. Fendler, J. H. Acc. Chem. Res. 1980, 13, 7.
44. Kunitake, T.; Sakamoto, T. J. Am. Chem. Soc. 1978, 100, 4615.
45. Whitten, D. G.; Russell, J. C.; Foreman, T. K.; Schmehl, R. H.; Bonilha, J.; Braun, A. M.; Sobol, W. In "Chemical Approaches to Understanding Enzyme Catalysis: Biomimetic Chemistry and Transition-State Analogs"; Green, B. S.; Ashani, Y.; Chipman, D., Eds.; Studies in Organic Chemistry, Vol. 10; Elsevier: Amsterdam, 1981.
46. Bonilha, J. B. S.; Foreman, T. K.; Whitten, D. G. J. Am. Chem. Soc. 1982, 104, 4215.
47. Semb, J. J. Am. Pharm. Assoc. 1935, 24, 547.
48. Emert, J.; Breslow, R. J. Am. Chem. Soc. 1975, 97, 670.
49. Mandelcorn, L., Ed. "Non-Stoichiometric Compounds"; Academic Press: New York, 1964.
50. Allcock, H. R. Acc. Chem. Res. 1978, 11, 81.
51. Lehn, J.-M. Acc. Chem. Res. 1978, 11, 49.
52. Russell, J. C.; Costa, S. B.; Seiders, R. P.; Whitten, D. G. J. Am. Chem. Soc. 1980, 102, 5678.
53. Turro, N. J.; Liu, K. C.; Chow, M. F. Photochem. Photobiol. 1977, 26, 413.
54. Worsham, P. R.; Eaker, D. W.; Whitten, D. G. J. Am. Chem. Soc. 1978, 100, 7091.
55. Nishimura, N.; Janado, M. J. Biochem. 1975, 77, 421.
56. Nakatani, H.; Shibata, K.-I.; Kondo, H.; Hiromi, K. Biopolymers 1977, 16, 2363.

57. Hui, Y.; Gu, J. Acta Chim. Sin. 1981, 39, 309.
58. Hui, Y.; Gu, J.; Jiang, X. Acta Chim. Sin. 1981, 39, 376.
59. Hui, Y.; Chen, X.; Gu, J.; Jiang, X. Sci. Sin. (Engl. Ed.) Ser. B 1982, 25, 698.
60. Hui, Y.; Wang, S.; Jiang, X. J. Am. Chem. Soc. 1982, 104, 347.
61. Bender, M. L.; Komiyama, M. "Cyclodextrin Chemistry"; Springer-Verlag: New York, 1978.
62. Harata, S. Bull. Chem. Soc. Jpn. 1975, 48, 2409.
63. VanEtten, R. L.; Sebastian, J. F.; Clowes, G. A.; Bender, M. L. J. Am. Chem. Soc. 1967, 89, 3242.
64. Manor, P. C.; Saenger, W. J. Am. Chem. Soc. 1974, 96, 3630.
65. Tabushi, I.; Kiyosuke, Y.; Yamamuna, K. J. Am. Chem. Soc. 1978, 100, 916.
66. Tabushi, I.; Yuan, L. C. J. Am. Chem. Soc. 1981, 103, 3574.
67. Suddaby, B. R.; Dominey, R. N.; Hui, Y.; Whitten, D. G. Can. J. Chem. accepted for publication.

RECEIVED January 10, 1985

12

Photosensitized Electron-Transfer Reactions in Organized Systems

The Role of Synthetic Catalysts and Natural Enzymes in Fixation Processes

I. WILLNER

Department of Organic Chemistry, The Hebrew University of Jerusalem, Jerusalem 91904, Israel

Charged colloids and water-in-oil microemulsions provide organized environments that control photosensitized electron transfer reactions. Effective charge separation of the primary encounter cage complex, and subsequent stabilization of the photoproducts against back electron transfer reactions is achieved by means of electrostatic and hydrophobic interactions of the photoproducts and the organized media.

Chemical utilization of the photoproducts has been accomplished by the introduction of synthetic catalysts or natural enzymes into the photochemical systems. With $Ru(NH_3)_5Cl^{2+}$ as electron acceptor and catalyst the photocleavage of acetylene (C_2H_2) to methane is observed. This process is a 6-electron reduction process and offers a model for the N_2-fixation reaction. Also, by the introduction of the enzyme ferredoxin reductase, a photochemical NADPH regeneration cycle has been established. The NADPH has been utilized in the reduction of ketones to alcohols in the presence of a secondary enzyme, alcohol dehydrogenase.

Induced disproportionation of photochemical single electron transfer products to two electron charge relays occurs in water-oil two phase systems. This process is a result of opposite solubility properties of the comproportionation products in the two phases.

Mimicking photosynthesis by means of artificial systems seems a promising route for solar energy conversion and storage (1-2). One possible cycle that is being extensively examined in recent years (3-4) is displayed in Figure 1. It involves a light absorbent, S, that upon excitation induces a transfer of an electron to an electron acceptor, A, leading to the photoproducts S^+ and A^-. Subsequent oxidation of an electron donor, D, recycles the sensitizer, and results in the conversion of light energy to chemical potential, stored in the products A^- and D^+ (eq. 1).

0097-6156/85/0278-0191$06.00/0

$$A + D \underset{\text{back reaction}}{\overset{h\nu}{\rightleftharpoons}} A^- + D^+ \qquad (1)$$

The photoproducts A^- and D^+ can then be utilized in chemical routes, e.g. the reduced photoproduct A^- can be used for the reduction of water to hydrogen, and fixation of carbon dioxide or nitrogen to organic fuels or ammonia. The oxidized photoproduct D^+ might be utilized in oxidation processes such as evolution of oxygen from water. Thus, one might envisage a variety of coupled photochemical-chemical processes that drive endoergic reactions converting abundant materials to fuels or useful chemicals (eq. 2-4).

$$H_2O \rightarrow H_2 + \tfrac{1}{2}O_2 \qquad (2)$$

$$CO_2 + H_2O \rightarrow CH_2O + O_2 \qquad (3)$$

$$N_2 + 3H_2O \rightarrow 2NH_3 + 1\tfrac{1}{2}O_2 \qquad (4)$$

The design of such artificial photosynthetic systems suffers from some basic limitations: a) The recombination of the photoproducts A^- and S^+ or D^+ is a thermodynamically favoured process. These degradative pathways prevent effective utilization of the photoproducts in chemical routes. b) The processes outlined in eq. 2-4 are multi electron transfer reactions, while the photochemical reactions are single electron transformations. Thus, the design of catalysts acting as charge relays is crucial for the accomplishment of subsequent chemical fixation processes.

Significant progress in the development of such artificial photosynthetic systems, particularly aimed at the photolysis of water, has been reported in recent years. Several approaches to resolve the problems involved in controlling the photoinduced electron transfer process as well as the development of catalysts for multi-electron fixation processes will be discussed in this paper.

Control of charge separation of the photoproducts.

The photosensitized electron transfer process involves two successive steps (eq. 5): In the primary event an encounter cage complex of the photoproducts is formed. This can either recombine to yield the original reactants or dissociate into separated photoproducts. The separated photoproducts can then recombine by a diffusional back electron transfer reaction to form the original reactants. We have introduced two conceptional approaches as a means for assisting the separation of the encounter cage complex and for the stabilization

$$S + A \overset{h\nu}{\rightleftharpoons} [S^+ \cdots A^-] \rightarrow S^+ + A^- \qquad (5)$$

of the photoproducts against the degradative recombination processes (5-6). These two approaches involve the organization of the photochemical system in interfacial systems that control charge separation by means of electrostatic or hydrophobic interactions of the interface with the photoproducts (Figure 2). Electrostatic interactions that control charge separation are exemplified in Figure 2(a) using a negatively charged interface as organization medium. In this system, a positively charged sensitizer, S^+, that is adsorbed onto the interface, and a neutral electron acceptor, serve as the photoreactants. Photoinduced electron transfer results in the encounter cage complex of the photoproducts. Although the photoproducts have opposite electric charges and exhibit mutual attractions, the highly charged interface is expected to repel the negative counterpart of the cage complex. Consequently, charge separation of the intermediate encounter complex is assisted. The recombination of the separated photoproducts via the diffusional mechanism is also retarded owing to the repulsion of the negative photoproduct, A^- from the oxidized species which is associated with the charged interface. Similarly, hydrophobic-hydrophylic boundaries capable of controlling the charge separation process are exemplified in Figure 2(b). Using this approach, the photosystem is solubilized in the aqueous media, while the reduced photoproduct is designed to exhibit hydrophobic character. Consequently, extraction of the reduced photoproduct from the water phase into the oil medium is anticipated to assist charge separation and to retard the recombination processes.

We have examined two types of organized media that effectively control the charge separation and back reactions of the intermediate photoproducts. These include, (a) charged colloids i.e. SiO_2 and ZrO_2 colloids that introduce electrostatic interactions between the photoproducts and interface (7-10), and (b) water-in-oil microemulsions that provide aqueous-oil two phase systems capable of controlling the reactions by proper design of the hydrophobic-hydrophilic balance of the photoproducts (6).

The silanol groups of the SiO_2 colloid are ionized in basic media (pH > 7.5). Consequently, a diffuse double layer is produced in the vicinity of the colloid particles, and the negatively charged colloid is characterized (11) by an electrical surface potential of ca. -170 mV. Similarly, ZrO_2 colloids are positively charged in aqueous acidic environments (pH = 4.2-4.5). We have synthesized two neutral, water-soluble, electron acceptors: di-(3-propylsulfonate)4,4'-bipyridinium, PVS^o, (1), and di-(2-propylsulfonato)-2,2'-bipyridinium, DQS^o, (2). The photosensitized reduction of these two electron acceptors with $Ru(bipy)_3^{2+}$ as sensitizer and triethanolamine, TEOA, as electron donor, has been examined in aqueous SiO_2 colloids (pH = 9.8) and compared to the similar reactions in a homogeneous aqueous phase (Table I) (7,9). The quantum yield of PVS^o reduction

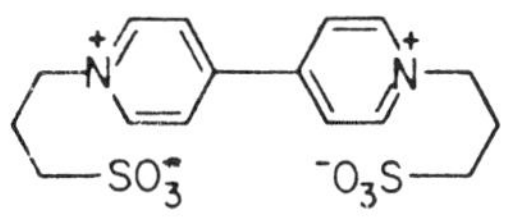

(1)

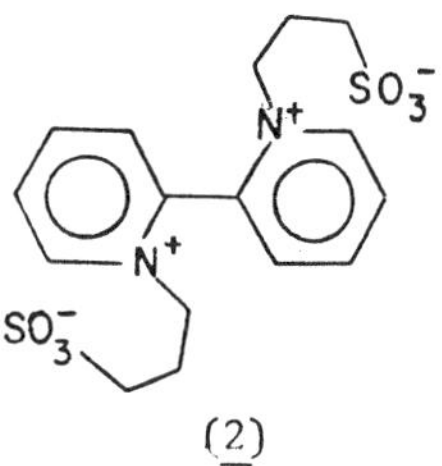

(2)

in the presence of the electron donor is 8-fold increased in the SiO_2 colloid as compared to that in a homogeneous phase. Similarly, the reduction of DQS^o proceeds in the SiO_2 colloid in the presence of TEOA ($2x10^{-3}$ M) with a quantum yield of $\phi=2.4x10^{-2}$ while no reduction

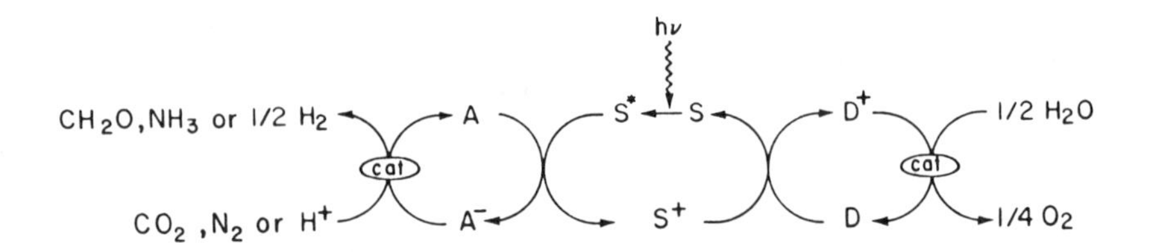

Figure 1. Conversion of light energy to chemical potential by means of photosensitized electron transfer reactions.

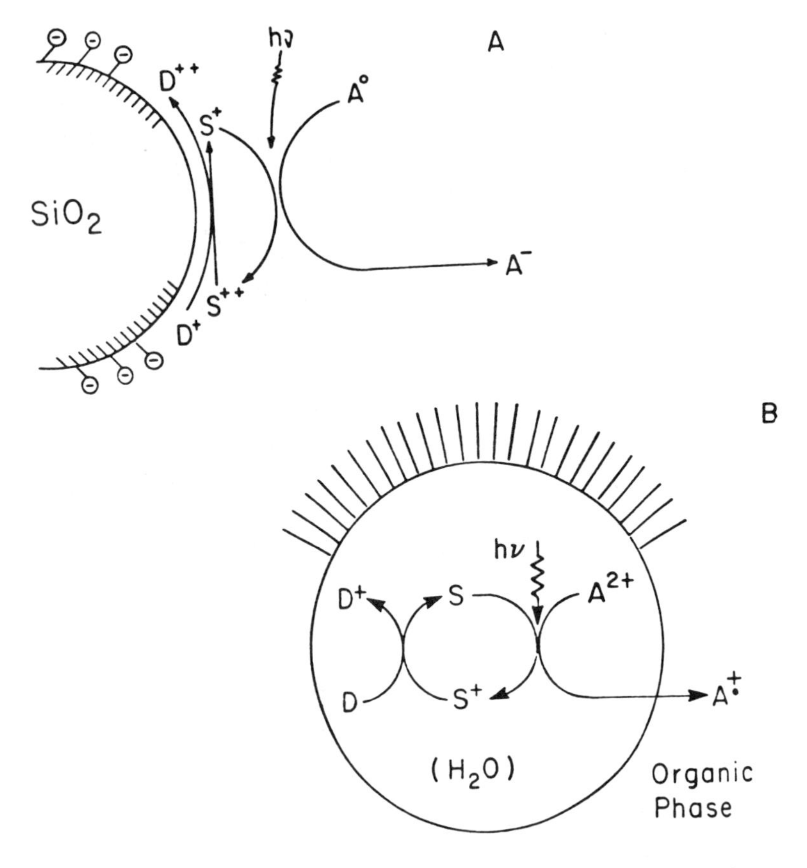

Figure 2. Charge separation and stabilization of photoproducts in organized environments: a) Application of the electrostatic interactions with charged SiO_2 colloids. b) Use of hydrophobic-hydrophilic interactions in water-in-oil microemulsions.

Table I. Effects of SiO_2 colloid on charge separation of the primary encounter cage complex and recombination rates of $PVS^{\bar{\cdot}}$ and $DQS^{\bar{\cdot}}$

	$PVS^{\bar{\cdot}}$ (SiO_2)				$DQS^{\bar{\cdot}}$ (SiO_2)			
	Hom.	pH=9.8	pH=8.2	[NaCl]= 0.1M pH=9.8	Hom.[a]	pH= 9.8	pH= 7.8	[NaCl]= 0.4M pH=9.8
ϕ_s^b	0.38	0.38	0.37	0.38	–[c]	0.26	0.20	0.16
$k_b(M^{-1}.s^{-1})^d$	$7.9x10^9$	$5.7x10^7$	$3.9x10^8$	$4.3x10^8$	–[c]	$1x10^7$	$2x10^8$	$8x10^8$

a. Homogeneous phase. No SiO_2 colloid included, pH=9.8.

b. Quantum yield for the separation of the encounter cage complex of photoproducts (equation 7). Determined by the absorbance of $PVS^{\bar{\cdot}}$ at λ=602 nm and bleaching of $Ru(bpy)_3^{2+}$ at λ=460 nm for $DQS^{\bar{\cdot}}$.

c. No separation of the encounter cage complex could be observed in a homogeneous aqueous phase. Therefore, no direct diffusional recombination rate constant could be estimated. Note however, that at high ionic strength and low pH media, where the SiO_2 colloid electric surface potential is low, the recombination rate constant is almost diffusion controlled ($k_b \simeq 10^9\ M^{-1}.s^{-1}$).

d. Determined by following the bleaching of photoproduced $PVS^{\bar{\cdot}}$ at λ=602 nm (ε_{602}= 12500 $M^{-1}\ cm^{-1}$) and following the recovery of bleached $Ru(bpy)_3^{2+}$ at λ=452 nm, for $DQS^{\bar{\cdot}}$ (ε_{452}= 14500 $M^{-1}\ cm^{-1}$).

of DQS^o occurs in homogeneous media (9). To account for the increased quantum yields under steady state illumination in the presence of SiO_2 colloid, we have characterized the processes involved in the photosensitized electron transfer reactions: i) quenching of the excited species (equation 6), ii) quantum yields for the separation of the primary encounter cage complex (equation 7) and iii) the recombination rate of the separated photoproducts (equation 8). The physical data for these reactions are summarized in Table 1.

$$*Ru(bpy)_3^{2+} + A \xrightarrow{k_q} [Ru(bpy)_3^{3+} \cdots A^{\bar{\cdot}}] \qquad (6)$$

$$[Ru(bpy)_3^{3+} \cdots A^{\bar{\cdot}}] \longrightarrow Ru(bpy)_3^{3+} + A^- \qquad (7)$$

$$Ru(bpy)_3^{3+} + A^{\bar{\cdot}} \xrightarrow{k_b} Ru(bpy)_3^{2+} + A \qquad (8)$$

The quenching rate constants (k_q) (7,9) of the excited sensitizer by PVS^o and DQS^o in the SiO_2 colloid are $1.5x10^9$ and $4x10^8\ M^{-1}.s^{-1}$ respectively. In a homogeneous aqueous phase the quenching rate constants correspond to $1.5x10^9\ M^{-1}.s^{-1}$ for PVS^o and $5.9x10^8 M^{-1}.s^{-1}$ for DQS^o. Thus, the higher quantum yields in the SiO_2 colloid cannot be attributed to the quenching process. Table 1 shows that the escape yields of photoproducts from the primary cage structure as well as the recombination rates of the separated photoproducts are strongly affected by the SiO_2 colloid.

It is evident that the initial encounter complex of $DQS^{\bar{\cdot}}$ is non-separable in a homogeneous aqueous phase and the photoproducts degrade in the cage structure. However, the SiO_2 colloid assists the separation of photoproducts from the cage structure (ϕ_s=0.26).

A second significant effect of the SiO_2 colloid is observed on the back reaction rate constants. In a homogeneous phase the recombination reaction of $PVS^{\overline{\cdot}}$ and $DQS^{\overline{\cdot}}$ with the oxidized species, $Ru(bpy)_3^{3+}$, is diffusion controlled. In the presence of the SiO_2 colloid this back electron transfer process is substantially retarded and ca. 200-fold slower than in the homogeneous phase. The functions of the SiO_2 colloid in charge separation and retardation of back reactions are attributed to electrostatic interactions of the photoproducts and the charged colloid interface (Figure 3). The positively charged sensitizer is bound to the colloid interface. The initial encounter cate complex formed upon electron transfer is still positively charged and associated with the particle. Yet, the negatively charged component, $DQS^{\overline{\cdot}}$, is repelled by the interface and ejected from the cage structure. The separated photoproducts are stabilized against diffusional back electron transfer reactions since $DQS^{\overline{\cdot}}$ or $PVS^{\overline{\cdot}}$ is repelled by the charged colloid to which the oxidized photoproduct, $Ru(bpy)_3^{3+}$, is bound. The effective control of the degradative pathways of the photoproducts, in the SiO_2 colloid allows the efficient subsequent oxidation of TEOA, by $Ru(bpy)_3^{3+}$, and consequently high quantum yields are observed under steady state illumination. The electrostatic functions of the SiO_2 colloid in these reactions have been confirmed by altering the pH and ionic strength of the colloid media. Acidification of the colloid environment results in partial neutralization of the silanol groups, and decrease of the colloid surface potential. Similarly, increase of the ionic strength reduces the diffusional double layer surface potential (7,9). Accordingly, reduction in the charge separation yields of the cage complex as well as enhanced back reaction rates are observed at lower pH values and high ionic strength conditions of the SiO_2 colloid (Table I). The application of the SiO_2 colloid in controlling the photosensitized electron transfer process is limited to basic aqueous solution. We should, however, note that we have used positively charged ZrO_2 colloids for effecting similar electrostatic interactions in acidic environments (pH=4.5) (10).

Hydrophobic interactions as a means for controlling charge separation and back electron transfer reactions have been demonstrated in a two phase system of a water-in-oil microemulsion (6). We have examined the photosensitized reduction of a series of 4,4'-bipyridinium salts, C_nV^{2+}, (3), (where n=1-16) with $Ru(bpy)_3^{2+}$ as sensitizer and $(NH_4)_3EDTA$ as electron donor in a water-in-toluene microemulsion media. Under steady state illumination the quantum yield of $C_nV^{\dot{+}}$ formation strongly depends on the alkyl chain length of the electron acceptor (Figure 4). It improves as the hydro-

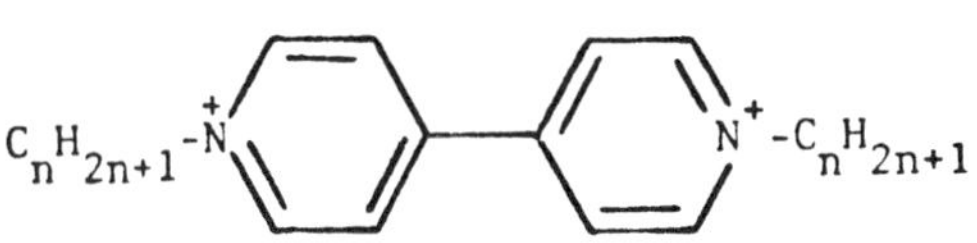

$$C_nH_{2n+1}-\overset{+}{N}\langle\bigcirc\rangle-\langle\bigcirc\rangle\overset{+}{N}-C_nH_{2n+1}$$

(a) R = CH_3 (d) R = C_8H_{17}
(b) R = C_4H_9 (e) R = $C_{14}H_{29}$
(c) R = C_6H_{13} (f) R = $C_{18}H_{37}$

3

phobicity of $C_nV^{\dot{+}}$ is increased and reaches an optimal value for n=8-16. The functions of the water-in-toluene microemulsion in controlling the process were verified by following the quenching reaction, cage structure separation and back electron transfer (Table II). It can be seen that for n=1 no charge separation occurs. For n=4 charge separation is inefficient and the recombination rate is very rapid. With the long chain electron acceptor ($n \geq 8$) the separation of the cage structure is effective and the back reaction rate constant is ca. 10-fold slower as compared to $C_4V^{\dot{+}}$. The solubility properties of the reduced photoproduct $C_nV^{\dot{+}}$ in water-organic two phase systems depends on the alkyl chain length: while C_1-$C_4V^{\dot{+}}$

Table II. Charge separation yields and recombination rates in the photosensitized reduction of C_nV^{2+} in water-in-toluene microemulsions (6)

C_nV^{2+}	1	4	6	8	14	18
$\phi_{separation} \times 10^3$	0	6	36	40	50	54
$k_b^{(a)} \times 10^{-9}$ sec^{-1} mole^{-1}	-	26	8	0.7	0.33	1.2
$\phi^{(b)}_{steady\ state} \times 10^3$	10^{-5}	0.8	2.5	7.5	8.1	7.2

a. Determined by following the disappearance of $C_nV^{\dot{+}}$ at λ=602 nm (ε=12.500 M^{-1} cm^{-1}).
b. Light intensity 7.56×10^{-3} einsteins.ℓ^{-1} $\cdot$ min^{-1}.

are rather soluble in water and insoluble in toluene, the amphiphilic electron acceptors $C_8V^{\dot{+}}$ -$C_{16}V^{\dot{+}}$ are extracted from the aqueous environment into the organic phases. Therefore, the enhanced quantum yields of the long chain photoproducts, $C_nV^{\dot{+}}$ (n=8-16), are assigned to hydrophobic interactions of the intermediate photoproducts with the water-oil microemulsion medium (Figure 4). The primary encounter cage complex for $C_nV^{\dot{+}}$ where $n \geq 8$ is associated with the hydrophobic interface boundary. Extraction of the hydrophobic component, $C_nV^{\dot{+}}$, into the oil phase assists the separation of the cage complex and the separated photoproducts are subsequently stabilized against the diffusional back reactions by means of the two phases. These two effects result in high quantum yields under steady state illumination.

Multi-Electron Charge Relays

Transformation of single electron transfer products into multi-electron charge relays is a basic requirement for accomplishing complex fixation reaction processes (equations 3 and 4). A possible way to achieve such transformations is the disproportionation of a single electron transfer product to the corresponding doubly reduced species (equation 9). The comproportionation equilibrium constant (K_d) is determined by the reduction potentials of the two species involved in the process (equation 10). Usually, $E_2 < E_1$ and consequently the disproportionation equilibrium lies overwhelmingly towards the single electron transfer product. Yet, this situation is valid in a homogeneous phase only, and might be rather altered in a two phase system (Figure 5). Assuming that the electron

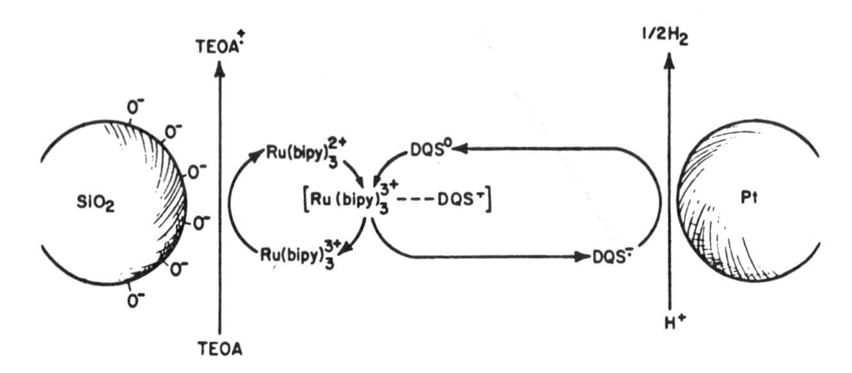

Figure 3. Functions of the SiO_2 colloid in controlling the photosensitized electron transfer process and H_2-evolution.

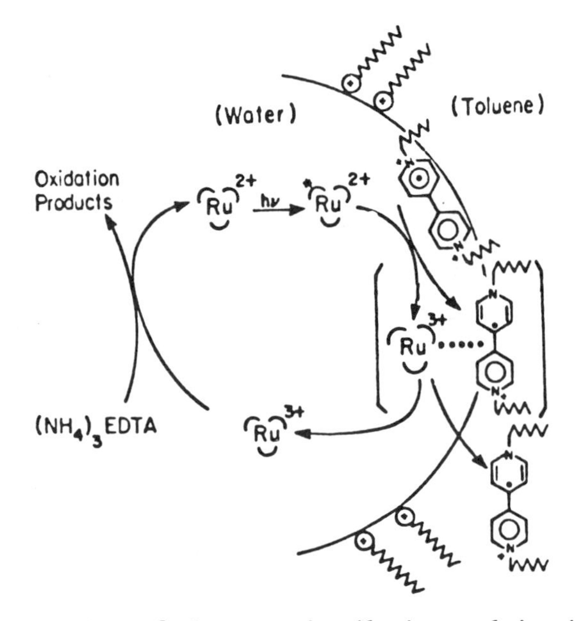

Figure 4. Functions of the water-in-oil microemulsion in charge separation and stabilization of photoproducts against back-reactions.

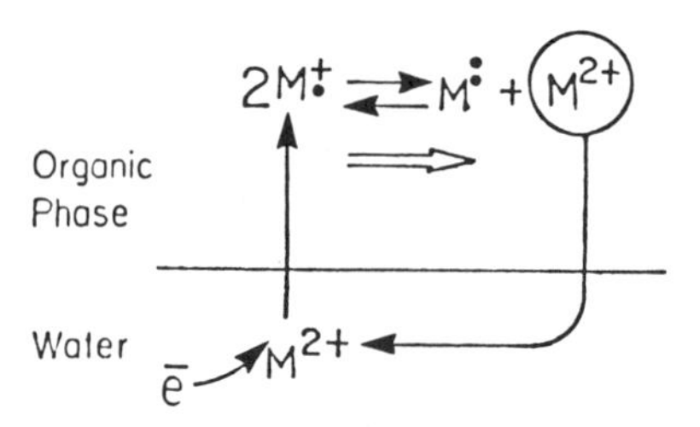

Figure 5. Induced comproportionation of an amphiphilic electron acceptor in a water-oil two phase system.

acceptor exhibits a delicate hydrophilic-hydrophobic balance, where the oxidized form A^{2+} is soluble in an aqueous media, while the mono-reduced product $A^{\dot{+}}$, is hydrophobic in nature and extracted from the aqueous solution into organic phases. Under such conditions disproportionation of $A^{\dot{+}}$ in the organic phase is accompanied by reextraction of A^{2+} to the aqueous phase. Consequently, a two phase system and the proper design of hydrophilic-hydrophobic balance of the disproportionation products provides an organized environment for induced comproportionation of a single electron transfer product to the doubly reduced charge relay. It is also evident from this cycle, that continuous reduction of A^{2+} will ultimately form the doubly reduced species, $A^{:}$, that functions as an electron sink.

$$2A^{\dot{+}} \underset{}{\overset{K_d}{\rightleftharpoons}} A + A^{2+} \qquad (9)$$

$$K_d = \frac{[A][A^{2+}]}{[A^{\dot{+}}]^2} \quad ; \quad K_d = 10^{-\frac{nF\Delta E^o}{RT}} \qquad (10)$$

$$(\text{where } \Delta E^o = E_1^o - E_2^o)$$

We have found (12) that the electron acceptor, N,N'-dioctyl-4,4'-bipyridinium, (3d), C_8V^{2+}, and its reduction products meet the correct hydrophobic-hydrophilic balance for an induced disproportionation of $C_8V^{\dot{+}}$, (4) to N,N'-dioctylbipyridinylidene (5), in organic-water two phase systems (equation 11). The electron acceptor (3d), undergoes two successive one-electron reduction processes to N,N'-dioctyl-4,4'-bipyridinium radical cation, $C_8V^{\dot{+}}$, and N,N'-dioctyl-4,4'-bipyridinylidene, C_8V, (5), (equation 12), (E_1=-0.47V and E_2=-0.90V, vs. NHE respectively). The comproportionation constant for $C_8V^{\dot{+}}$ in homogeneous aqueous phase is $K_d = 5.5\times10^{-8}$. Thus, the disproportionation equilibrium of $C_8V^{\dot{+}}$ in a homogeneous aqueous phase lies overwhelmingly towards the single electron reduction product. The electron acceptor, C_8V^{2+}, is soluble in water and insoluble in organic media such as toluene or ethylacetate. In turn, the one electron reduction product, $C_8V^{\dot{+}}$, is hydrophobic in nature and extracted into organic phases from aqueous environments. The consequence of the opposite solubility properties of C_8V^{2+} and its reduction products in the organized two phase system on the disproportionation of $C_8V^{\dot{+}}$ is shown in Figure 6. The electron acceptor, C_8V^{2+}, is photoreduced in water using $Ru(bpy)_3^{2+}$ as sensitizer and $(NH_4)_3EDTA$ as electron donor. The photoproduct, $C_8V^{\dot{+}}$, is extracted from the aqueous phase to the organic phase and the absorption spectra of the photoproducts in the organic phase is displayed in Figure 6. After a short illumination time of the aqueous phase, the absorption spectrum of the photoproduct in the organic phase resembles that of $C_8V^{\dot{+}}$. However, when the absorption spectrum of $C_8V^{\dot{+}}$ in a homogeneous phase, where disproportionation is negligible, is subtracted from the experimental absorption spectrum of the photoproducts present in the organic phase, an absorption pattern of a second component at λ=400 nm is observed (Figure 6(b)). This absorption band is identical to that of N,N'-dioctylbipyridinylidene, C_8V, that is produced electrochemically (13). (The subtraction procedure is based on the fact that C_8V does not absorb at λ=602 nm (ε=12.500 M^{-1} cm^{-1}). Therefore, the absorption of $C_8V^{\dot{+}}$

can be subtracted from the composite spectrum using the respective concentration factor) (14).

$$2C_8V^{\dot{+}} \rightleftharpoons C_8V + C_8V^{2+} \qquad (11)$$

$$C_8V^{2+}\ (\underline{\underline{3}}) \xrightarrow{\bar{e}} C_8V^{\dot{+}}\ (\underline{\underline{4}}) \xrightarrow{\bar{e}} C_8V\ (\underline{\underline{5}}) \qquad (12)$$

Thus it is evident that in the two phase system the doubly reduced photoproduct, C_8V, is formed in conjunction with the photosensitized one electron transfer process (12, 14). In addition, prolonged illumination results in the accumulation of the doubly reduced product, C_8V, at the expense of the single electron transfer product, $C_8V^{\dot{+}}$ (Figure 6(c)). These results are consistent with the induced disproportionation mechanism outlined in Figure 5. Due to the opposite solubility properties of the disproportionation products in the two phases, C_8V^{2+} is reextracted into the aqueous phase and the doubly reduced comproportionation product, C_8V, is accumulated in the organic phase. The quantitative spectroscopic estimation of $[C_8V^{\dot{+}}]$ and $[C_8V]$ in the organic phase allowed us to estimate the comproportionation constants of $C_8V^{\dot{+}}$ in various organic-water two phase systems (14). For example, in ethylacetate we have estimated a value of $K_d=3\text{x}10^{-1}$ M for the disproportionation equilibrium of $C_8V^{\dot{+}}$. This value is ca. 10^7 higher than the comproportionation constant of the similar process in an homogeneous aqueous medium.

It is evident that a single electron transfer photoproduct is transformed into a doubly reduced charge relay in two phase systems. The primary processes in the natural photosynthetic apparatus involve single electron transfer reactions that proceed in hydrophobic-hydrophilic cellular microenvironment. Thus, we suggest similar induced disproportionation mechanisms as possible routes for the formation of multi-electron charge relays, effective in the fixation of CO_2 or N_2.

The subsequent chemical utilization of the two electron charge relay has also been accomplished (12, 14). The electrochemical reduction of C_8V^{2+}, (3d) by means of cyclic voltammetry shows two reversible, one electron, reduction waves at $E_1=-0.47V$ and $E_2=-0.90V$ (vs. NHE) corresponding to the formation of $C_8V^{\dot{+}}$ and C_8V respectively (equation 12). Addition of meso-1,2-dibromostilbene (6) does not affect the reversibility of the first reduction wave. Yet, the reoxidation wave of C_8V is depleted upon addition of 6 implying a chemical reaction of C_8V with the dibromides. Introduction of meso-1,2-dibromostilbene, (6), into the organic phase of an ethylacetate-water two phase system, that includes $Ru(bpy)_3^{2+}$ as sensitizer, C_8V^{2+} as electron acceptor and $(NH_4)_3EDTA$ as electron donor, results upon illumination in the quantitative formation of trans-stilbene (7) in the organic phase (equation 13).

$$\underset{\underline{\underline{6}}}{\text{H}(\text{Ph})\text{C(Br)}-\text{C(Br)(H)Ph}} + 2\bar{e} \longrightarrow \underset{\underline{\underline{7}}}{\text{PhCH}=\text{CHPh}} + 2Br^- \qquad (13)$$

The electrochemical studies reveal that the active species in debromination is the two electron reduction product, C_8V. Yet, the primary photochemical process is a single electron transfer reaction that yields $C_8V^{\dot{+}}$. In view of our previous discussion, we suggest the cycle presented in Figure 7 as the mechanistic route for debromination of 6. In this cycle photoreduction of C_8V^{2+} in the aqueous solution is accompanied by extraction of $C_8V^{\dot{+}}$ from the aqueous phase into the organic phase. Induced disproportionation of $C_8V^{\dot{+}}$ yields the doubly reduced species, C_8V, that is the active reductant in the debromination process. The discussion suggests that a similar process should be prevented in a homogeneous phase since the formation of C_8V is not favoured. Indeed, illumination of 6 and the previously described photosystem in a homogeneous dimethylformamide solution does not lead to the formation of trans-stilbene (7), despite the effective formation of $C_8V^{\dot{+}}$.

Catalysts for Chemical Utilization of the Photoproducts

Heterogeneous and Homogeneous Catalysis

For chemical utilization of the electron transfer photoproducts inclusion of catalysts seems to be essential. These catalysts might function as charge storage sites for the complex multi-electron transfer fixation processes, and/or might participate in the activation of the substrates towards the chemical reactions. In the natural photosynthetic system enzymes function as catalytic sites for the complex fixation processes. We might envisage two alternative approaches in developing catalysts for the chemical utilization of the photoinduced electron transfer products: (i) One possibility involves the development of synthetic catalysts that mimic the functions of enzymes with respect to charge storage and substrate activation capabilities. (ii) The second approach might involve the introduction of natural enzymes into artificial chemical systems (provided that the enzymes are stable in this artificial environment).

Most of the past efforts for the utilization of the photoproducts in subsequent chemical reactions were directed towards the photolysis of water to hydrogen and oxygen (15-19). In the course of our studies using charged colloids as a means for controlling the charge separation and recombination reactions, we have designed an additional H_2-evolution system (9,20). As stated previously the negatively charged SiO_2 colloid assists the separation of the photoproduced N,N'-dipropyl-2,2'-bipyridinium radical anion , $DQS^{\dot{-}}$, and stabilizes the intermediate photoproducts, $DQS^{\dot{-}}$, and $Ru(bpy)_3^{3+}$

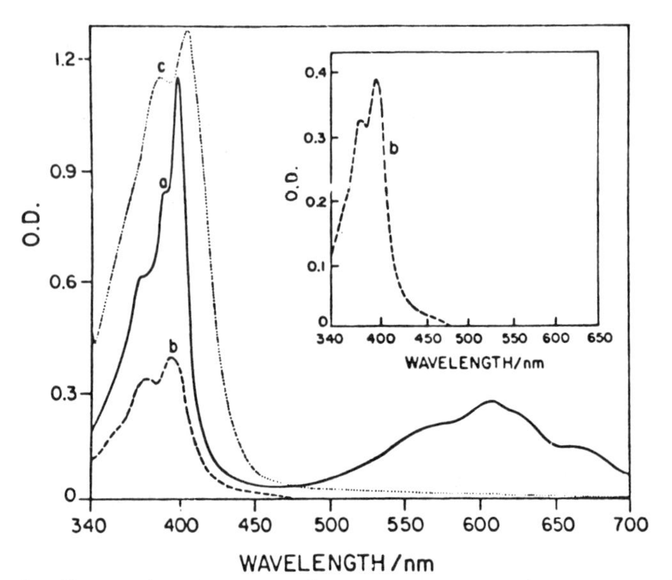

Figure 6. Absorption spectra of the components in the organic phase obtained after extraction of photogenerated C_8V^+ from the aqueous phase: a) Composite spectrum of C_8V^+ and C_8V in ethylacetate after illumination of the system for 10 minutes; b) Spectrum of C_8V in ethylacetate obtained after subtraction of C_8V^+.spectrum from the composite spectrum (a). Insert is enlarged absorption spectrum of C_8V. c) Spectrum of the photoproducts when toluene is used as the organic phase in the two phase system. Spectrum recorded after illumination for 15 minutes corresponds to C_8V as major product.

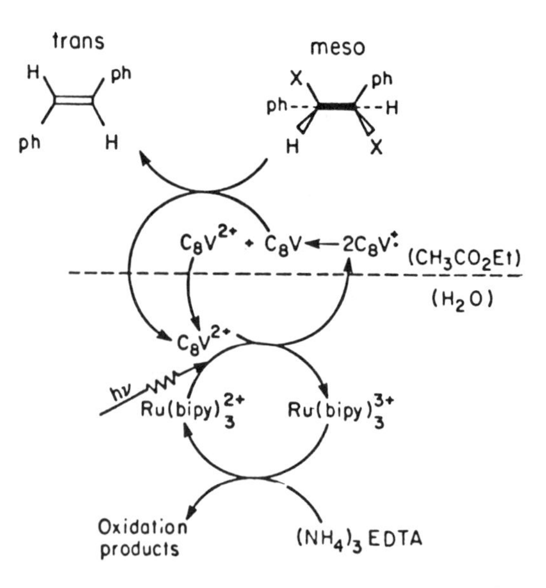

Figure 7. Photosensitized debromination of meso-1,2-dibromo-stilbene, (6), in a two phase system.

against the back-electron transfer process (Figure 2). The reduced photoproduct, $DQS^{\cdot-}$, (E^o=-0.75V vs. NHE) mediates the evolution of hydrogen in the presence of Pt colloid. It should be noted that in this system H_2 is evolved from the basic aqueous SiO_2 colloid (pH=8.5-10.2). This might facilitate the complementary O_2-evolution process. The similar H_2-evolution cannot be accomplished in a homogeneous phase since no separated photoproducts are obtained. Thus, the SiO_2-colloid provides an organized medium that stabilizes the electron transfer products and allows the subsequent evolution of hydrogen (Figure 2).

Recently we have attempted to pursue multi-electron fixation processes as models for N_2 or CO_2 fixation. In nature, the N_2-fixation enzyme, nitrogenase, exhibits non-specificity properties, and acetylene competes for nitrogen as the fixation substrate (21). The fixation process of acetylene to methane and of nitrogen to ammonia (euqations 14 and 15) have several common features: (i) both involve the cleavage of a triple bond; (ii) the two reactions involve 6 electrons in the fixation mechanism. Thus, it seems that the photocleavage of acetylene to methane might offer a good model for development of N_2-fixation cycles (22).

$$N_2 + 6\bar{e} + 6H^+ \longrightarrow 2NH_3 \quad (14)$$

$$HC{=}CH + 6\bar{e} + 6H^+ \longrightarrow 2CH_4 \quad (15)$$

Ruthenium (II) pentamine, $Ru(NH_3)_5H_2O^{2+}$ is known to bind acetylene (23) as well as nitrogen (24). We have therefore examined the photosensitized reduction of acetylene using $Ru(bpy)_3^{2+}$ as sensitizer, triethanolamine, TEOA, as electron donor and $Ru(NH_3)_5Cl^{2+}$ as electron acceptor and catalyst for the fixation process (22). Illumination of this system with visible light results in the formation of methane (Figure 8). Similarly, methylacetylene ($CH_3C{\equiv}CH$) and ethylacetylene (CH_3CH_2-$C{\equiv}CH$) are photocleaved to methane and ethane or propane respectively. In turn doubly substituted acetylene substrates e.g. dimethylacetylene are neither photocleaved nor reduced.

Several mechanistic aspects involved in the photocleavage of acetylene to methane have been elucidated (Scheme 1). The primary event involves the photoreduction of $Ru(NH_3)_5Cl^{2+}$ by $^*Ru(bpy)_3^{2+}$ followed by aquation of the reduced product to $Ru(NH_3)_5(H_2O)^{2+}$ (equations 16 and 17). This photoproduct adds acetylene in a "side on" complex that has been isolated and characterized (equation 18). Nevertheless, the π-acetylene complex appears to be inert towards reduction or cleavage to methane in the presence of reducing agents. Yet, upon illumination of the π-acetylene coordination compound it undergoes a transformation to a new complex that is active in methane evolution in the presence of reducing agents including $Ru(NH_3)_5(H_2O)^{2+}$. The photoinduced activation of the π-bonded acetylene complex has been followed spectroscopically and is attributed to a π-σ acetylene ligand rearrangement (equation 19). The σ-acetylene complex undergoes subsequent reductive cleavage to methane (equation 20). The photoinduced σ-π acetylene bond rearrangement clearly explains the lack of reductive cleavage of dimethylacetylene since this transformation is not possible. The

Scheme I.

$$Ru(bipy)_3^{2+} + Ru(NH_3)_5Cl^{2+} \xrightarrow{h\nu} Ru(bpy)_3^{3+} + Ru(NH_3)_5Cl^{+} \quad (16)$$

$$Ru(NH_3)_5Cl^{+} + H_2O \rightarrow Ru(NH_3)_5(H_2O)^{2+} + Cl^{-} \quad (17)$$

$$Ru(NH_3)_5(H_2O)^{2+} + HC{\equiv}CH \rightarrow Ru(NH_3)_5(HC{\equiv}CH)^{2+} \quad (18)$$

$$Ru(NH_3)_5(HC{\equiv}CH)^{2+} \xrightarrow{h\nu} (NH_3)_5Ru{-}C{\equiv}CH^{1+} + H^{+} \quad (19)$$

$$(NH_3)_5Ru{-}C{\equiv}CH^{1+} \rightleftarrows [(H_3N)_5Ru{=}C{=}CH_2]^{2+}$$

$$5Ru(NH_3)_5(H_2O)^{2+} + (NH_3)_5{-}Ru{-}C{\equiv}CH^{+1} + 6H^{+} \rightarrow 6Ru(NH_3)_5(H_2O)^{3+} + 2CH_4 \quad (20)$$

photocleavage of acetylene to methane involves 6 electrons and protons. Thus, the feasibility of accomplishing the isoelectronic fixation of nitrogen to ammonia by similar means is conceivable and remains a challenge.

Enzymatic catalysis

A different approach for utilization of the photoproducts in chemical routes involves the introduction of natural enzymes as catalysts in the photochemical system. In nature, dihydronicotinamide adenine dinucleotide (NADH) and dihydronicotinamide dinucleotide phosphate (NADPH) participate as reducing cofactors in a variety of enzymatic reduction processes. Thus, the development of photochemical NADH and NADPH regeneration cycles is anticipated to allow a variety of reduction processes by inclusion of substrate specific NAD(P)H dependent enzymes.

Several chemical routes have been developed for the regeneration of NADH and NADPH (25, 26). We have developed (27) a photochemical system for regeneration of NADPH (Figure 9). 4,4'-Bipyridinium radical cations reduce $NADP^{+}$ to NADPH in the presence of the enzyme ferredoxin reductase, FDR. Illumination of a system composed of the sensitizer $Ru(bpy)_3^{2+}$, ethylenediamine tetraacetic acid, the electron acceptors dimethyl-4,4'-bipyridinium, (methyl viologen, MV^{2+}) and $NADP^{+}$ and the enzyme ferredoxin reductase (FDR) leads to the quantitative formation of NADPH. Addition of 2-butanone and the second enzyme alcohol dehydrogenase (from T. Brockii), ALDH, results in the formation of (-)2-butanol and $NADP^{+}$. Continuous illumination of this system results in the accumulation

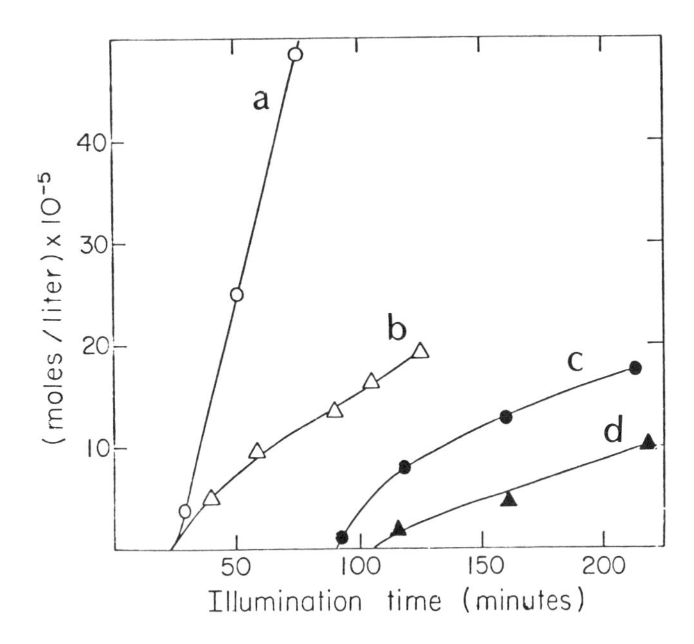

Figure 8. Rate of photocleavage of acetylene (C_2H_2) and methylacetylene (C_3H_4) as a function of illumination time. a) Methane (CH_4) from acetylene and c) ethane (C_2H_6) from methylacetylene using meso-Zn-tetramethylpyridinium porphyrin as sensitizer. b) Methane (CH_4) from acetylene and d) ethane (C_2H_6) from methylacetylene with $Ru(bpy)_3^{2+}$ as sensitizer.

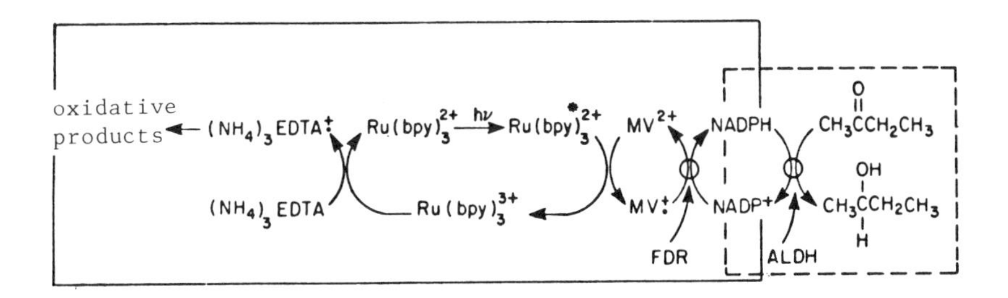

Figure 9. Photoreduction of 2-butanone using two coupled enzymes and the NADPH regeneration cycle.

of (-)2-butanol at the expense of the sacrificial electron donor EDTA (Figure 10). In this process two enzyme catalyzed sub-cycles are operative: The primary cycle represents the photo-regeneration of NADPH, while the subsequent cycle describes the chemical use of NADPH in the fixation of a ketone to an alcohol. The net reaction accomplished in this cycle is the reduction of 2-butanone by EDTA (equation 20). This process is endoergic by ca. 33 kcal mol^{-1} of EDTA consumed.

$$R{-}N\begin{matrix} CH_2CO_2^- \\ CH_2CO_2^- \end{matrix} + CH_3\overset{O}{\overset{\|}{C}}CH_2CH_3 + 2H_2O \rightarrow RNHCH_2CO_2^- + CH_2O + HCO_3^- + CH_3{-}\overset{OH}{\underset{H}{C}}{-}CH_2CH_3 \quad (20)$$

A major aspect to consider in such enzyme-catalyzed photochemical systems is the stability of enzymes in the artificial chemical environments. Table III summarizes the turnover (TN) numbers for the different enzymes and cofactors involved in the reduction of 2-butanone. It is evident that the enzymes exhibit high stability

Table III. Turnover Numbers (TN) of the components involved in the photosensitized reduction of 2-butanone.

	$Ru(bpy)_3^{2+}$	MV^{2+}	FDR[b]	$NADP^+$	ALDH[c]
TN	530	40	24000	40	6000

a. No loss of system activity could be detected after determination of these turnover numbers.
b. F.W. ≃ 40,000; Cf. M. Shin, Methods in Enzymology 23, 441 (1971).
c. F.W. ≃ 40,000; Cf.R.J. Lamed and G. Zeikus, Biochem. J., 195, 183 (1981).

towards denaturation and that the rate of product formation is not affected even after prolonged illumination times. The stability of the system and the optical purity of the product (-)2-butanol (100%) demonstrate an effective method for production of optically active alcohols. Certainly, such photosensitized fixation cycles based on the regeneration of NAD(P)H- might be generalized. By proper substitution of the secondary NAD(P)H-dependent enzyme reduction of various other substrates is conceivable. Some of these possibilities e.g. production of amino acids and CO_2-fixation are now being examined in our laboratory.

Conclusions

Different aspects involved in the design of artificial photosynthetic systems have been discussed. Charged colloids and water-oil microemulsions provide effective organized media for controlling photosensitized electron transfer processes. Development of catalysts capable of utilizing the photoproducts in chemical routes, particularly in multi-electron fixation processes is of major

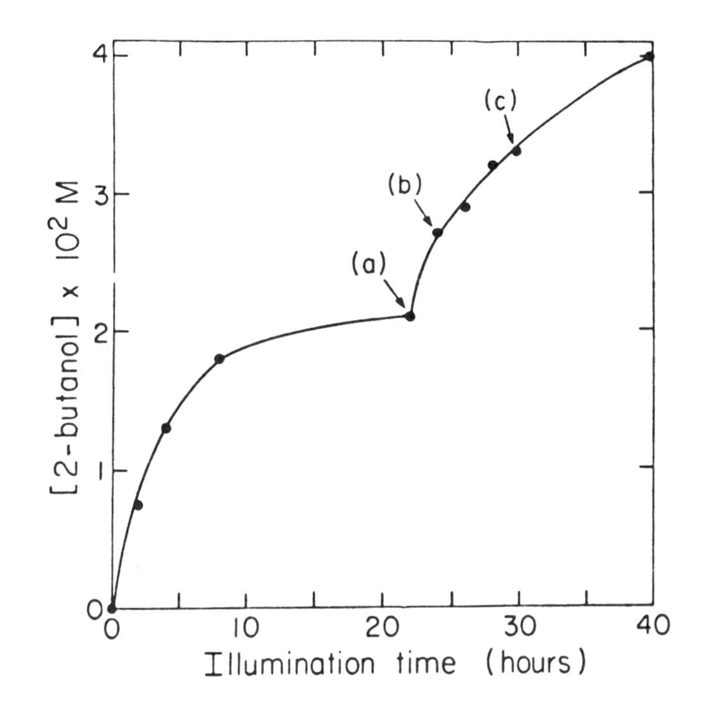

Figure 10. Rate of 2-butanol formation at different time intervals of illumination. Initial $(NH_4)_3EDTA$ concentration $2x10^{-2}M$. a) Addition of $(NH_4)_3EDTA$, $2x10^{-2}M$, b) and c) Addition of $(NH_4)_3EDTA$, $1.7x10^{-2}M$.

importance. The induced disproportionation of a single electron transfer product to a two electron charge relay in a water-oil two phase system seems a general principle worth developing. Two different approaches in developing catalysts have been discussed. One approach involves the introduction of homogeneous and heterogeneous synthetic catalysts. The other alternative suggested the introduction of natural enzymes into the photochemical systems. The photocleavage of acetylene to methane using the homogeneous catalyst $Ru(NH_3)_5Cl^{2+}$ implies the feasibility of designing homogeneous catalysts for multielectron fixation processes. Other reactions, such as N_2 and CO_2-fixations remain future challenges. Similarly, introduction of natural enzymes as specific catalysts in artificial photosynthetic systems seems to be of broad applicability. Introduction of NADH and NADPH dependent enzymes for the preparation of amino acids, CO_2-fixation into sugars as well as reduction of keto substituted natural products seems feasible.

We thus realize that the duplication of functions of natural photosynthesis by means of artificial processes is conceivable. The progress in recent years is encouraging us to continue pursuing these challenging goals.

Acknowledgment

The author wishes to express his deep gratitude to his collaborators: Y. Degani, Z. Goren, P. Dan, D. Mandler, R. Maidan, and E. Por whose inspired and enthusiastic efforts have made possible the success of this work.

Literature Cited

1. Sutin, N.; Creutz, C. Pure Appl. Chem., 1980, 52, 2717-38.
2. Gratzel, M. Acc. Chem. Res., 1981, 14, 276-84.
3. Kalyanasundaram, K. Coord. Chem. Rev., 1982, 46, 159-244.
4. Whitten, D.G.; Russel, J.C.; Schnell, R.H. Tetrahedron, 1982, 38, 2455-87; Acc. Chem. Res., 1980, 13, 83-90.
5. Willner, I.; Laane, C.; Otvos, J.W.; Calvin, M. in "Inorganic Reactions in Organized Media"; Holt, S.L., Ed.; Advances in Chemistry Series, No. 177, American Chemical Society, Washington, D.C. 1982, p. 71.
6. Mandler, D.; Degani, Y.; Willner, I. J. Phys. Chem. in press 1984.
7. Willner, I.; Yang, J.-M.; Otvos, J.W.; Calvin, M. J. Phys. Chem. 1981, 85, 3277.
8. Willner, I.; Otvos, J.W.; Calvin, M. J. Am. Chem. Soc., 1981, 103, 3203.
9. Degani, Y.; Willner, I. J. Am. Chem. Soc., 1983, 105, 6228.
10. Willner, I.; Degani, Y. Isr. J. Chem., 1982, 22, 163.
11. Laane, C.; Willner, I.; Otvos, J.W.; Calvin, M. Proc. Natl. Acad. Sci., U.S.A., 1981, 78, 5829.
12. Goren, Z.; Willner, I. J. Am. Chem. Soc., 1983, 105, 7764.
13. Watanabe, T.; Honda, K. J. Phys. Chem., 1982, 86, 2617.
14. Maidan, R.; Goren, Z.; Becker, J.V.; Willner, I. J. Am. Chem. Soc., 1984, in press.
15. Keller, P.; Moradpour, A.; Amouyal, E., Kagan, H.B. Nouv. J. Chim., 1980, 4, 377-84; J. Am. Chem. Soc., 1980, 102, 7193-96.

16. Harriman, A.; Porter, G.; Richoux, M.-C. J. Chem. Soc. Faraday Trans. 2, 1981, 77, 1939.
17. Kalyanasundaram, K.; Gratzel, M., Helv. Chim. Acta, 1980, 63, 478.
18. Kirch, M.; Lehn, J.-M.; Sauvage, J.-P. Helv. Chim. Acta, 1979, 62, 1345.
19. Lehn, J.-M.; Sauvage, J.-P.; Zeissel, R. Nouv. J. Chim., 1980, 4, 623-27.
20. Degani, Y.; Willner, I. J. Chem. Soc. Chem. Commun., 1983, 710-12.
21. Schrauzer, G.N. in "New Trends in the Chemistry of Nitrogen Fixation", Chatt, J.; da Camara Pina,L.M.; Richards, R.L. Eds., Academic Press, London, 1980, p. 105.
22. Degani, Y.; Willner, I., submitted for publication.
23. Lehman, H.; Schenk, K.J.; Chapuis, G., Ludi, A. J. Am. Chem. Soc., 1979, 101, 6197.
24. Harrison, D.F.; Weissberger, E.; Taube, H., Science, 1968, 159, 320.
25. Wong, C.-H.; Daniels, L.; Orme-Johnson, W.H.; Whitesides, G.M.; J. Am. Chem. Soc., 1981, 103, 6227-28.
26. Wong, C.-H.; Whitesides, G.M. J. Org. Chem., 1982, 104,1552-54.
27. Mandler, D.; Willner, I. J. Am. Chem. Soc., 1984, 106,5352-53.

RECEIVED January 10, 1985

13

Intrazeolite Photochemistry: Use of β-Phenylpropiophenone and Its Derivatives as Probes for Cavity Dimensions and Mobility

J. C. SCAIANO, H. L. CASAL, and J. C. NETTO-FERREIRA[1]

Division of Chemistry, National Research Council of Canada, Ottawa, Canada K1A 0R6

The photochemistry of β-arylpropiophenones has been examined in Silicalite and other zeolites. The channel structure of Silicalite prevents intramolecular quenching by the β-phenyl group in β-phenylpropiophenone and leads to a dramatic enhancement of the phosphorescence. Ring substitution can have a very different effect on the phosphorescent properties depending on its size, polarity and ring position. Oxygen quenching studies are consistent with the presence of at least two types of inclusion sites.

Zeolites are crystalline aluminosilicates that have found a wide variety of applications in industry and in the laboratory (1-6). Different compositions and crystallization procedures yield zeolites with pores or channels. A new class of zeolites with low Al_2O_3 content has been developed recently (7,8); these are being used extensively as catalysts. Silicalite (>99% SiO_2) belongs to this new class of dealuminized zeolites and has been reported to contain no Al (8). The framework structure of dealuminized zeolites consists of a tetrahedral arrangement of five SiO_2 units. The channel system of Silicalite is formed by near-circular zig-zag channels defined by 10 oxygen centers cross-linked by ellipical straight channels (see Figure 1 in ref. 8); the average free diameter of these channels is ca. 6Å.

As a result of the absence of AlO_2^- units in the crystalline lattice, Silicalite has no ion exchange properties. This characteristic makes Silicalite hydrophobic; while zeolites are commonly used to extract water from organic substances, Silicalite selectively extracts organic materials from aqueous solutions (9).

Photochemistry on surfaces, solids, inclusion compounds and other organized media continue to be the subject of considerable attention (10-15). Zeolites offer a new possibility in the study of photoprocesses in ordered media, and the study of guest-host

---

NOTE: This chapter is Part III in a series. Issued as NRCC-23767

[1]Current address: Departamento de Quimica, Universidade Federal Rural do Rio de Janeiro, Rio de Janeiro, CEP 23460, Brazil

0097-6156/85/0278-0211$06.00/0

relationships in these systems can help in the understanding of the characteristics of the active sites in these powerful catalysts.

In earlier work from this laboratory (14,15), we have shown that several aromatic ketones show phosphorescence when included in the channel structure of Silicalite. In particular, the behaviour of β-phenylpropiophenone is virtually identical to that of acetophenone; by contrast, in homogeneous solution in benzene at room temperature the two ketones show very different behaviour (16). While acetophenone has a triplet lifetime of a few microseconds, the triplet of β-phenylpropiophenone lives only ca. 1 ns (16-19). This difference has been demonstrated to reflect quenching of the carbonyl triplet by the β-phenyl ring. The geometry of the Silicalite channels does not allow for molecular motions involving large side groups. In general, the predominant motions of included molecules involve rotation along a molecular axis roughly parallel to the channel axis (20). This restriction to molecular motion is reflected in the long triplet lifetime observed for β-phenylpropiophenone in Silicalite ($\tau \sim 0.14$ ms) (14). This lifetime enhancement, in excess of five orders of magnitude from the value in solution, is believed to reflect motional restrictions that prevent the triplet state from achieving the conformation required for intramolecular quenching. We have now examined this question in more detail by using a variety of substituted β-phenylpropiophenones (Chart I). In this report we review our earlier findings in the case of β-phenylpropiophenone and analyze our new results for the various molecules in Chart I. In addition, a few preliminary results in zeolites other than Silicalite are also presented.

Experimental

The compounds studied (I-IX) have been described previously (19).

Silicalite (S-115), with particle sizes of 1-2 μm, was obtained from Union Carbide Canada and calcined at 450-550°C for 12 hours. The chemical composition, according to X-ray fluorescence measurements showed Si (>99.5%), Al (<0.5%) and traces (<0.01%) of Ca and K.

Inclusion of the different ketones in Silicalite was achieved as described previously (14,15), using 2,2,4-trimethylpentane (isooctane) as a solvent for the ketones. In our study of substituent effects we have used 120 mg of ketone per gram of Silicalite in all inclusion attempts. The quantities used would yield samples in which up to 60% of the void volume of Silicalite (0.19 mL/g) could be occupied. The actual yields (and therefore the efficiency of inclusion) were determined by microanalysis, and the results are listed in Table I.

Phosphorescence spectra (uncorrected, front face) were recorded on a Perkin-Elmer LS-5 fluorescence spectrometer using a pulsed excitation source (~10 μs) and gated detection. The instrument was controlled by a P-E 3600 data station. The samples were typically excited at 313 nm using the instrument's monochromator and an additional interference filter. Excitation and emission bandpasses were 2 nm. Typically the emission spectra were recorded using a 50 μs delay following excitation and a 20 μs gate. The samples were contained in cells made of 3×7 mm^2 Suprasil tubing, under a continuous stream of N_2, O_2 or O_2/N_2 mixtures of known composition.

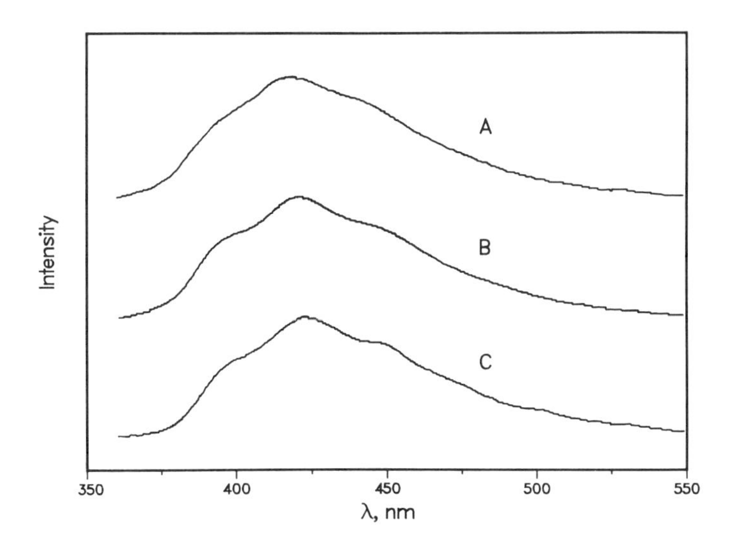

Figure 1. Phosphorescence emission spectra of:
A) β-phenylpropiophenone (I); B) β-(4-methylphenyl)propiophenone (IV); and C) β-(4-cyanophenyl)propiophenone in Silicalite.

CHART I

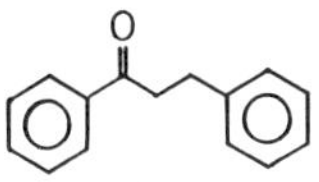

I 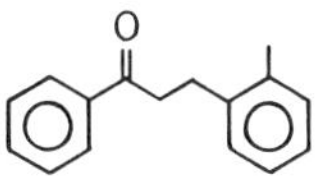 II

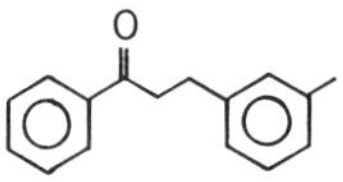

III 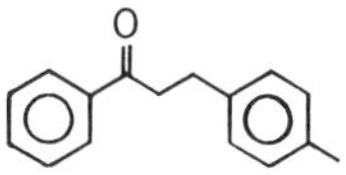 IV

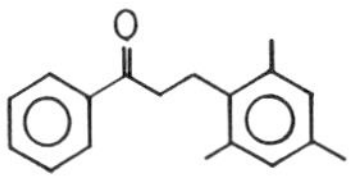

V 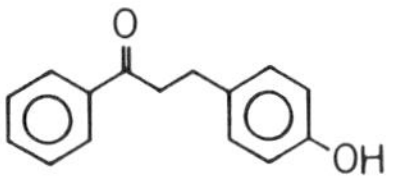 VI

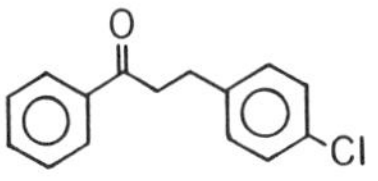

VII 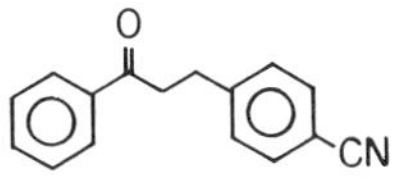 VIII

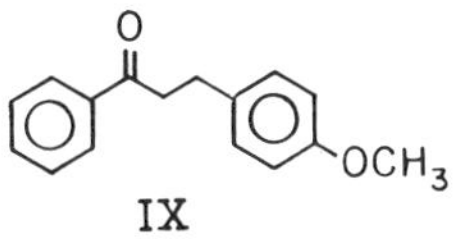

IX

Results and Discussion

Photochemistry of β-Phenylpropiophenone. The short lifetime of triplet β-phenylpropiophenone in solution has been recognized for a number of years (16-19); this efficient deactivation makes this ketone particularly photostable (21). In earlier studies we have shown that the efficiency of intramolecular quenching is controlled by the ability of the ketone to achieve conformation **Ia** (16).

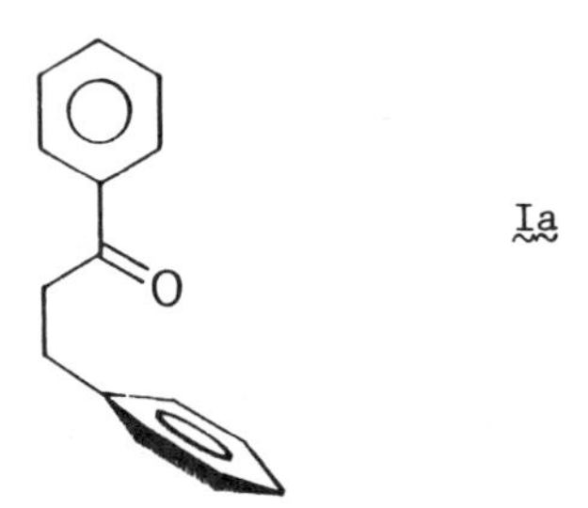

Ia

Intramolecular quenching seeems to require an n,π* triplet, since the lifetime of β-phenyl-p-methoxypropiophenone in non polar solvents (which has a low lying π,π* state) is ~60 ns, compared to ca. 1 ns for **I** (19).

Crystalline β-phenylpropiophenone is not phosphorescent at room temperature. Similarly, adsorption on silicagel does not lead to significant phosphorescence. It would thus appear that a channel structure is essential. In fact, channel (or pore) dimensions also appear to be quite important; for example, when β-phenylpropiophenone is adsorbed on Mordernite, a zeolite with ~9Å pores, no luminescence can be detected (15).

Examination of molecular models indicates that β-phenylpropiophenone can readily adopt a conformation whose kinetic diameter is comparable to that of benzene, which is known to be included in the channels of Silicalite.

In the next Section we explore the possibility of using substituted β-phenylpropiophenones as 'molecular calipers'.

Substituted β-phenylpropiophenones on Silicalite. We have recently shown that the lifetime of β-phenylpropiophenone triplets is rather insensitive to substitution in the β-phenyl ring (19). We have suggested that this insensitivity reflects the fact that triplet decay is predominantly controlled by molecular motion; i.e. by how rapidly a molecule can achieve a conformation suitable for intramolecular deactivation.

When we applied ketones **I**-**IX** to Silicalite using the inclusion technique described earlier we observed phosphorescence in only a few examples, see Table I. Elemental analysis of the samples revealed a clear correlation between their carbon content and the relative phosphorescent intensity. The behaviour observed can be adequately analyzed by separating the ketones in three main groups:

(a) **I** and **IV** show high luminescent intensity and their carbon content suggests that at least 60% of the ketone applied has been incorporated to the Silicalite substrate.

Table I. Composition and Phosphorescence of the samples examined.

Guest	% C		Relative	τ,ms
	expected	found[a]	Intensity[b]	
I	10.3	7.6	41	0.14
II	10.3	2.3	0	
III	10.3	1.8	0	
IV	10.3	6.2	105	0.14
V	10.3	1.9	0	
VI	9.6	2.5	0	
VII	8.8	2.9	1	0.18
VIII	9.8	2.2	6	0.15
IX	9.6	2.8	3	0.18
None	-	-	0[c]	-

a.- % C determined by microanalysis
b.- Calculated as I_{max}/instrument scale factor
c.- The relative emission intensity of empty, calcined Silicalite is ~0.5 at 405 nm and 0 at 420 nm.

(b) II, III and V show no phosphorescence and very poor inclusion (note the low carbon content). These three molecules are sufficiently large that we believe they do not enter the channel structure of silicalite. The fraction that does incorporate presumably remains on the surface, or in sites where the conformation required for decay is readily achieved.

(c) VI, VII, VIII and IX show weak or no phosphorescence and generally poor inclusion; however, these ketones can be expected to fit in the channel of Silicalite (IX may be a borderline case). Why do they fail to include and phosphoresce efficiently? We believe that this is determined by a combination of the size and hydrophilic properties of each substrate. Thus, the 4-hydroxy derivative (VI) is probably too polar to have much affinity for the hydrophobic Silicalite. Polarity is clearly a factor for all the members of this group, while size may play a role in at least the case of IX. Similar size effects (as to inclusion) have also been observed for ortho- and para-xylene (22).

Figure 1 shows representative phosphorescence spectra for I, IV and VIII in Silicalite. In all cases emission peaks at ~420 nm, and, except for the small differences in the degree of vibrational resolution the three spectra are identical.

Site Inhomogeneity. In all cases where we observed phosphorescence emission from ketones in Silicalite the decay of this luminescence was non-exponential. In this respect, it should be noted that 'lifetimes' given in Table I and in earlier papers in this series refer to the first observable lifetime following a 10 μs excitation pulse and a 20 μs delay. All values are comparable and probably reflect a similar degree of immobilization. Non-exponential decays are common for other luminescent probes adsorbed on surfaces (23).

It is known that Silicalite in particular, and zeolites in general contain several different sites (24); the inhomogeneity of adsorption sites is largely responsible for the non-exponential decay law. Figure 2 illustrates the decay observed in the case of β-phenylpropiophenone in Silicalite.

Quenching experiments have also demonstrated that some classes of sites are more readily accessible than others to oxygen and to the ketones; the latter seem to show some degree of cooperative effect in their binding to the zeolite (15).

In several examples we have studied the effect of adding oxygen to the sample while monitoring the emission intensity. Figure 3 illustrates this effect for β-phenylpropiophenone on Silicalite at ca. 40% loading. Quite clearly the efficiency of quenching is a function not only of the oxygen concentration in the gas in equilibrium with the zeolite, but also of the value of the delay time (t_d) used for monitoring the emission.

It is probably appropriate to note at this point that Stern-Volmer type of analysis of luminescence data obtained using modern gated spectrometers (such as the Perkin-Elmer LS5) follows different mathematical expressions than those used for data obtained under continuous irradiation conditions. This is true even for homogeneous systems where the excited state decays with simple mono-exponential behaviour.

If we assume that the time gate (t_g) used when monitoring the luminescence is considerably shorter than the delay (t_d) used, and than the triplet lifetime (τ_T), then

$$I_t^0 = I_o^0 \, e^{-t_d/\tau_T} \qquad (1)$$

where I_o^0 is the intensity at time zero in the absence of quencher and I_t^0 at time 't_d', also in the absence of quencher. If we add a quencher in concentration [Q], and which quenches with a rate constant $\underline{k}_q$, then

$$I_t = I_o^0 \, e^{-(\underline{k}_q[Q]+\tau_T^{-1})t_d} \qquad (2)$$

Thus, the equivalent of the conventional Stern-Volmer equation becomes:

$$\frac{I_t^0}{I_t} = e^{\underline{k}_q[Q]t_d} \qquad (3)$$

or, in logarithmic form

$$\ell n\left(\frac{I_t^0}{I_t}\right) = \underline{k}_q[Q]t_d \qquad (4)$$

We note that the expressions above have been derived for a homogeneous system, where only one triplet lifetime is involved; even here a Stern-Volmer type plot based on data from gated spectrometers is not expected to follow "conventional" Stern-Volmer behaviour.

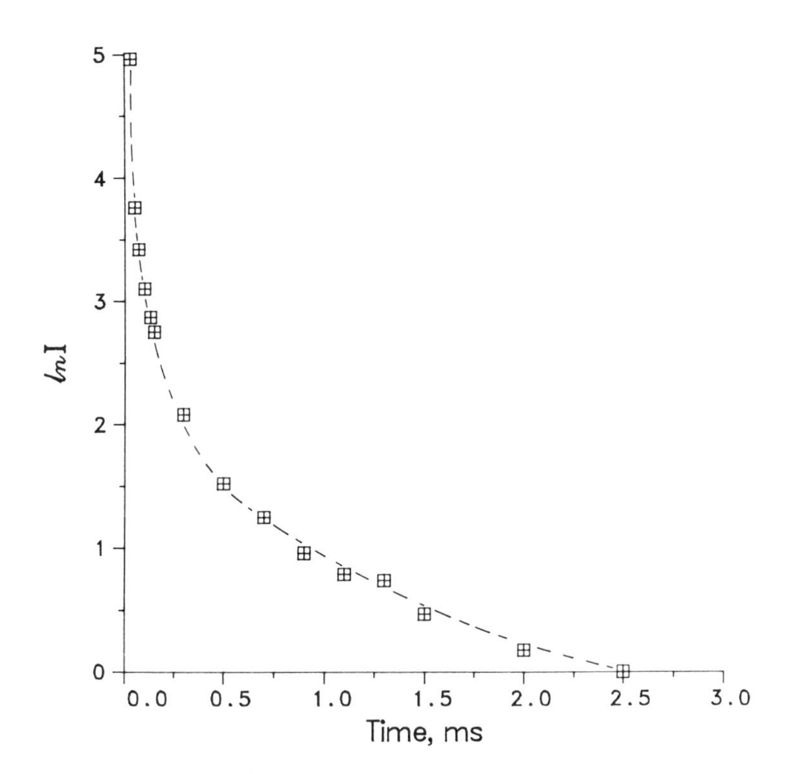

Figure 2. Luminescence (ℓn I) intensity of β-phenylpropiophenone in Silicalite as function of time.

Figure 4 shows three representative plots for β-phenylpropiophenone of $\ln(I_t^0/I_t)$ vs. the oxygen mole fraction for different values of the time delay, t_d. The abscissa represents the mole fraction of oxygen in oxygen-nitrogen mixtures in equilibrium with the sample at a total pressure of 1 atm., and is a convenient variable in experiments of this type. Quite clearly the plots are not linear; further, even the initial slopes are not linearly proportional to the value of t_d. This seems fully consistent with our earlier suggestions from co-inclusion and other oxygen quenching experiments that Silicalite has several classes of inclusion sites (14,15), and that these differ in the lifetime of the triplet (see Fig. 2), as well as in their accessibility to oxygen. Similar arguments have been used to explain quenching plots in uranyl-exchanged zeolites (25). Equation 4 can be modified replacing the concentration term for $\underline{a}.X(O_2)$, where $X(O_2)$ is the mole fraction of oxygen, and $\underline{a}$ is a parameter that measures the accessibility of oxygen molecules to the various sites in the zeolite.

$$\ln\left(\frac{I_t^0}{I_t}\right) = -\underline{k_q}\ \underline{a}\ X(O_2)t_d \tag{5}$$

If we further assume that $\underline{k_q}$ has the same value for the different sites, the differences and variations in slopes in plots of $t_d^{-1}.\ln(I_t^0/I_t)$ vs. $X(O_2)$ must be due to changes in the accessibility parameter $\underline{a}$. Figure 5 shows a plot of the same data as in Figure 4, in the form indicated above.

Quenching by butadiene. Dienes are known to be excellent quenchers of carbonyl triplets (25); in homogeneous solution the rate constants for triplet quenching frequently approach the diffusion controlled limit.

We have observed that butadiene quenches the phosphorescence from β-phenylpropiophenone very efficiently. For example for t_d=20 μs we observed:

- Oxygen $I_t^0/I_t = 9$ at $X(O_2) = 1.0$
- Oxygen $I_t^0/I_t = 3$ at $X(O_2) = 0.1$
- Butadiene $I_t^0/I_t \geqslant 30$ at $X(C_4H_6) = 0.1$

Thus, since oxygen and butadiene are quenchers of similar efficiency in homogeneous solution, we conclude that the local concentration of butadiene at the individual adsorption sites is much higher than that of oxygen under similar gas phase concentrations. Accessibility is naturally a function of the properties of the molecule in addition to its size.

Comparison with **I** in other zeolites. The same procedure for inclusion was used with β-phenylpropiophenone on Silicalite and on molecular sieves (MS) 3A, 5A, 10X and 13X. No emission was observed on MS-3A or MS-5A, while the emission on MS-10X and MS-13X (under nitrogen) was 9% and 54% of that in Silicalite.

The intensity of emission was independent of the observation time for Silicalite and MS-10X (for at least a few hours). However, in the case of MS-13X the phosphorescence intensity decays quickly

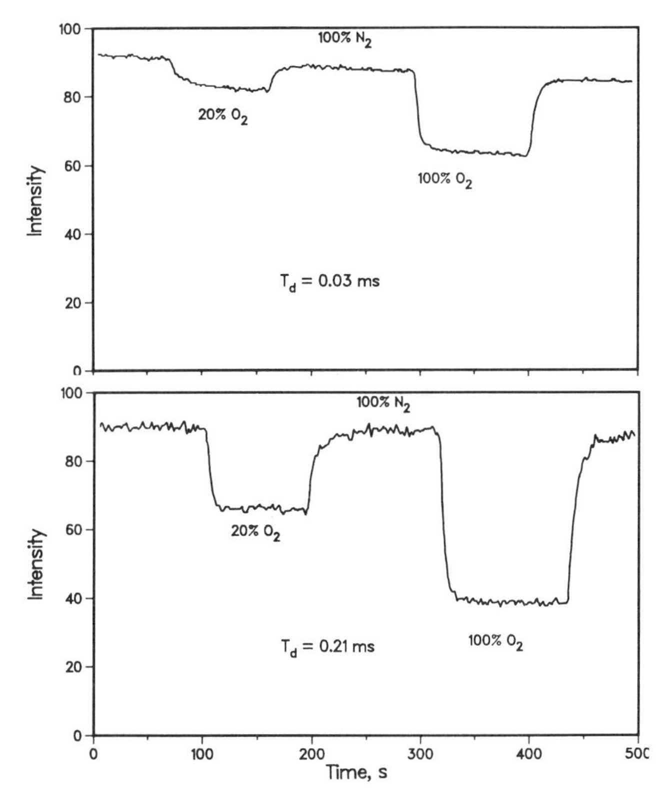

Figure 3. Time dependence of the phosphorescence intensity of β-phenylpropiophenone (λ_{em}=450 nm) with varying conditions.

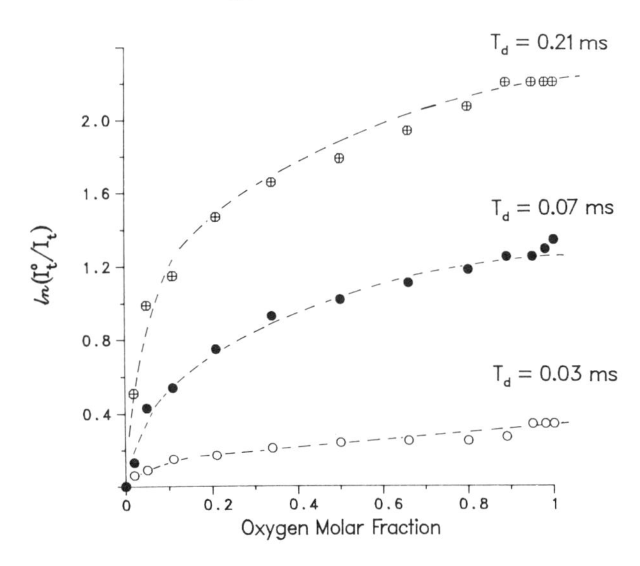

Figure 4. "Stern-Volmer" quenching plot of the phosphorescence emission of β-phenylpropiophenone according to eq. 4.

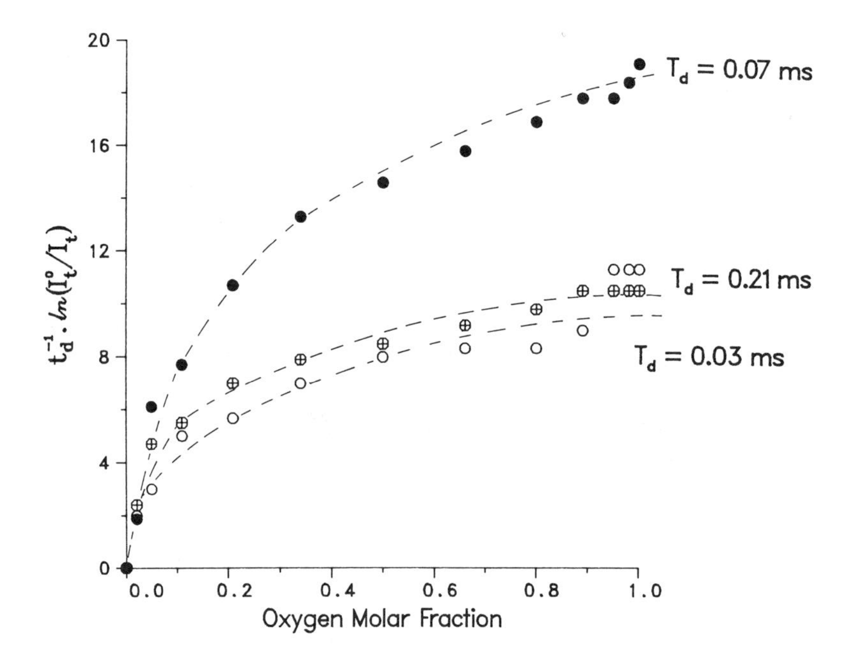

Figure 5. Same data as Figure 4 but plotted following eq. 5.

with time. In one particular observation the half-life was ~80 seconds. A similar, but even more dramatic effect could be achieved by exposing the sample to the unfiltered beam from a 200 Watt Xenon lamp. Quite clearly, even the weak beam from the excitation source in the spectrometer is enough to induce sufficient change for the luminescence to disappear. Whether these changes result from substrate relocation, quencher generation or photodecomposition of I in the active sites is unclear at this point (although the last possibility is favored) and is the subject of current work.

No luminescence was observed from ketones II, III, IV and V on MS-13X.

Conclusion

The results reported here and in earlier publications in this series suggest that cavity size and limitations to molecular motion play a dominant role in the photochemistry and photophysics of alkyl aryl ketones included in zeolites. In the case of Silicalite the size and polarity of various substituted β-phenylpropiophenones seem to determine the efficiency of inclusion and ultimately of luminescence. The same factors, relating to size and mobility can be expected to play an important role in the use of zeolites as catalysts for other reactions, whether these are photochemical or thermal processes. In this sense studies with β-phenylpropiophenones may lead to considerable information on adsorption sites and on the freedom (or lack of it) of molecular motion as well as on the accessibility of these sites to other reactants. Recent work from Turro's laboratory has shown that pyrene aldehyde can be used to probe the nature of inclusion sites in various zeolites (27); dibenzyl-ketones were also used as probes on porous silica (28).

Preliminary studies with other types of zeolites (such as MS-13X) seem to indicate that photochemical change may play a far more important role in these systems as compared with Silicalite.

Acknowledgments. Thanks are due to S.E. Sugamori for technical assistance and to H. Séguin for elemental analysis. One of us (J.C.N.F.) would like to thank CNPq (Brazil) and NSERC (Canada) for support.

Literature Cited

1. Barrer, R.M. "Zeolites and Clay Minerals as Sorbents and Molecular Sieves," Academic Press, London, 1978.
2. Breck, D.W. "Zeolite Molecular Sieves," Wiley-Interscience, New York, 1974.
3. "The Properties and Applications of Zeolites" Edited by R.P. Townsend. The Chemical Society Special Publication No. 33, London, 1980.
4. "Zeolite Chemistry and Catalysis" Edited by J.A. Rabo, ACS Monograph 171. American Chemical Society, Washington, DC, 1976.
5. "Catalysis by Zeolites" Edited by B. Imelik, Elsevier, Amsterdam, 1981.
6. "Natural Zeolites: Occurrence, Properties, Use" Edited by L.B. Sand and F.A. Mumpton, Pergamon Press, 1978.

7. Kokotailo; G.T.; Lawton, S.L.; Olson, D.H.; Meier, W.M. Nature 1978, 272, 437; US Patent 3, 702, 886 (1972).
8. Flanigen, E.M.; Bennett, J.M.; Grose, R.W.; Cohen, J.P.; Patton, R.L.; Kirchner, R.M.; Smith, J.V. Nature 1978, 271, 512.
9. Milestone, N.B.; Bibby, D.M. J. Chem. Tech. Biotechnol. 1981, 31, 732; Shultz-Sibbel, G.M.W.; Gjerde, D.T.; Chriswell, C.D.; Fritz, J.S.; Coleman, W.E. Talanta 1982, 29, 447.
10. de Mayo, P. Pure & Appl. Chem. 1982, 54, 1623.
11. Scheffer, J.R. Acc. Chem. Res. 1980, 13, 283; Trotter, J. Acta Cryst. 1983, B39, 373.
12. Casal, H.L.; de Mayo, P.; Miranda, J.F.; Scaiano, J.C. J. Am. Chem. Soc. 1983, 105, 5155; Hui, Y.; Winkle, J.R.; Whitten, D.G. J. Phys. Chem. 1983, 87, 23.
13. Ozin, G.A.; Hugues, F., Mattav, S.M., McIntosh, D.F. J. Phys. Chem. 1983, 87, 3445.
14. Casal, H.L.; Scaiano, J.C. Can. J. Chem. 1984, 62, 628.
15. Casal, H.L.; Scaiano, J.C. Can. J. Chem. in press.
16. Scaiano, J.C.; Perkins, M.J., Sheppard, J.W.; Platz, M.S.; Barcus, R.L. J. Photochem. 1983, 21, 137.
17. Wagner, P.J.; Kelso, P.A.; Kemppainen, A.E.; Haug, A.; Graber, D.R. Mol. Photochem. 1970, 2, 81.
18. Stermitz, F.R.; Nicodem, D.E.; Muralidharan, V.P.; O'Donnell, C.M. Mol. Photochem. 1970, 2, 87.
19. Netto-Ferreira, J.C.; Leigh, W.J.; Scaiano, J.C. J. Am. Chem. Soc. in press.
20. Eckman, R.; Vega, A.J. J. Am. Chem. Soc. 1983, 105, 4841.
21. Bergmann, F.; Hirshberg, Y. J. Am. Chem. Soc. 1943, 65, 1429.
22. Kelusky, E.C.; Fyfe, C.A.; 67th Annual Conference, Chemical Institute of Canada, Montreal, June 1984. Abstract Ph 12-06.
23. Habti, A.; Keravis, D.; Levitz, P.; van Damme, H. J. Chem. Soc. Faraday Trans. 2, 1984, 80, 67.
24. Fyfe, C.A.; Gobbi, G.C.; Klinowski, J.; Thomas, J.M.; Randas, S. Nature 1982, 296, 530.
25. Suib, S.L.; Bordeianv, O.G.; McMahon, K.C.; Psaras, D., A.C.S. Symp. Ser. 1982, 177, 225.
26. Wagner, P.J.; Kochevar, I. J. Am. Chem. Soc. 1968, 90, 2232.
27. Baretz, B.H.; Turro, N.J. J. Photochem. 1984, 24, 201.
28. Turro, N.J.; Cheng, C.-C.; Mahler, W. J. Am. Chem. Soc. 1984, 106, 5022.

RECEIVED January 10, 1985

14

Polymer-Based Sensitizers for the Formation of Singlet Oxygen

JERZY PACZKOWSKI and D. C. NECKERS[1]

Department of Chemistry, Bowling Green State University, Bowling Green, OH 43403

We outline herein the synthesis and photochemical properties of a new series of polymer-based rose bengals. These derivatives, soluble in non-polar solvents in contrast to Ⓟ -rose bengal,[1,2*] are based on the recently described new chemistry of rose bengal.[3] The electronic absorption spectra of these polymeric derivatives suggest the possibility of aggregation of the rose bengal (RB) molecules. This is indicated by changes in the absorption spectra with the loading of rose bengal on the polymer. Singlet oxygen formation from these polymeric sensitizers in methylene chloride solution (the polymers are soluble in this solvent) suggest that $\Phi_{^1O_2}$ is dependent on the number of RB units attached to the polymer chain. At low concentrations of RB the quantum yield of singlet oxygen formation increases with concentration of RB and reaches a maximum value when 1 RB is attached per 30 units of the polymeric chain. Subsequent increases in the RB content cause a sharp decrease in $\Phi_{^1O_2}$. Electronic absorption spectra and quantum yields of singlet oxygen formation by monomeric RB benzyl ester (the form of rose bengal in the polymer) suggest that hydrogen bonding between rose bengal moieties and protic solvents lead to a blue shift in the absorption maximum and an increase in the quantum yield of singlet oxygen formation.

*As a matter of convention, Ⓟ -rose bengal refers to rose bengal immobilized on insoluble crosslinked styrene-co-divinylbenzene beads. All other polymers referred to are not crosslinked and hence soluble in various solvents.

[1] Author to whom correspondence should be directed.

0097-6156/85/0278-0223$06.00/0

The synthesis, use, and basic photochemical properties of the first heterogeneous sensitizer for singlet oxygen formation were outlined in 1973. This sensitizer, called Ⓟ- rose bengal by Neckers and Schaap, was important because it led to singlet oxygen formation in solvents in which rose bengal was not soluble.[1,2] Polymer bound photosensitizers have a number of practical advantages over soluble sensitizers. There is decreased interaction or self-quenching of one dye molecule with another when the dyes are bound to a polymeric backbone, and the chromophores may be "site-isolated". Polymeric sensitizers are also functional in solvents where the dye itself is non-functional, and there is greater stability of the dye bound sensitizer (it does not bleach). An added advantage is that the polymeric sensitizer can be easily separated from the reaction mixture by either filtration or sedimentation.

There are three general ways to design a polymer-based photosensitizer. In the first, the sensitizer is covalently bound to crosslinked polymer beads or specially functionalized silica gels.[1-2, 4-5] These polymeric sensitizers are generally prepared via chemical reaction of the sensitizer with functionalized polymers. In the second, the sensitizer is immobilized by absorption of the sensitizer on the solid support. This technique is generally applied with silica gel as the support.[6,7] It is a convenient method, but the sensitizer is easily eluted by polar solvents. In the third approach, the sensitizer is incorporated into a polymeric thin film. This is accomplished by dissolving the sensitizer and the polymer in a solvent and depositing the polymer/sensitizer solution on a flat surface. The number of sensitizers which may be used in heterogeneous form is quite large.[8]

Various compounds are used as singlet oxygen sensitizers. These include the xanthene dyes like rose bengal (RB), eosin—a xanthene analog, (rhodamine B), victoria blue, 1-amino-4-hydroxyanthraquinone, hematoporphyrin, tetraphenylporphin, tris(2,2'-bipyridine) ruthenium (II), malachite green, methylene blue, chlorophyllin and many other compounds.[2,8-11] The only thing these compounds have in common is that they all absorb in the long wavelength region of the visible spectrum. The ability of excimers and exciplexes to decay to yield triplets, quench oxygen, and generate singlet oxygen are important considerations in sensitizer design.[12]

Rose bengal is a uranine analog and was first synthesized by Gnehm 11 years after Baeyer discovered fluorescein.[13] It is sold at dye grade as 90% the di-sodium salt of 3,4,5,6-tetrachloro-2,4,5,7-tetraiodouranine and in purified form by DyeTel, Inc.[14], Scheme 1. It is nucleophilic at C-2', but has no nucleophilicity at the C-6 phenoxide.[15] Consequently, (1) was reacted with chloromethylated polystyrene beads in a Merrifield manner[16] to produce a polymeric analog of the dyestuff.

Scheme I. Structure and Properties of Rose Bengal Derivatives

		$\underline{R}^1$	$\underline{R}^2$	λmax(nm)	Φ_F[9]	Φ_{ST}[9]	Φ_{1O_2}[9]	Φ_{ST}[29]
$\underline{1}$	uranine*	H	H	491	0.93	0.03	0.1	0.03
$\underline{2}$	eosin	Br	H	514	0.63	0.3	0.4	0.32
$\underline{3}$	erythrosin	I	H	525	0.08	0.6	0.6	0.69
$\underline{4}$	rose bengal	I	Cl	548	0.08	0.76	0.76	0.86

*Uranine is the disodium salt of fluorescein.
The systematic name of fluorescein in its quinoid form is Benzoic acid, 2-(6-hydroxy-3-oxo-3H-xanthen-9-yl)-. The systematic name of fluorescein in its lactonic form is Spiro[isobenzofuran-1(3H),9'-[9H]xanthen]-3-one, 3',6'-dihydroxy-.

Polymer rose bengal, or Sensitox I, was derived from this concept. It was, in the words of an immortal reviewer, a "citation classic".[17]

The original polymer-based xanthene dyes were prepared from a commercial chloromethylated polystyrene/divinylbenzene copolymer trade-named "BioBeads". BioBeads were originally manufactured for gel permeation chromatography applications. But with the appearance and general acclaim of Merrifield's publications on "Solid Phase Synthesis"[18] workers at BioRad, obviously familiar with the ion exchange literature, chloromethylated these crosslinked beads with chloromethyl ether and stannic chloride to obtain the precursor polymer for both Merrifield's studies and our own.

Though Ⓟ-rose bengal was first reported in 1973 and Sensitox I appeared on the market in 1976, essentially nothing was known about the fundamental chemistry of rose bengal until the studies of Lamberts and Neckers.[3,13,15] These studies were important in establishing that rose bengal could be converted using standard organic chemistry to derivatives of rose bengal whose photochemistry was previously unknown but which behaved, more or less, in a normal photochemical fashion.

Until the beginning of this work, little was known about the effect of the polymer on the behavior of the sensitizer, rose bengal. In this paper we shall establish that the polymer is indeed an extremely important component of the chemistry of polymeric derivatives of rose bengal.

The initial polymer rose bengal was prepared from chloromethylated polystyrene/divinylbenzene beads using the following reaction:

$$Ⓟ\text{-}CH_2Cl + RB \longrightarrow Ⓟ\text{-}RB + Cl^-$$

The results of several control experiments made it clear that polymer rose bengal was an authentic singlet oxygen sensitizer.[1,2] Photooxidation of typical singlet oxygen acceptors in non-polar solvents (a condition where rose bengal itself is not soluble or effective) showed that polymer rose bengal produced the same products as did other known sources of singlet oxygen. These reactions were inhibited by singlet oxygen quenchers.[19] Nevertheless, the quantum yield of singlet oxygen formation from these polymer beads in methylene chloride was low (0.43) when compared with the quantum yield of singlet oxygen formation from rose bengal dissolved in MeOH (0.76). Other studies[20] indicated that Sensitox I does not form singlet oxygen as efficiently as do other polymer-based sensitizers, though Sensitox I does have a major advantage - it does not bleach.

The first detailed kinetic study of sensitized photooxidation was reported by Schenck[21] in 1951. His group developed a procedure which was based on the quenching of all formed dye triplets by oxygen and the trapping of all the formed singlet oxygen with a very reactive acceptor.[22] Halofluoroscein dyes like rose bengal and eosin are efficient photosensitizers because they have high quantum yields of triplet formation. Triplet yield formation is directly related in these systems to the number and atomic weight of heavy atoms present in the molecule, and rose bengal is by far the most efficient and most common sensitizer in the halofluoroscein series.

Because rose bengal itself is soluble only in polar solvents, its effectiveness as a singlet oxygen sensitizer is quite restricted. Polymer rose bengal, which took advantage of the hydrophobicity of the polystyrene backbone to carry rose bengal into non-polar solvents, is a very convenient sensitizer because it circumvents this solubility problem. It is also stable to oxidative bleaching, though the yield of singlet oxygen is somewhat lower than is the yield of singlet oxygen from the soluble sensitizer.

The principle spectroscopic and photochemical properties of rose bengal derivatives recently synthesized by Lamberts and Neckers[3,13,15] are shown in Table I. These new derivatives cover the gamut of solubility, and are soluble in every solvent from water to pentane. In methylene chloride and chloroform the quantum yields of singlet oxygen are somewhat lower than they are in methanol, e.g., Table I.

It was our intention when this new work began to outline, more fully, the effect of the polymeric structure on the effectiveness of rose bengal as a photosensitizer and to compare the behavior of the polymer-bound dye to the behavior of the dye - free - in solution. In the present paper, therefore, we report on the photochemical and spectral properties of new singlet oxygen sensitizers based on soluble polystyrenes. These new derivatives are referred to as P -RB.

Polymer-bound rose bengals were obtained in soluble form from the reaction of poly(styrene-vinylbenzyl chloride) with rose bengal in DMF. The ratio of styrene units to vinylbenzyl chloride units (60% meta, 40% para) was 3:1 in the copolymer as originally synthesized. In order to obtain polymer-based rose bengals containing differing concentrations of rose bengal on the chain backbone, differing amounts of rose bengal were allowed to react with equivalent amounts of the copolymer. The yields of the obtained polymers, the percent of chloromethyl groups which corresponds to the number of RB moieties in the reaction mixture, and the wavenumber of the carbonyl ester absorption of the obtained polymers are given in Table II.

Because of the substantial excess of chloromethyl groups in the polymer relative to the quantity of added rose bengal (Table II), it could be assumed that all of the RB becomes attached to the polymer chain. The previous elegant work of Lamberts and Neckers[3] demonstrated clearly that the esterification of RB with benzyl chloride formed the benzyl ester only. There was no reaction at the C-6 phenoxide. Thus, we assumed that only polymeric ester derivatives of rose bengal were formed in this reaction.

The resultant polymeric rose bengals are soluble in solvents like methylene chloride and chloroform, but are not soluble in MeOH with the exception of P -RB- 1520 which is soluble in a mixture of $MeOH/CH_2Cl_2$ (1:1). In addition all of the polymers show the typical carbonyl absorption of an ester between 1725 and 1741 cm^{-1}.

In order to obtain a soluble polymer-based RB copolymer of styrene and vinylbenzyl chloride, a mixture of 60% meta isomer and 40% para isomer was used for the polymerization. The molecular weight of this polymer was 107,000 and the molecular weight distribution of the starting polymer is shown in Figure 1a.

All of the rose bengal polymers were synthesized without protection from light and oxygen. The molecular weight distribution for P-RB-51 obtained after reaction with RB is shown in Figure 1b. Examination of both gps curves 1a and 1b indicated that partial decomposition of the starting copolymer had occurred in its reaction with RB. The molecular weight distribution curves show that only part of the P-RB-51 had a molecular weight approximate to that of the starting copolymer.

In order to compare the influence of light and oxygen on the final P-RB, subsequent polymer syntheses were carried out in the dark under nitrogen. The molecular weight distribution curve in this latter case was very close to the molecular weight distribution curve of the starting copolymer (Figure 1d). In other words, when the reaction of chloromethylated polystyrene was carried out with rose bengal in the presence of both light and oxygen, the polymer decomposed during the chemical process (we assume by a RB photoinitiated reaction). This degradation is likely the result of known polystyrene photooxidations.[23,24]

Both the polymer which had been prepared in the absence of oxygen and light, and the polymer which had been functionalized in the presence of oxygen and light were irradiated at 566 nm at room temperature (5 hr). The P -RB-51 obtained by reaction in the absence of light was irradiated an additional 10 hours at refluxing methylene chloride temperature (40°). The molecular weight distribution curve for P-RB-51 (which was not protected from light during preparation) is shown after irradiation for this length of time in Figure 1c. P-RB-51 prepared under anaerobic conditions in the absence of light was stable over the entire time of radiation. This indicated that the polymer degradation process was likely sensitized by monomeric rather than polymeric rose bengal.

Table I. Visible Absorption Spectra and Quantum Yields of Singlet Oxygen Formation of Rose Bengal Derivatives

R_1	R_2	λ_{max} [nm]	Solvent	ϕ_{AO_2}
Na	Na	558	MeOH	0.76
CH_2Ph	Na	564	MeOH	--
CH_2Ph	H	564	MeOH	--
		496;407	CH_2Cl_2	--
CH_2Ph	TEA*	563	MeOH	0.74
		569	CH_2Cl_2	0.67
TEA	TEA	557	MeOH	0.72
		556	CH_2Cl_2	0.48
Et	TEA	563	MeOH	0.74
		563	CH_2Cl_2	0.71
Et	H	564	MeOH	0.76
		496;407	CH_2Cl_2	0.61
Et	Ac	494;400	MeOH	--
		494;395	CH_2Cl_2	0.61

TEA* = triethylammonium

Table II. The Yields, %-CH_2Cl which Correspond to the Number of Rose Bengals Added to the Reaction Mixture of P -CH_2Cl, and the cm^{-1} of C=O in P-RB.

Polymer	Yield [g]	% of CH_2Cl group	C=O
P-RB-51	0.85	1.65	1725.0
P-RB-102	0.95	3.30	1725.0
P-RB-152	0.77	5.00	1725.0
P-RB-305	1.03	10.0	1741.4
P-RB-450	1.15	15.0	1728.1
P-RB-610	0.90	20.0	1738.3
P-RB-1520	1.05	50.0	1739.8

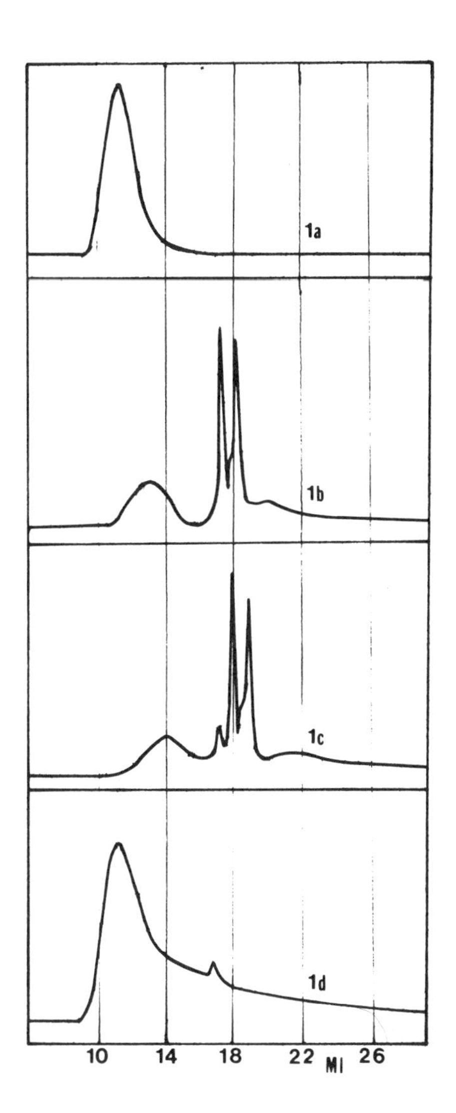

Figure 1. High-Pressure GPS (μ -styragel columns) of Polymers
1a-starting polymer-poly(styrene vinylbenzyl chloride) copolymer
1b-P-RB-51 obtained without protection from light and oxygen
1c-P-RB-51 after irradiation at 566 nm (5 hr)
1d-P-RB-51 obtained with protection from light and oxygen

Electronic Absorption Spectra

Figure 2 shows the electronic spectra in the visible region of the analyzed polymers in methylene chloride solution, and the spectral data is shown in Table III.

These polymers separate into three groups which have different electronic absorption properties. The first is P-RB-51, the lowest in rose bengal content. This polymer has an $A_{\lambda 1}/A_{\lambda 2}$ ratio which is very similar to the ratio of $A_{\lambda 1}/A_{\lambda 2}$ for RB (Figure 2). The second group of polymers P-RB-102, P-RB-152, and P-RB-305 has a ratio of the two absorption maxima, $A_{\lambda 1}/A_{\lambda 2}$, which is very similar to, but lower than that of either P-RB-51 or RB. The last group of polymers shows a reduction in the ratio of $A_{\lambda 1}/A_{\lambda 2}$. In this case, the amount of RB attached to the polymer is much larger by comparison.

In order to measure the quantitative properties of the electronic absorption spectra, the absorption of a 1mg/l polymer solution at the λ_{max} was calculated. These data were plotted against the amount of RB used in the synthesis, and this is shown in Figure 3.

There are two different types of behavior apparent from Figure 3. At low concentrations of RB on the polymer support, the absorption of a 1mg/l polymer solution increases with the content of RB regularly. The second area is that between P-RB-305 and P-RB-451 where the absorption for a 1mg/l solution dramatically decreases. Finally at the higher concentrations of RB on the polymer backbone, the increase in absorption of a 1mg/l solution is apparent again, though it does not grow as rapidly as in the lower RB cases. In the case of the first four polymers, the distance between the rose bengals is large - these are effectively "site isolated" rose bengals (statistically P-RB-51 has one RB for every 100 styrene units and P-RB-305 has one rose bengal for every 30 styrenes). We can assume at the low concentrations of dye on the polymer chain that the absorption spectra would resemble that of rose bengal in dilute solution, e.g., Figure 3.

The situation is completely different when the distance between the rose bengal chromophores on the polymeric backbone is diminished. There is then the possibility of aggregation of the rose bengals along the polymer chain allowing one dye molecule to influence another. In the higher concentration cases, both the form of the absorption curve and the molar absorptivity (as measured for a 1 mg/l solution) no longer appear as that of simple rose bengal.[25,26]

Every polymer, no matter what the rose bengal concentration, shows a maximum absorption for the rose bengal moiety at 571-2 nm when the spectrum is taken in a non-polar solvent such as methylene chloride. RB benzyl ester, on the other hand, shows a maximum absorption at 564 nm in MeOH. This phenomenon is explained by a relatively large influence of hydrogen bonding solvents on the absorption maxima. The absorption maximum of RB benzyl ester (essentially a monomeric rose bengal polymer unit) in different solvents was measured. The positions of the absorption maxima, as a function of solvent, are shown in Table IV.

Data in Table IV suggests that the red shift of the P-RB maximum is a typical solvent effect, and that the hydrogen bonding between

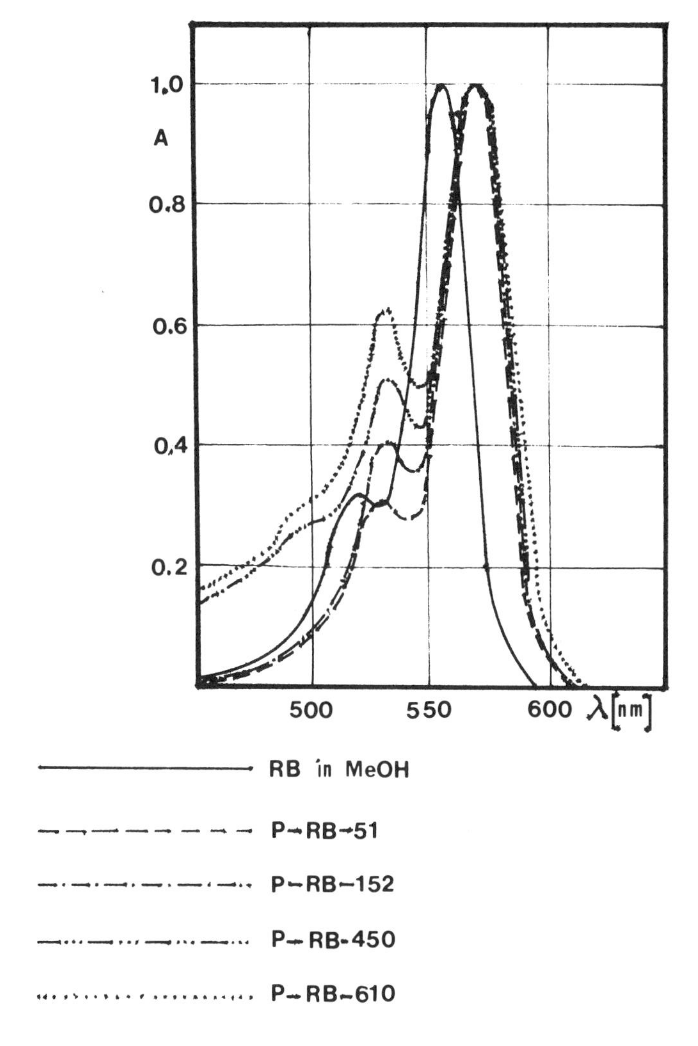

Figure 2. The Electronic Spectra in the Visible Region of P-RB Samples

Table III. The Electronic Absorption Data in the Visible Region of Analyzed Polymers

Polymer	$A^{**}_{1mg/1}$	$\lambda 1_{max}$	$\lambda 2_{max}$	$A_{\lambda 1}/A_{\lambda 2}$	$A3_{min}$
P-RB-51	$4.63 . 10^{-3}$	571.0	529.0	3.165	541.0
P-RB-102	$6.33 . 10^{-3}$	572.0	532.0	2.685	544.0
P-RB-152	$9.11 . 10^{-3}$	572.0	532.0	2.418	544.0
P-RB-305	$22.35. 10^{-3}$	572.0	532.0	2.674	544.0
P-RB-450	$11.0 . 10^{-3}$	572.5	532.0	1.935	546.0
P-RB-610	$11.5 . 10^{-3}$	572.5	532.0	1.593	547.0
P-RB-1520*	$18.2 . 10^{-3}$	570.0	531.0	2.057	541.0

*The electronic absorption in mixture of methanol and methylene chloride (1:1).

**This is the absorption of a 1 mg sample at λ_{max}.

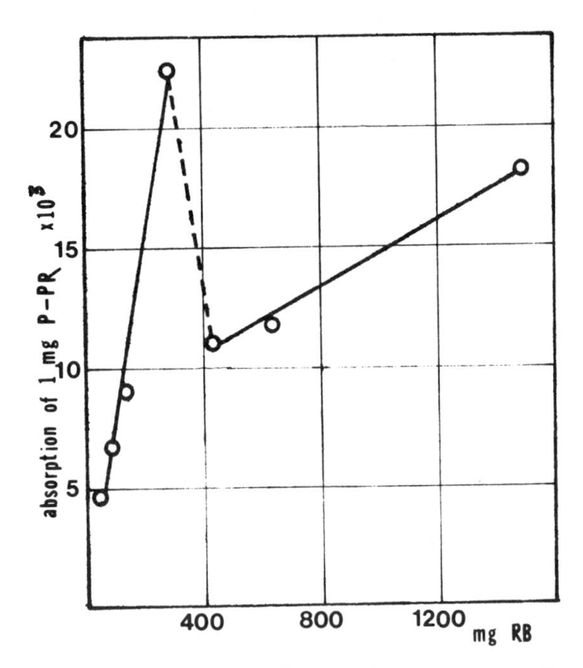

Figure 3. Calculated Absorption of 1mg P-RB as a Function of the Amount of RB Attached to Polymer

Table IV. λ_{max} Positions for RB Benzyl Ester in Different Solvents and Mixture of Solvents

Solvents	λ_{max}	Solvents	λ_{max}	Solvents	λ_{max}	Solvents	λ_{max}
DMF	572	DMF-NMF (90:10)	571	Acetone	570	CH_2Cl_2-MeOH (96:4)	
DMF-MeOH (50:50)	568			Acetone-H_2O (50:50)	566		568**
		DMF-NMF (50:50)	568			CH_2Cl_2-MeOH (50:50)	566
DMF-MeOH (25:75)	565	NMF***	566	Acetone-MeOH (50:50)	565		
MeOH*	564						

*From [19]; ** saturated solution, RB benzyl ester is insoluble pure CH_2Cl_2; *** N-Methylformamide.

the RB benzyl ester and protic solvents leads to a blue shift of the maximum absorption of the benzyl ester[27,28]. Small differences between the position of the maximum of P-RB (571-2nm) and a saturated solution of RB benzyl ester in methylene chloride-MeOH (568 nm) result from the presence of methanol, the effect of polymer microenvironment[27], and RB-RB interactions.

Quantum Yields of Singlet Oxygen Formation of Soluble Polymeric Rose Bengals in Solution

The intermediacy of singlet oxygen in photosensitized oxidations was originally postulated by Kautsky.[30,31] The sequence of events involves excitation of the sensitizer, intersystem crossing, energy transfer from the triplet to molecular oxygen, and reaction of the formed singlet oxygen with the substrate.

$$\text{Sens} \xrightarrow{h} {}^{1}\text{Sens} \qquad (1)$$

$$\text{Sens} \xrightarrow{k_2} {}^{3}\text{Sens} \qquad (2)$$

$${}^{3}\text{Sens} + O_2 \xrightarrow{k_3} O_2 + \text{Sens} \qquad (3)$$

$$A + {}^{1}O_2 \xrightarrow{k_4} AO_2 \qquad (4)$$

$${}^{1}O_2 \xrightarrow{k_5} {}^{3}O_2 \qquad (5)$$

The quantum yields of singlet oxygen formation from Polymeric Rose Bengals (P-RB) in solution were obtained using the actinometric method described by Schaap, Thayer, Blossey and Neckers[2]. The steady state treatment of the kinetic scheme given (Equations 1-5) yields the following result (6).

$$\Phi(AO_2) = \Phi({}^{1}O_2)\,\frac{k_4\,[A]}{k_5 + k_4\,[A]} \qquad (6)$$

where $\Phi_{^1O_2}$ is the quantum yield for singlet oxygen formation, and $\Phi(AO_2)$ is the quantum yield of product formation. If a relatively reactive acceptor is used in high concentrations, then $k_4\,[A] >> k_5$ and the reaction is zero order in [A], i.e., $\Phi_{AO_2} = \Phi(O_2{}^1)$. Since all photooxidation reactions were carried out under similar experimental conditions in solution both for oxidations sensitized by RB and reactions sensitized by P -RB, the ratio of the rate of photooxidation with RB and the rate of photooxidation with P-RB is equal to the ratio of the quantum yield for singlet oxygen formation with RB to that of the quantum yield for singlet oxygen formation with P-RB; (7).

$$\frac{V_{(AO_2)RB}}{V_{(AO_2)P\text{-}RB}} = \frac{\Phi_{(AO_2)RB}}{\Phi_{(AO_2)P\text{-}RB}} \tag{7}$$

The quantum yield of singlet oxygen formation is:

$$\Phi_{(AO_2)P\text{-}RB} = \Phi_{(AO_2)RB} \frac{V_{(AO_2)P\text{-}RB}}{V_{(AO_2)RB}} \tag{8}$$

Gollnick has reported that the quantum yield $\Phi_{^1O_2}$ for RB in MeOH is 0.76. Therefore the quantum yield of singlet oxygen formation for P -RB is:

$$\Phi_{(AO_2)P\text{-}RB} = 0.76 \frac{V_{(AO_2)P\text{-}RB}}{V_{(AO_2)RB}} \tag{9}$$

The photooxidation or 2,3-diphenyl-p-dioxene was used to define the quantum yield of singlet oxygen formation.

II + $^1O_2$ ⟶ IV

The conversion of II to IV was monitored by gas chromatography. Irradiation under identical conditions of a mechanically stirred methylene chloride solution of II with P -RB gives the results shown in Figure 4.

The data shown in Figure 4 indicated a different rate of singlet oxygen formation for the synthesized polymers. The calculated quantum yields of singlet oxygen formation are plotted against the amount of rose bengal used in the synthesis of the polymer in Figure 5.

It is clear that as the amount of rose bengal attached to the polymeric support increased in the regime where the polymer can be called "lightly functionalized", the quantum yield of singlet oxygen formation increases also, reaching a maximum value for the

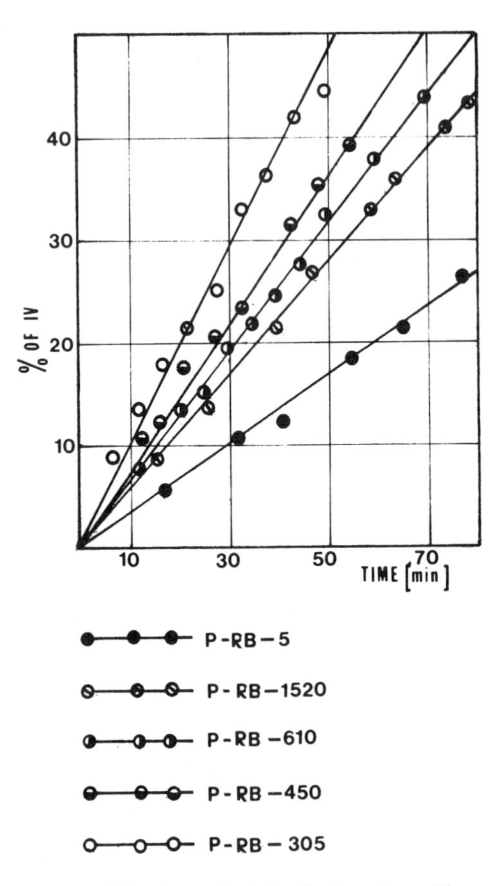

Figure 4. Photooxidation of 2,3-diphenyl-p-dioxene with P-RB (Photooxidation with P-RB-1520 was carried out in 50% MeOH and 50% CH_2Cl_2 Solution)

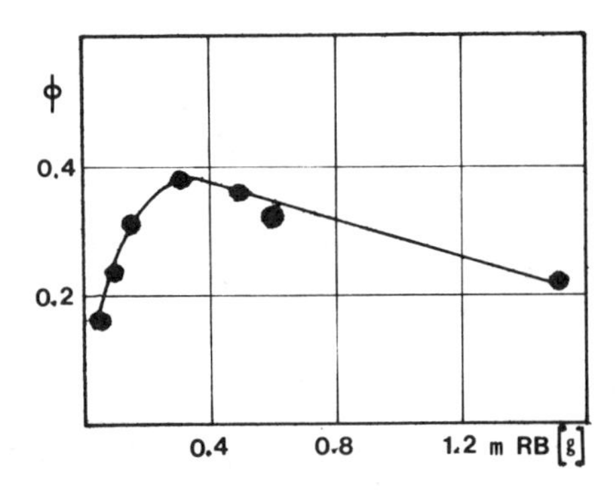

Figure 5. Quantum Yield of Singlet Oxygen Formation as a Function of the Amount of RB Attached to a Polymer

polymer functionalized with one rose bengal for every 30 styrene units (P-RB-305; $\Phi_{^1O_2} = 0.38$). The quantum yield for singlet oxygen formation decreases as additional rose bengals appear on the polymer chain.

There are two reasonable explanations for this behavior in the so-called lightly functionalized regime. First, it could be suggested that singlet oxygen might be quenched by the polystyrene. This quenching might either lead to a chemical product, as for example, the formation of the tertiary hydroperoxide on the polymer backbone, or be physical quenching of the singlet oxygen by the polymer. If polymer quenching were important, then it would be anticipated that a monomeric model of the polymeric backbone might also retard the reaction. To confirm the latter point, the quantum yield for singlet oxygen formation from rose bengal benzyl ester in MeOH in the presence and absence of added excess quantities of the polystyrene model, cumene, was measured. The results are clear - cumene does not quench the formation of singlet oxygen from RB-benzyl ester ($\Phi_{^1O_2} = 0.72$ in the absence of cumene under standard conditions and $\Phi_{^1O_2} \cong 0.71$ in the presence of a large excess of cumene).

A second explanation, and one we confirmed experimentally, is that there is a substantial effect of solution viscosity on the apparent quantum yield of singlet oxygen formation in the case of the polymeric sensitizers. Thus, the polymers used in our experiments were studied such that the apparent rose bengal concentration was kept constant in the cell. To do this, much more polymer containing a lower loading of rose bengal was required than was required for polymers of higher loading. Thus, an 8.2 mg equivalent of rose bengal in a 25 mL solution required 180 mg of P-RB-51; 94.2 mg of P-RB-102; 65.8 mg of P-RB-152 and 35.2 mg. of P-RB-305. As the loading of the polymer is increased, the actual amount of polymer required to achieve a fixed concentration of rose bengal in the solvent became less. More polymer is dissolved in the solution in the lower loading case, hence the solution is more viscous.

That the photooxidation is actually being controlled by solution viscosity was confirmed by measuring the quantum yield of singlet oxygen formation from solutions of RB-benzyl ester containing the same quantity of standard polystyrenes whose molecular weight is gradually increased in a mixed solvent system; methylene chloride/MeOH (4:1). The relative viscosity of P-RB and RB benzyl ester solutions containing polystyrene standards could be calculated. Results are shown in Figure 6.

Both curves have similar shapes and it is clear that the photooxidation process is controlled by the diffusion of oxygen into the rose bengal sites in the polymer solution. These results suggest that when a small amount of rose bengal is attached to the polymeric backbone (P-RB-51, 102; 152; 305) the quantum yield of of singlet oxygen formation is essentially the same (about 0.38).

Polymeric rose bengal derivatives are soluble in non-polar solvents such as methylene chloride. The maximum quantum yield of singlet oxygen formation is 0.38. Whereas with the rose bengal benzyl ester, which can be studied in methanol solution, the quantum yield for singlet oxygen formation reaches 0.72 in that solvent. In order to check how the solvent effects singlet oxygen formation, we measured the quantum yield for singlet oxygen formation from RB-benzyl ester in methylene chloride/MeOH mixtures. Results are shown in Figure 7. Hydrogen bonding between the rose bengal benzyl ester and MeOH influences the quantum yield. This effect is particularly distinct in the range of solvent concentrations between 40 and 80% MeOH. Since the RB benzyl ester is insoluble in pure methylene chloride, the direct measurement of the quantum yield in the pure solvent is impossible, but likely the quantum yield approximates the quantum yield in 80% methylene chloride - 20% MeOH ($\Phi_{^1O_2}$ = 0.37). Figure 5 shows that the quantum yield of singlet oxygen formation is affected differently at high loadings of rose bengal than at low loadings. The highest value of the quantum yield of singlet oxygen formation is observed for P-RB-305 (Φ = 0.38). The quantum yield then subsequently decreases as the amount of rose bengal attached to the polymer increases. This phenomenon suggests interesting self-quenching interactions are influenced by the structure of the polymer. Studies of these phenomena are currently in progress.

Experimental Section

Rose bengal, dye content 92%, was purchased from Aldrich and was used as received. Solvents used for preparation of solutions for quantum yield measurement were spectroscopic grade and were purchased from Aldrich.

Infrared spectra were obtained using a Nicolet 20DX Fourier Transform Infrared spectrometer and electronic absorption spectra using a Varian Cary 219 instrument. Quantum yields were measured with Bausch and Lomb high-intensity monochromator with an Osram HBO-200L2 super pressure mercury lamp. GLC analysis was performed on a Hewlett-Packard 5800 gas chromatograph fitted with a glass capillary column (0,20mm I.D., length 12 m) containing a crosslinked methyl silicone film (film thickness 0,33 mm) and a flame ionization detector. High-pressure gel permeation chromatograms (GPS) were run on a Waters Associates instrument with UV detection and μ-styragel columns having pore sizes of 15^5, 10^4, and 500 Å arranged in series.

Synthetic Procedures; Purification of Monomers

Commercial monomers were washed three times with 2% NaOH and four times with distilled water. The monomers were dried over anhydrous $CaCl_2$ and were distilled under vacuum before use.

Polymerization of Poly(styrene-vinylbenzyl chloride) Copolymer

In order to obtain poly(styrene-vinylbenzyl chloride) copolymer, a mixture of 10.4g (0.1 mol) styrene, 7.6 g (0.05 mol) vinylbenzyl chloride (mixture of 60% meta isomer and 40% para isomer) and 90 mg benzoyl peroxide was used. Polymerization was carried out in degassed (three cycles of freeze-thaw under high vacuum) sealed ampules at 60°C. The polymer was precipitated with excess methanol, filtered, and dried.

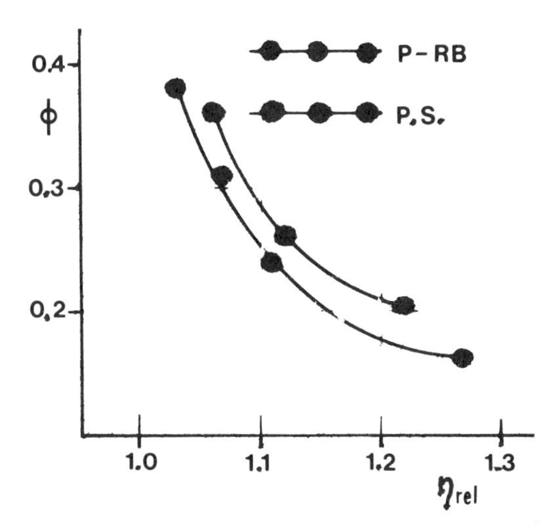

Figure 6. Quantum Yield of Singlet Oxygen Formation as a Function of Viscosity

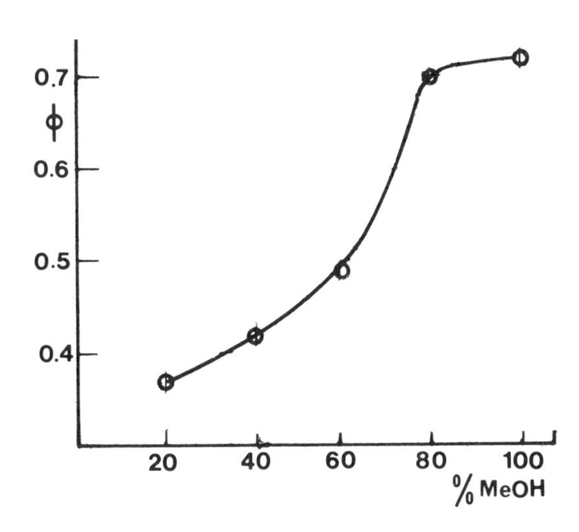

Figure 7. Quantum Yield of Singlet Oxygen Formation for RB Benzyl Ester in Mixture of MeOH and CH_2Cl_2 as a Function of MeOH Concentration

Polymer Bound Rose Bengal (P-RB)

To 1.07 g of poly(styrene vinylbenzyl chloride) copolymer in DMF was added respectively: 51 mg RB for P-RB-51, 102 mg RB for P-RB-102, 152 mg RB for P-RB-152, 305 mg RB for P-RB-305, 450 mg RB for P-RB-450, 610 mg RB for P-RB-610 and 152 mg RB for P-RB-1520. The mixtures were stirred magnetically and heated (80°) for 24 hours. The reaction mixtures were then cooled to ambient temperature. All the polymers were precipitated by addition of an excess of distilled water. They were then purified by precipitating from DMF solution by excess of methanol and washed continuously with methanol until the final filtrates were colorless. The polymers were dried in a vacuum.

2,3-Diphenyl-p-dioxene was prepared according to the method of Summerbell and Berger, and the solid thus obtained was recrystalized twice from ethanol, m.p. 93-94°C (uncorrected)[32].

Rose Bengal benzyl ester was prepared according to the method described by Lamberts and Neckers in a previous publication[17].

Quantum Yield Measurements of Singlet Oxygen Formation

For quantum yield measurement procedure was as follows: P-RB solutions in 25 ml CH_2Cl_2 were prepared so that the amount of RB attached to the polymer corresponded to 8.2 mg of RB in solution. In order to obtain these solutions we used: 180 mg P-RB-51; 94 mg P-RB-102; 65.8 mg P-RB-152; 35.2 mg P-RB-305; 27.6 mg P-RB-450; 22.6 mg P-RB-610 and 13.8 mg P-RB-1520 (solution of P-RB-1520 prepared in mixture 50% MeOH and 50% CH_2Cl_2). To the polymer solution was added 172 mg of 2,3 diphenyl-p-dioxene. The solution (2.5 ml) in a rectangular cell was irradiated with a high intensity monochromator at 566 nm. The solution was stirred, and a continuous flow of oxygen over the surface of the solution was maintained throughout the irradiation. The formation of the photooxidation product ethylene glycol dibenzoate (IV) was followed by analysis of the solution by GLC at 225°C. $\Phi(^1O_2)$ was calculated from the ratio of the rate of formation IV for P-RB compared to the rate for RB and using the known quantum yield for RB $\Phi(^1O_2)$ = 0.76) according to equation 9.

The same procedure was used to measure singlet oxygen formation from RB benzyl ester under different conditions. The following solutions (25 ml) were prepared:

1. 8.2 mg RB benzyl ester, 172 mg II in MeOH
2. 8.2 mg RB benzyl ester, 172 mg II, 172 mg cumene in MeOH
3. 8.2 mg RB benzyl ester, 172 mg II in 20% MeOH and 80% CH_2Cl_2
4. 8.2 mg RB benzyl ester, 172 mg II in 40% MeOH and 60% CH_2Cl_2
5. 8.2 mg RB benzyl ester, 172 mg II in 60% MeOH and 40% CH_2Cl_2
6. 8.2 mg RB benzyl ester, 172 mg II in 80% MeOh and 20% CH_2Cl_2

In order to describe the influence of viscosity, the following solutions (25 ml) were used:

1. 8.2 mg RB benzyl ester,172 mg II,100 mg polystyrene m.w.200,000
2. 8.2 mg RB benzyl ester,172 mg II,100 mg polystyrene m.w.110,000
3. 8.2 mg RB benzyl ester, 172 mg II,100 mg polystyrene m.w.35,000

Solutions were prepared in 20% MeOH and 80% CH_2Cl_2 mixture.

Acknowledgment

This work was supported by the National Science Foundation (DMR #8103100). We are very grateful for their support. Many helpful discussions with Dr. J. J. M. Lamberts are acknowledged with gratitude.

Literature Cited

1. Blossey, E. C.; Neckers, D. C.; Thayer, A. L.; Schaap, A. P. J. Amer. Chem. Soc., 1973, 95, 5820.
2. Schaap, A. P.; Thayer, A. L.; Blossey, E. C., Neckers, D. C. J. Amer. Chem. Soc., 1975, 97, 3741.
3. Lamberts, J. M. M., and Neckers, D. C. J. Amer. Chem. Soc., 1983, 105, 7465.
4. Schaap, A. P.; Thayer, A. L.; Taklika, K. A., Valenti, P. C. J. Amer. Chem. Soc., 1979, 101, 4016.
5. Tamagaki, S.; Liesner, C. E.; Neckers, D. C. J. Org. Chem., 1980, 45, 1573.
6. Nilsson, R.; Kearns, D. R.; Photochem. Photobiol, 1974, 19, 181.
7. Kautsky, H. Trans. Faraday, 1939, 35, 3795.
8. Kenley, R. A.; Kirshen, N. A.; Mill, T.; Macromolecules, 1980, 13, 808.
9. Buell, S. L.; Demas, J. N.; J. Phys. Chem., 1983, 87, 4675.
10. Gollnick, K.; Griesbeck, A.; Tet. Letters, 1984, 25, 725.
11. Denny, R. W.; Nickou, A. Org. Reactions, 1973, 20, 177.
12. Davidson, R. S.; Pratt, J. E. Tetrehedron, 1984, 40(6), 1004.
13. For an extensive review, see Lamberts, J. J. M. and Neckers, D. C. Zeit. F. Natur. 39b, 474-84 (1984).
14. P. O. Box 23, Perrysburg, Ohio 43551.
15. Lamberts, J. J. M., Schumacher, D. R., Neckers, D. C. J. Amer. Chem. Soc., in press.
16. Merrifield, R. B. J. Amer. Chem. Soc., 1963, 85, 2149.
17. Neckers, D. C. Review of rejected proposal to a funding agency, May 1984.
18. Blossey, E. C.; Neckers, D. C. Solid Phase Synthesis, Dowden, Hutchinson and Ross, E. Stroudsburg, Penn. (1975).
19. Schaap, A. P. Tet. Lett., 1970, 1757.
20. Midden, R. W., Wang, S. Y. J. Amer. Chem. Soc., 1983, 105, 4129.
21. Schenck, G. E. Z. Elektrochem, 1951, 55(6), 505.
22. Gollnick, K.; Schenck, G. O. Pure Appl. Chem., 1964, 8, 507.
23. Rabek, J. F.; Ranby, B.; J. Polym. Sci., 1974, A1, 12, 273.
24. Ranby, B.; Rabek, J. F. Photodegradation, Photooxidation and Photostabilization of Polymers, John Wiley, 1975.
25. Arbeloa, L. I., J. Chem. Soc., Faraday Trans., 1981, 77, 1725.
26. Yuzhakov, V. I.; Russ. Chem. Rev., 1979, 48, 1076.

27. Martin, M. Chem. Phy. Lett., 1975, 35, 105.
28. Tssa, R. M.; Ghoneim, M. M.; Idriss, K. A.; Harfoush, A. A. J. Physik. Chem., NF, 1975, 94, 135.
29. Kamat, P. V.; Fox, M. A. J. Photochem., 1984, 24, 285.
30. Kausky, H., deBruijn, H., Naturwissenschaften, 1931, 19, 1043.
31. Kautsky, H. Biochem. L., 1937, 291, 271.
32. Summerbell, R. K.; Berger, D. R. J. J. Amer. Chem. Soc., 1959, 81, 633.

RECEIVED January 10, 1985

15

Steric Compression Control

A Quantitative Approach to Reaction Selectivity in Solid State Chemistry

SARA ARIEL, SYED ASKARI, JOHN R. SCHEFFER[1], JAMES TROTTER, and LEUEEN WALSH

Department of Chemistry, University of British Columbia, Vancouver, Canada V6T 1Y6

The solid state photochemistry of eight closely related α,β-unsaturated cyclohexenones is correlated with their crystal and molecular structures as determined by X-ray diffraction methods. It is suggested that the observed changes in reactivity are caused by differences in the packing arrangements near the reaction site which either sterically impede or allow certain key atomic motions along the reaction coordinate. Computer simulation of these motions coupled with calculations of the resulting non-bonded steric compression energies support the theory. Steric compression also explains the case of a molecule which fails to undergo [2+2] photocycloaddition when irradiated in the solid state despite an almost perfect crystal lattice alignment of the potentially reactive double bonds. The packing diagram suggests that photodimerization would lead to unfavorable steric interactions between the reacting molecules and their stationary lattice neighbors. Computer simulation of the early stages of the photo-dimerization coupled with estimates of the resulting steric compression energies corroborate the theory.

It is well established, in a qualitative sense, that chemical reactions occurring in crystals are subject to restrictive forces, not found in solution, which limit the allowable range of atomic and molecular motions along the reaction coordinate. This often leads to differences, either in the product structures or the product ratios, in going from solution to the solid state. This was first demonstrated in a systematic way by Cohen and Schmidt in 1964 in their studies on the solid state photodimerization of cinnamic acid and its derivatives (1). This work led to the formulation of the famous topochemical principle which states, in

[1]Author to whom correspondence should be directed.

0097-6156/85/0278-0243$06.00/0

effect, that reactions in crystals tend to be least motion in character.

In 1975, Cohen introduced the concept of the reaction cavity in solid state chemistry (2). The reaction cavity was defined as the space occupied by the reacting species and bounded by the surrounding, stationary molecules. Cohen viewed the topochemical principle as resulting from the preference for chemical processes to occur with minimal distortion of the reaction cavity, either in the formation of voids within it or extrusions from it (Figure 1).

The next advance in the understanding of the forces which underlie the topochemical principle was due to McBride (3). He introduced the concept of local stress to explain the details of the mechanisms by which diacyl peroxides decompose in the solid state. McBride showed that least motion can be overcome in these cases by anisotropic stresses equivalent to many tens of kilobars of pressure exerted by the product carbon dioxide molecules trapped in unfavorable lattice poitions.

Most recently, Gavezzotti (4) has analyzed theoretically certain solid state processes in terms of the volume of the constituent molecules and the size and location of the empty and filled spaces in the crystal lattice. With a statement that will be seen to be directly pertinent to the results of our investigation, he concludes that "a prerequisite for crystal reactivity is the availability of free space around the reaction site".

What is lacking at this point in theories relating lattice restraints and chemical reactivity is the identification of *specific* steric interactions which alter reactivity and an estimation of their magnitude. This requires an extensive database of structure-reactivity information for a series of closely related compounds. This we have from our studies on the solid state photochemistry and X-ray crystallography of a large number of variously substituted bicyclic dienones of general structure 1 (5). In this series, we recently observed a photorearrangement

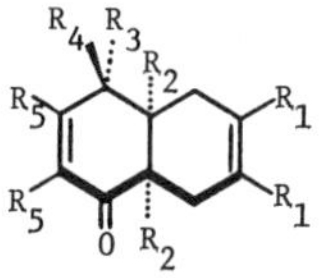

1

which did not conform to the normal reactivity observed for these systems in the solid state and which could not be accounted for using traditional stereoelectronic arguments. In this paper we demonstrate that in all likelihood, this change in reactivity is caused by specific crystal packing effects near the reaction site which are unique to the compound which behaves abnormally. We suggest the term steric compression control for this effect and estimate its magnitude using non-bonded repulsion energy calculations. We also demonstrate that the concept of steric compression control can be applied to bimolecular reactions ([2+2] photocycloadditions) in the solid state.

Enone Photorearrangements. As will be seen, the steric compression is associated with the substituents attached to the α, β-unsaturated double bond in bicyclic dienones of general structure **1**. We therefore select compounds for comparison in which R_5 is constant and equal to methyl throughout the series. Table I outlines the eight compounds, **1a**-**1h**, studied; each has had its molecular and crystal structure determined by X-ray diffraction methods (6-11).

Table I. Reactants, Hydrogen Abstraction Distances and Steric Compressions in Solid State Photorearrangements

Enone	R_1	R_2	R_3	R_4	$H..C_\alpha$ (Å)	$H..C_\beta$ (Å)	Steric Compression Accompanying Pyramidalization C_α	C_β
1a	CH_3	CH_3	H	OH	2.78	2.75	yes	yes
1b	CH_3	CH_3	OH	H	2.88[b]	2.92[b]	yes	yes
1c	CH_3	CH_3	OH	CH_3	2.86	2.81	yes	yes
1d	H	CH_3	H	OH	2.82	2.78	yes	no
1e	H	CH_3	OH	H	2.74	2.85	yes	yes
1f	H	H	H	OH	2.92	2.84	yes	yes
1g	CH_3	CH_3	H	OAc	2.74	2.70	no	yes
1h	CH_3	CH_3	OAc	H	2.79	2.84	yes	yes

[a]Yes indicates a hydrogen-hydrogen contact upon pyramidalization of <1.9 Å. In some cases, more than one contact is developed. No indicates no contacts <2.2 Å. The exact values are not given as they vary with methyl group rotation (see text). [b]Enone **1b** does not react when photolyzed in the solid state.

The solid state photochemical results are summarized in Figure 2. With two exceptions (**1b** and **1g**), the solid state photoreactivity consists of allylic hydrogen transfer to the β-carbon atom of the enone double bond followed by closure of the resulting biradical species **2** or **3** (whether path A leading to **2**, or path B leading to **3** is followed, depends upon the dienone conformation adopted in the solid state (5)). In contrast, crystals of dienone **1b** are photochemically inert and, most remarkably, irradiation of dienone **1g** in the solid state leads, via allylic hydrogen transfer to the enone α-carbon atom, to biradical **4g** (path C). That the photochemical reactions of enones **1a**-**1h** are true solid state processes and not the result of photochemistry occurring in liquid regions of the crystal was shown by the fact that irradiation of enones **1a**-**1h** in solution affords exclusively intramolecular [2+2] photocycloaddition. The reasons for the solid state/solution rectivity differences have been discussed (5).

There is considerable current interest over the question of what factors govern the α versus β regioselectivity of hydrogen atom abstraction by the carbon-carbon double bonds of photoexcited α, β-unsaturated ketones (12-14). In solution, abstraction by the β-carbon is normally preferred, and this is the regioselectivity followed in six of the eight enones studied in this work. This can be attributed to the preference for forming a resonance-stabilized radical center next to the carbonyl group (2 and 3) rather than a radical which is not conjugated with the carbonyl group (4).

Possible Origins of Abnormal Regioselectivity. What is the reason for the reversed regioselectivity in the case of enone 1g? One possible explanation is that the acetate group (R_4) of enone 1g favors the formation of biradical 4 (path C) through interaction with the radical center produced on the adjacent carbon atom. However, it is now generally accepted that bridged 1,3-dioxolan-2-yl radicals are not formed in 1,2-acyloxy radical rearrangements (15-16). Also, if such a species were dictating the preferred formation of biradical 4 in the case of enone 1g, then enone 1h should behave similarly, but it does not. In addition, enone 1i (R_1=CH_3, R_2=R_3=H and R_4=OAc, crystal structure not complete) behaves normally, i.e., reacts via initial β-carbon abstraction when photolyzed in the solid state. Using similar logic, other explanations which rely on the difference between acetate and hydroxyl (e.g., hydrogen bonding capability) can be dismissed.

A second possible explanation for the anomalous behavior of compound 1g is that the allylic hydrogen being abstracted in this case is much closer to the enone α-carbon atom than to the β-carbon. That this is not so is evident from the carbon-hydrogen distances summarized in Table I. These values, which are taken from the X-ray crystallographic work (6-11), show no trend between ground state abstraction distance and preferred reactivity. For example, enone 1g reacts at C_α even though the hydrogen atom being abstracted is closer to C_β.

Steric Compression Control. A clue to the explanation we favor came from an inspection of the crystal packing for enones 1a-1h. It appeared that the change in hybridization of C_α or C_β from sp^2 to sp^3, which necessarily accompanies hydrogen transfer to these atoms, would force the methyl groups at these centers into close contacts with certain hydrogen atoms on neighboring molecules and thus sterically impede the reaction. For the same reason, twisting about the carbon-carbon double bond, which is believed to accompany photoexcitation of α,β-unsaturated ketones in solution (13, 17-18), is unlikely to be important in the solid state for enones 1a-1h. The steric hindrance to pyramidalization is represented schematically in Figure 3. The packing diagrams indicated that steric hindrance to pyramidalization (steric compression) was present in all eight compounds studied. For enones 1a, 1b, 1c, 1e, 1f and 1h, steric compression occurs upon pyramidalization at both C_α and C_β. The only exceptions to this trend were the C_α methyl group of compound 1g and the C_β methyl group of enone 1d whose pyramidalization appeared to be unimpeded. This formed the basis of our working hypothesis, namely that it is the void space surrounding the C_α methyl group of enone 1g which allows reaction and pyramidalization at this center in contrast to the steric compression which would attend reaction and pyramidalization at C_β.

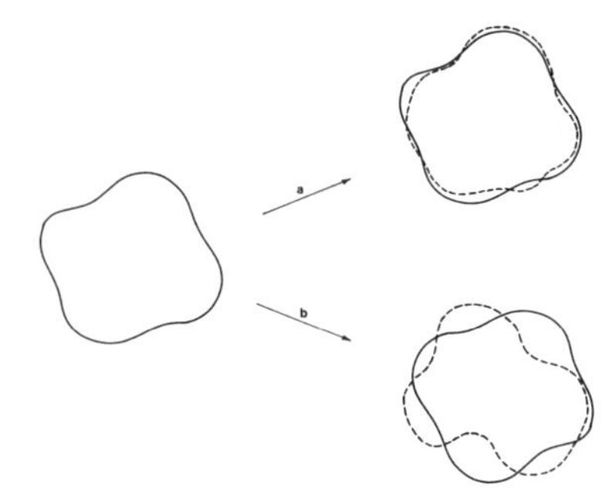

Figure 1. The reaction cavity before reaction (full line) and in the transition state (broken line) for energetically favorable (a) and unfavorable (b) reactions. Adapted with permission from Ref. 2. Copyright 1981, Verlag Chemie.

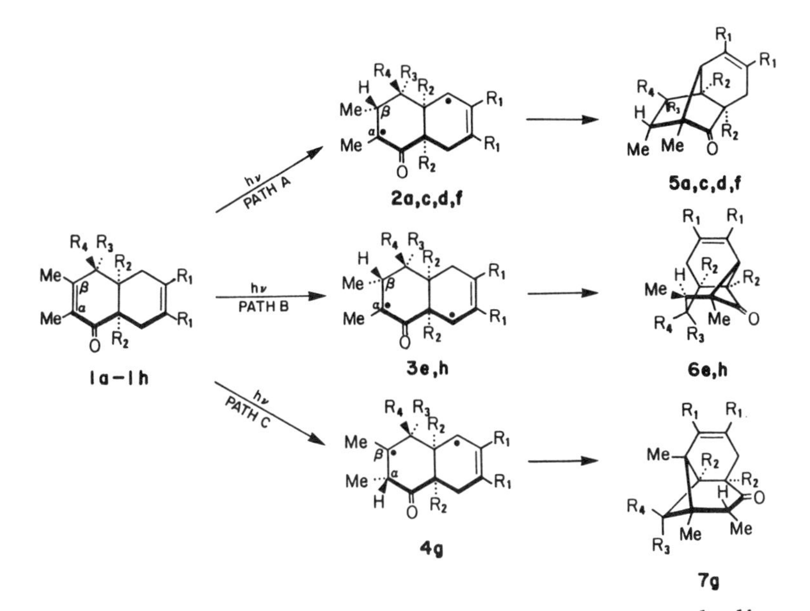

Figure 2. Photochemical reaction pathways for enones 1a–1h. Refer to Table I for list of substituents.

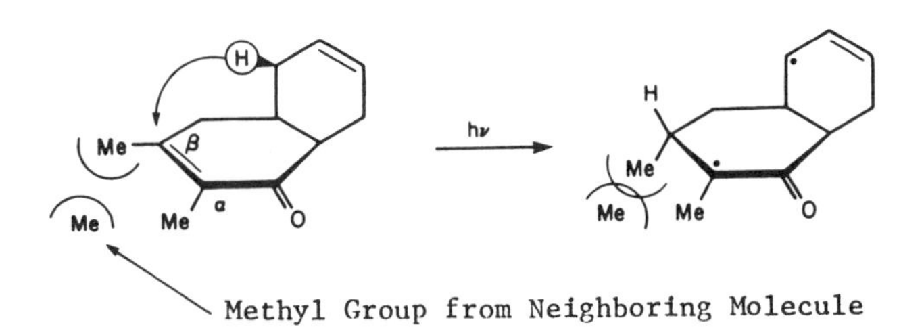

Figure 3. Steric compression resulting from pyramidalization at the β-carbon atom. The substituents have been omitted for clarity.

Computer Simulation of Pyramidalization. This hypothesis was tested by computer simulation of the pyramidalization at either C_α or C_β of a single molecule surrounded by its stationary lattice neighbors. Using the X-ray crystal structure-derived coordinates of enones 1a-1h as a starting point, the methyl groups attached to C_α and C_β were rotated downwards by intervals of 11°, 22°, 33°, 44° and 55° keeping all other coordinates unchanged. At each interval, any new intermolecular hydrogen-hydrogen contacts of less than 2.2 Å involving the methyl groups undergoing pyramidalization were noted. The distance of 2.2 Å was selected because this is the distance below which the non-bonded repulsion energy betweeen hydrogen atoms becomes significant (vide infra).

The results confirmed our working hypothesis. Pyramidalization at C_α of enone 1g and C_β of 1d led to no significant steric compression (contacts >2.2 Å), whereas pyramidalization in all other cases led to new contacts averaging (at their minimum) 1.6±0.3 Å. These results are summarized in Table I. As an example, Figure 4 shows stereodiagrams of enone 1g before and after pyramidalization at C_β. The steric compression accompanying full 55° pyramidalization is indicated by the dotted lines and consists of hydrogen-hydrogen contacts of 1.71 and 1.87 Å; pyramidalization at C_α is unimpeded.

Steric Compression Energies. An estimate of the steric compression energies accompanying pyramidalization may be reached using one or more of the several semi-empirical equations which relate interatomic distance and non-bonded repulsion energy. Two of the better known equations are the Lennard-Jones 6-12 potential function (19) and the Buckingham potential as parameterized by Allinger for his MM2 force field program (20). These two functions are plotted graphically in Figure 5 for interactions involving hydrogen atoms. Using this plot we can estimate the steric compression energy for enone 1g fully pyramidalized at C_β (contacts of 1.71 and 1.87 Å). This amounts to 11.7 kcal/mole (MM2) or 12.7 kcal/mole (6-12).

Methyl Group Rotation. Methyl group rotation, which can be rapid in the solid state, can obviously alter the steric compression contacts. Initially, the computer simulation of pyramidalization was carried out keeping the methyl groups in their original, ground state rotational orientations. Having determined the minimum intermolecular contacts attending pyramidalization, we then rotated the interacting methyl groups in both directions by 30° and redetermined the contacts. The results of such a computer experiment for rotation of the pyramidalized C_β methyl group of enone 1g are given in Figure 6 which is a plot of intermolecular hydrogen-hydrogen contact versus angle of rotation. This shows that rotation of this methyl group in either direction does not relieve the steric compression caused by pyramidalization. For example, rotation in the positive direction, while slightly relieving the 1.71 Å contact, strongly decreases the 1.87 Å contact. Rotation in the opposite direction is no better; the 1.71 Å contact is decreased slightly while the 1.87 Å is relieved. In addition, a third contact, 2.20 Å, which is unimportant at 0°, becomes a significant at approximately -20°. A similar rotation

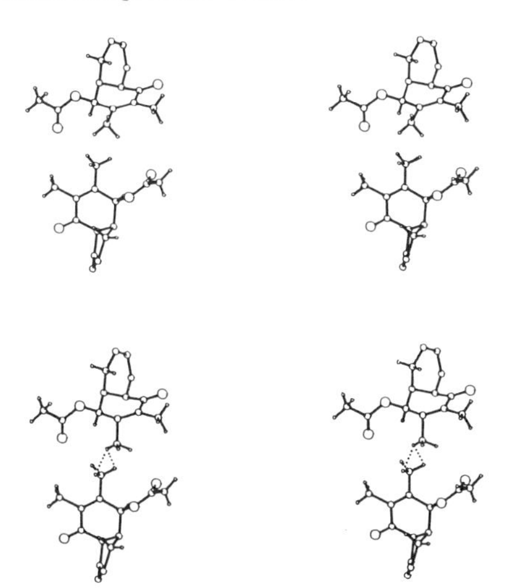

Figure 4. Stereodiagrams of enone **1g** before (above) and after (below) pyramidalization at the β-carbon atom. The steric compression contacts which develop are shown by the dotted lines.

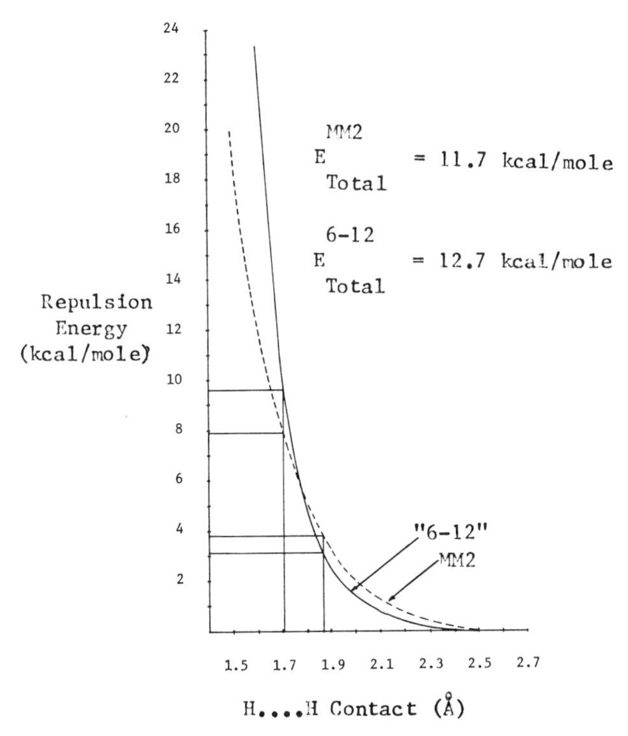

Figure 5. Hydrogen-hydrogen non-bonded repulsion energies versus distance for MM2 (dotted line) and 6-12 (solid line) potential functions. The energies at 1.71 and 1.87 Å are shown.

simulation was carried out for the non-pyramidalized methyl group for the interacting pair in the case of enone 1g. Again, rotation was found to be ineffective in relieving the hydrogen-hydrogen steric compression contacts. Methyl rotation was tested for all eight enones studied. Although the hydrogen-hydrogen contacts varied with rotation (as above), in no case did rotation alter the conclusions reached on the basis of the 0° rotation contacts.

Mechanistic Interpretation. We interpret the steric compression results kinetically in terms of the relative activation energies for hydrogen atom transfer. It is well established that hydrogen abstraction is the rate determining step in other hydrogen transfer-initiated photorearrangements such as the Norrish type II reaction (21-25). The situation is summarized in Figure 7. In the absence of any steric compression, hydrogen abstraction by the β-carbon has a lower activation energy than hydrogen transfer to the α-carbon atom for reasons already discussed. Steric compression accompanying hydrogen transfer to both C_α and C_β raises both activation energies, but maintains the ordering of C_β below C_α (enones 1a, 1b, 1c, 1e, 1f and 1h). Steric compression at C_α but not C_β (enone 1d) increases the normal activation energy difference resulting in abstraction by C_β being even more favored than before. In the anomalous case of enone 1g, however, steric compression occurs only at C_β with the result that abstraction by C_β has a higher activation energy than abstraction by C_α thus accounting for the observed change in regioselectivity. An additional interesting point concerns the photochemical inertness of enone 1b. Originally this unreactivity was ascribed solely to the unusually long hydrogen abstraction distances involved (Table I), but it now can be seen to be due in addition to the steric compression which would accompany abstraction at either carbon.

At this point, while the main features of the theory are clear, it is not worthwhile to try to calculate the actual activation energy differences for each enone based on the hydrogen-hydrogen contacts accompanying pyramidalization in each case. The main reasen for this is the relatively large (but normal) experimental errors in determining the hydrogen atom coordinates from the room temperature crystallographic data. As is apparent from Figure 5, compression energy is a very sensitive function of distance below 2 Å. Neutron diffraction studies would permit more accurate quantification of the theory.

Steric Compression Inhibition of [2+2] Photocycloaddition. Following the pioneering work of Schmidt and co-workers on the solid state photodimerization reactions of the cinnamic acids (26), a very large body of evidence has accumulated which demonstrates that intramolecular [2+2] photocycloaddition is the virtually inevitable result of a crystal packing arrangement which orients lattice neighbors so that the potentially reactive double bonds lie

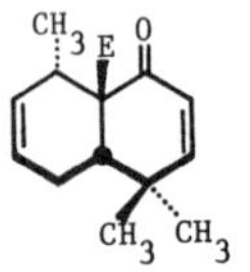

CH_3 E O

8 (E = CO_2CH_3)

CH_3 CH_3

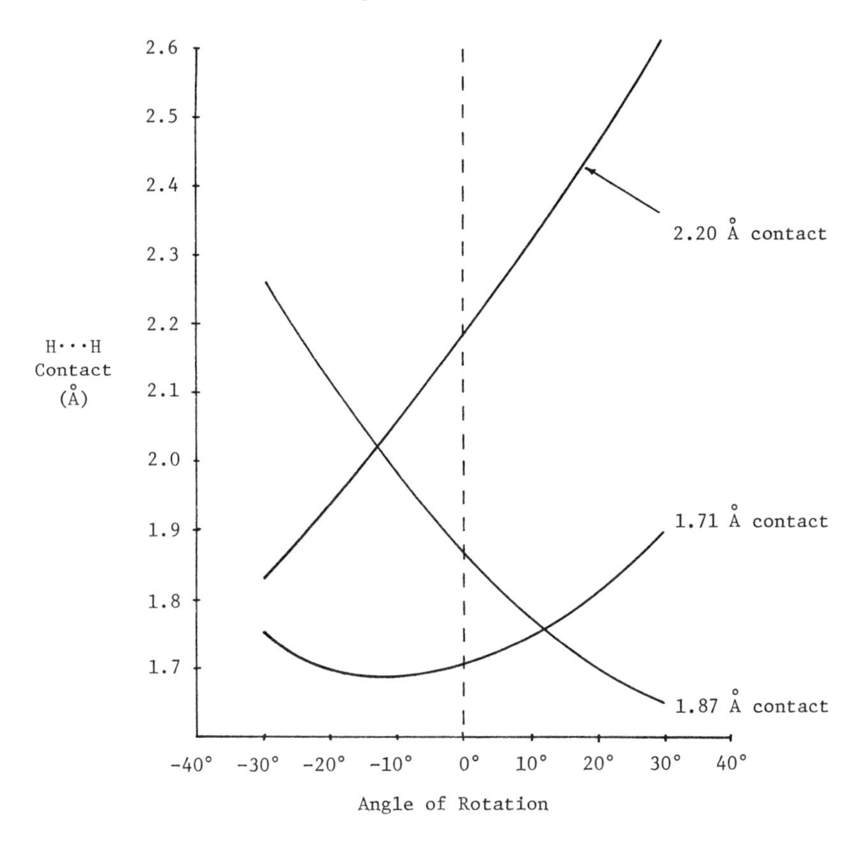

Figure 6. Hydrogen-hydrogen contacts versus angle of rotation for the pyramidalized methyl group of enone **1g**. Of the nine total contacts, only the three shown are below 2.2 Å.

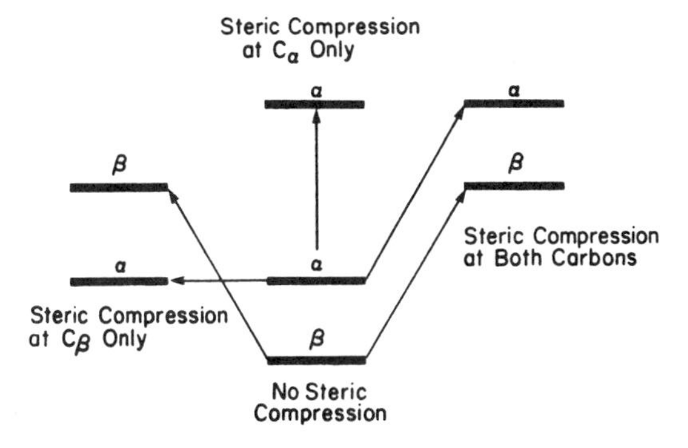

Figure 7. Relative activation energies for hydrogen transfer to C_α or C_β as a function of steric compression accompanying pyramidalization.

in a parallel arrangement at center to center distances of 4.1 Å or less (27-34). We were thus very surprised to observe the complete lack of photochemical reactivity of enone **8** when irradiated in the solid state. Compound **8** crystallizes in a lattice arrangement which is ideal for intermolecular [2+2] photocycloaddition (Figure 8) (35). The potentially reactive double bonds are oriented in a head to tail fashion and are parallel, directly above one another and only slightly offset along the double bond axis (0.52 Å); the center to center distance is 3.79 Å. Nevertheless, photolysis of single crystals of **8** for up to 40 hours at -16° to -18°C (to prevent melting) with a Liconix Helium-Cadmium 325 nm CW laser showed less than 1% reaction by capillary gas chromatography. That this lack of reactivity is not an intrinsic property of enone **8** was shown by the fact that its solution phase irradiation at the same wavelength leads to essentially quantitative yields of the cage compound resulting from intramolecular [2+2] cycloaddition.

What is the source of this remarkable lack of reactivity in the solid state? The packing diagram shown in Figure 8 reveals the probable answer. As the potentially reactive molecules X and $\bar{X}$ start to move towards one another in the initial stages of [2+2] photocycloaddition, each experiences increasingly severe steric compression of two of its methyl groups (dotted lines). The key feature of this steric compression is that it is developed, not between the potential reactants X and $\bar{X}$ (after all, a certain amount of steric compression between reactants must always accompany bond formation between them), but between X and $\bar{Y}$ and $\bar{X}$ and Y. Thus molecules Y and $\bar{Y}$ act as stationary impediments to photodimerization in exactly the same way as the stationary lattice neighbors inhibited hydrogen abstraction in the case of enones **1a**-**1h**.

Computer Simulation of Photodimerization. These ideas were tested by computer simulation of the solid state [2+2] photocycloaddition. Two mechanisms were considered: (1) Molecules X and $\bar{X}$ move toward each other in 0.24 Å increments along the double bond center to center vector (dual motion mechanism) and (2) molecule $\bar{X}$ remains fixed while molecule X moves toward it in 0.48 Å increments along the center to center vector (single motion mechanism). In both cases, the coordinates of molecules Y and $\bar{Y}$ were left unchanged during the hypothetical dimerization. The new hydrogen-hydrogen contacts were then determined at each stage of the dimerizations. By virtue of the symmetry of the system, all four hydrogen-hydrogen contacts are identical, and the contacts developed by moving X toward $\bar{X}$ are the same as those developed by moving $\bar{X}$ toward X.

The results are shown graphically in Figure 9. This is a plot of the total steric compression energy (MM2) versus double bond center to center distance for both the dual motion and single motion dimerization pathways. In both cases, the steric compression accompanying dimerization is of sufficient magnitude to account reasonably for the lack of dimerization. For example, at a center to center distance of 2.35 Å (dual motion mechanism), the hydrogen-hydrogen contact is 1.9 Å and the total MM2 repulsion

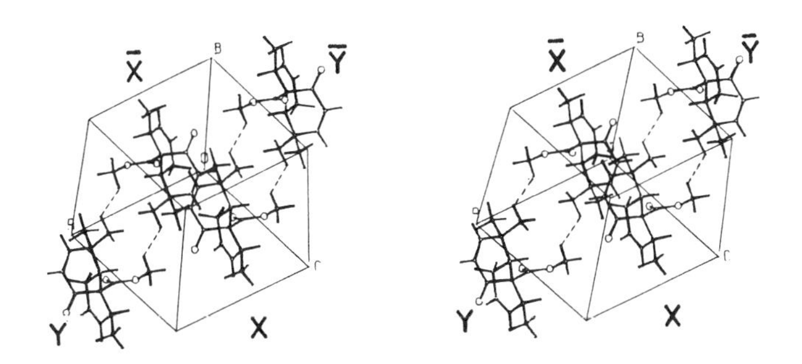

Figure 8. Stereo packing diagram of compound 8. Molecules X and X̄ are related through a center of symmetry. Translation of X along a generates Y, and translation of X̄ along -a generates Ȳ.

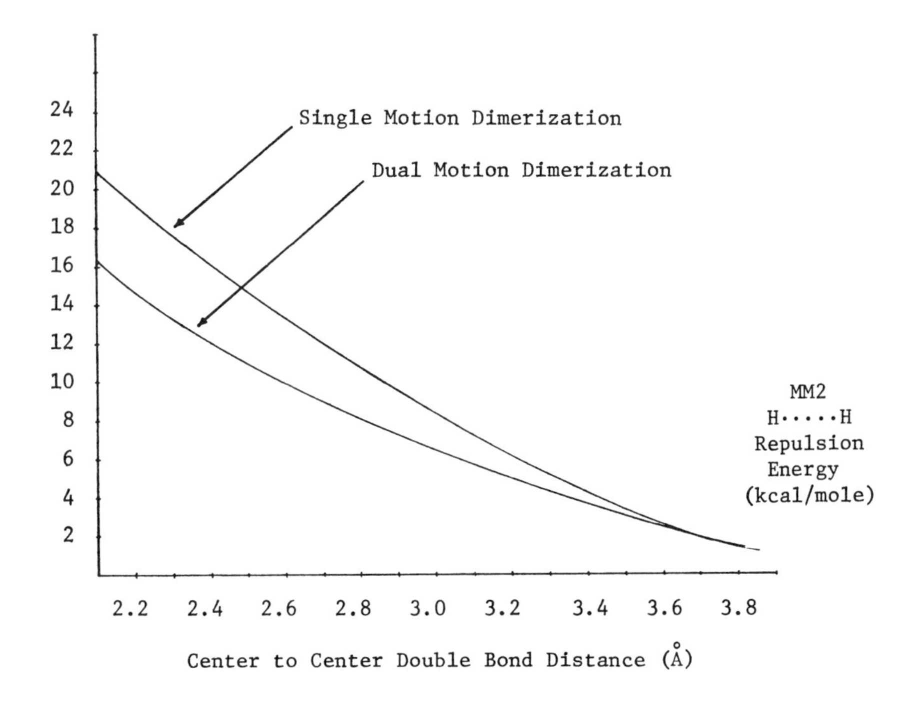

Figure 9. MM2 steric compression energy versus double bond center to center distance for single and dual motion photodimerization pathways.

energy is 13.2 kcal/mole. This corresponds to a pre-dimerization geometry in which 2p - 2p orbital overlap is small. This is based on Roberts' calculations of the overlap integral S_{ij} versus distance for 2p-σ and 2p-π bonding (36). Using these data, we estimate that at a center to center separation of 2.35 Å (offset 0.33 Å), the p-orbital overlap between molecules X and $\bar{X}$ is less than 20% of maximum. Further movement of X and $\bar{X}$ towards each other becomes prohibitively expensive owing to the fact that the hydrogen-hydrogen repulsion energy rises very steeply below 1.9 Å (Figure 5). As in the case of enones 1a-1h, these overall conclusions are not altered when rotation of the interacting methyl groups is taken into account.

A final point concerns the interesting prediction that at center to center distances above 2.1 Å the dual motion photodimerization is less sterically hindered than the single motion pathway. This can be explained with the aid of Figure 10 which is an idealized drawing of the packing arrangement for compound 8 showing the methyl-methyl interactions. Simply put, the sum of the four interactions developed in moving both reactants toward each other by a distance d is less than the sum of the two much more severe interactions which result when one of the reactants is moved toward the other by a distance 2d.

Summary. We have shown that the course of both unimolecular and bimolecular solid state chemical reactions can be influenced profoundly by certain specific steric interactions which develop between he reacting molecules and their stationary lattice beighbors. We suggest the term steric compression control for this effect and predict that it will find general utility in understanding chemical processes in the solid state. Our results provide strong support for Cohen's reaction cavity (2) and Gavezzotti's volume analysis (4) view of solid state specificity. We are in the process of testing the concept of steric compression control using a wide variety of solid state systems.

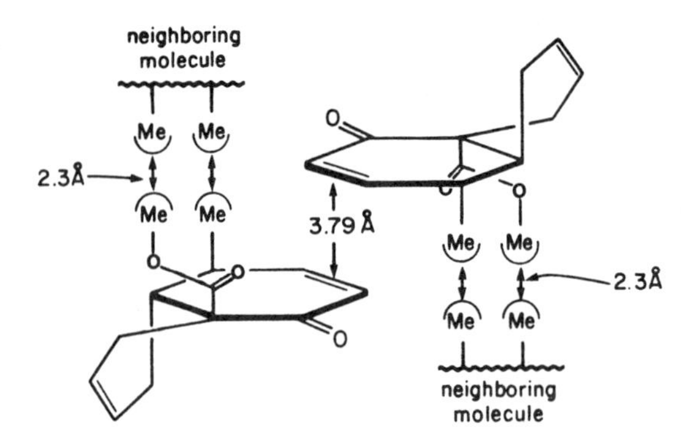

Figure 10. Idealized packing diagram for compound 8. Steric compression between four methyl groups develops as [2+2] photodimerization proceeds. All substituents other than those involved in the hydrogen-hydrogen interactions have been omitted for clarity.

Acknowledgment. We thank the Natural Sciences and Engineering Research Council of Canada for financial support.

Literature Cited

1. Cohen, M.D.; Schmidt, G.M.J. J. Chem. Soc. 1964, 1996-2000.
2. Cohen, M.D. Angew. Chem., Int. Ed. Engl. 1975, 14, 386-93.
3. McBride, J.M. Acc. Chem. Res. 1983, 16, 304-12.
4. Gavezzotti, A. J. Am. Chem. Soc. 1983, 105, 5220-5.
5. Appel, W.K.; Jiang, Z.Q.; Scheffer, J.R.; Walsh, L. J. Am. Chem. Soc. 1983, 105, 5354-63.
6. Compounds 1a and 1b: Greenhough, T.J.; Trotter, J. Acta Cryst. 1980, B36, 1831-5.
7. Compound 1c: Secco, A.S.; Trotter, J. Acta Cryst. 1982, B38, 2190-6.
8. Compound 1d: Greenhough, T.J.; Trotter, J. Acta Cryst. 1980, B36, 2843-6.
9. Compound 1e: Greenhough, T.J.; Trotter, J. Acta Cryst. 1980, B36, 1835-9.
10. Compound 1f: Secco, A.S.; Trotter, J. Acta Cryst. 1982, B38, 1233-7.
11. Compounds 1g and 1h: Ariel, S.; Trotter, J. Acta Cryst. 1984, in press.
12. Byrne, B.; Wilson, C.A. II; Wolff, S.; Agosta, W.C. J. Chem. Soc., Perkin I, 1978, 1550-60.
13. Schuster, D.I. In "Rearrangements in Ground and Excited States"; de Mayo, P., Ed.; Academic: New York, 1980; Vol. 3, pp 199-204.
14. Chan, C.B.; Schuster, D.I. J. Am. Chem. Soc. 1982, 104, 2928-9.
15. Beckwith, A.L.J.; Ingold, K.U. In "Rearrangements in Ground and Excited States"; de Mayo, P., Ed.; Academic: New York, 1980; Vol. 1, pp 242-4.
16. Barclay, L.R.C.; Lusztyk, J.; Ingold, K.U. J. Am. Chem. Soc. 1984, 106, 1793-6.
17. Pienta, N.J. J. Am. Chem. Soc. 1984, 106, 2704-5.
18. Schuster, D.I.; Bonneau, R.; Dunn, D.A.; Rao, J.M.; Joussot-Dubien, J. J. Am. Chem. Soc. 1984, 106, 2706-7.
19. Warshel, A.; Lifson, S. J. Chem. Phys. 1970, 53, 582-94.
20. Allinger, N.L. J. Am. Chem. Soc. 1977, 99, 8127-34.
21. Wagner, P.J. Top. Curr. Chem. 1976, 66, 1-52.
22. Wagner, P.J. In "Rearrangements in Ground and Excited States"; de Mayo, P., Ed.; Academic: New York, 1980; Vol 3, pp 381-444.
23. Scaiano, J.C.; Lissi, E.A.; Encina, M.V. Rev. Chem. Intermed. 1978, 2, 139-96.
24. Scaiano, J.C. Acc. Chem. Res. 1982, 15, 252-8.
25. Wagner, P.J. Acc. Chem. Res. 1971, 4, 168-77.
26. Schmidt, G.M.J. Pure Appl. Chem. 1971, 27, 647-78.
27. Schmidt, G.M.J. "Solid State Photochemistry"; Ginsburg, D., Ed.; Verlag Chemie, New York, 1976.
28. Thomas, J.M.; Morsi, S.E.; Desvergne, J.P. In "Advances in Physical Organic Chemistry"; Gold, V.; Bethell, D., Eds.; Academic: New York, 1977; Vol. 15.

29. Thomas, J.M. Pure Appl. Chem. 1979, 51, 1065-82.
30. Lahav, M.; Green, B.S.; Rabinovich, D. Acc. Chem. Res. 1979, 12, 191-7.
31. Scheffer, J.R. Acc. Chem. Res. 1980, 13, 283-90.
32. Addadi, L.; Ariel, S.; Lahav, M.; Leiserowitz, L.; Popovitz-Biro, R.; Tang, C.P. In "Chemical Physics of Solids and their Surfaces", Roberts, M.W.; Thomas, J.M., Eds.; The Royal Society of Chemistry: London, 1980; Specialist Periodical Reports, Vol. 8; Ch. 7.
33. Byrn, S.R. "The Solid State Chemistry of Drugs"; Academic: New York, 1982.
34. Hasegawa, M. Chem. Rev. 1983, 83, 507-18.
35. Ariel, S.; Trotter, J. Acta Cryst. 1984, submitted for publication.
36. Roberts, J.D. "Molecular Orbital Calculations"; Benjamin, New York, 1961; p. 30.

RECEIVED January 10, 1985

16

Recent Advances in Organic Materials

FRED WUDL

Department of Physics, Institute for Polymers and Organic Solids, University of California, Santa Barbara, CA 93106

Organic conductors can naturally be subdivided into two types: single crystals and polymers. The former have experienced a rapid development from semiconductors (1950's - 1960's) (1) to metals (1970's) (2) to superconductors (1980's) (3). Highly conducting organic polymers (except "pyropolymers" and composites) have had a shorter history.

Whereas the mechanisms responsible for metallic conductivity as well as the loss of conductivity at lower temperatures in single crystals are now understood (4), the same cannot be said for polymeric conductors because there is still some controversy centered around the nature of the transport process and change carriers (5) For that reason and due to spatial requirements, this article will deal only with polymeric conductors and will emphasize poly(heterocycles), particularly poly(thiophene).

The only certainty, insofar as requirements to observe high conductivities in polymers is concerned, is that the polymers must be partially oxidized or reduced ("p-doped" or "n-doped", respectively). It is therefore not an unreasonable assumption that any polymer with an extended π system will be a good candidate for the achievement of activated (semiconductor) or non-activated (metal) conductivity upon doping (6). Clearly, the simplest π-conjugated polymer is poly(acetylene) (Figure 1) and has therefore been the most extensively studied. The reasons for popularity of a polymer in this highly specialized area of research are ease of preparation and "form" (7) (film, fiber, or powder). The latter can be overwhelming; for example, polyacetylene has existed for decades as black, amorphous powder and was, in fact, doped in the 1960's (8) but attracted little attention because not much could be done with an intractable (non-fibricable) "brick dust". When Shirakawa discovered the process to produce the same polymer in film form, the attention it attracted was phenomenal.

0097–6156/85/0278–0257$06.00/0

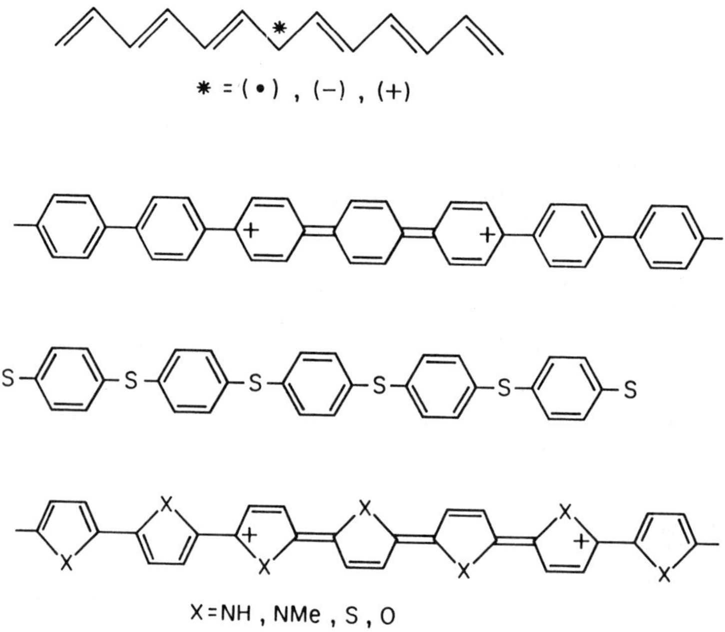

* = (•) , (−) , (+)

X = NH , NMe , S , O

Figure 1. Examples of conducting polymers and species responsible for charge storage. Top, poly(acetylene) and soliton: 2nd, poly-p-phenylene and bipolaron: 3rd, poly-p-phenylene sulfide; 4th, poly(heterocycles) and bipolaron. Bipolarons in poly(furan) have not yet been established.

Since poly(pyrrole) and poly(thiophene) (PT) can be prepared in film form by a simple procedure, they also fulfill the above requirements for popularity. However, it is clear, particularly from spectroscopic investigations, that the nature of the charge storage (and carriers) in the poly(heterocycles) is different from poly(acetylene) (9).

The topics to be covered in this article are: (a) brief review of poly(heterocycles), (b) description of the latest results on poly(thiophene), and (c) description of poly(isothianaphthene), (PITN).

Brief Review of Poly(heterocycles)

These polymers, particularly poly(pyrrole), are most conveniently prepared from the parent molecule *via* electrolysis. So far, furan, pyrrole, thiophene, and various methylated derivatives have been polymerized by this procedure (10). The anodic polymerization apparently also works for relatively electron rich aromatic compounds such as aniline and azulene (11).

In the case of the five-membered heterocycles, polymerization occurs predominantly through the 2,5 positions and when the 3,4 positions of pyrrole are blocked by methyl groups, number average molecular weights on the order of 1000 have been obtained (12). Of all the heterocycles, unsubstituted pyrrole yields the most highly conducting polymer [≅ 200 S/cm ($S = \Omega^{-1}$)]. Since the polymerization is performed under oxidative conditions, the as-formed polymer is p-doped with the counter ion corresponding to the anion of the supporting electrolyte. These doped films are considerably more stable to the atmosphere than their poly(acetylene) counterparts. However, when they are connected to the cathode of a battery, they can be "dedoped" (brought to a neutral state) and converted to an unstable form. This conversion usually occurs with a concomitant change in color (from blue-black to brown or orange-red), indicating that these polymers are also electrochromic (13).

The bandgap for poly(pyrrole) is ≅ 3eV (≅ 410nm), for poly(thiophene) it is ≅ 2eV (≅ 620nm) and for poly(acetylene) it is ≅ 1.6eV (≅ 780nm). From these numbers one could imagine that poly(acetylene) and PT would be well suited for solar energy conversion devices. However, while the bandgap of poly(acetylene) is the best of the three polymers to match the solar spectrum, from a theoretical point of view, it appears as though this material may be doomed to have very low solar energy conversion efficiency because the photogenerated carriers can form intrinsic localized gap states (soliton - antisoliton pair) at a calculated rate of ≅ 1013 sec^{-1}. In fact, the observed efficiency of a liquid junction poly(acetylene) solar cell is very low (14).

At the molecular level, studies of the doping mechanism in the poly(heterocycles) reveal that very short lived radical cations ("polarons") immediately decay (combine) to form dications ("bipolarons"); i.e., charge storage appears to involve dications rather than radical cations (9,15). Whether bipolarons will be as detrimental to the efficiency of solar cells based on poly(heterocycles) as are solitons to solar energy conversion devices based on poly-(acetylene) remains to be established.

Recent Results on Poly(thiophene)

Unlike pyrrole, thiophene can be coupled through the 2,5 positions by a non-electrochemical approach involving Grignard intermediates (16). Careful purification of starting materials (in this case 2,5-diiodothiophene) allowed the isolation of an electrically insulating sample of PT with a molecular weight of *ca* 4000 and clean ir and uv-vis spectra as well as elemental analysis corresponding to the empirical formula $C_{188}H_{97}IS_{46}$. This composition indicates that, on the average, for every 46 thiophene units there is one butadiene (17). Electronic and infrared spectroscopy revealed that the "chemically coupled" PT was cleaner (pratically no absorption below the bandgap) than a sample of dedoped electrochemically polymerized thiophene.

Since the ionization potential of thiophene is relatively high, the electric fields required for its anodic polymerization are rather steep (≅ 20V *vs* SCE). In addition, the simplest supporting electrolyte for this operation is $LiBF_4$ and deposition of Li at the cathode (usually Pt) is also energetically unfavorable. Recently, Druy (13) reported that substitution of 2,2'-bithiophene for thiophene gave better quality films, probably due to the lower ionization potential of the dimer relative to thiophene. An additional improvement consisted in replacing the Pt counter electrode by Al (9). Spectroscopy revealed that dedoped PT films produced with the above improvements were indistinguishable in quality from the chemically coupled PT.

With these excellent films on hand we were able to do highly sophisticated experiments of *in situ* doping and dedoping while performing another measurement such as electronic spectroscopy (9). The results of such experiments showed that charges, in PT, are stored as dications; a finding that parallels observations on poly-(pyrrole (15).

Electron spin resonance experiments revealed that samples of the chemically coupled polymer had very few spin defects (≅ 65 ppm per carbon). Preliminary results of epr experiments during *in situ* doping seem to corroborate the proposal of charge storage in the form of weakly confined bipolarons because doping, particularly at high dopant concentrations, shows almost no paramagnetism (spinless carriers) (18).

Solid State Photoeffects

The technique of photoinduced spectroscopy has been used very effectively to probe the lifetime and nature of states in the gap of polyacetylene (19-25); i.e., both the lattice distortion and the associated electronic structure of these photogenerated species were investigated by recording the small changes in spectroscopy (ir, epr) that occur during photoexcitation (19-25). Frequency dependence of the photogenerated signals revealed that solitons could be generated by photons with $\hbar\omega < E_g$ (E_g is the semiconductor energy gap). Similarly, electron spin resonance measurements during photoexcitation revealed the spinless nature of the photogenerated species (25). Because the characteristic features of the infrared spectrum of the charged photogenerated species were found to be in a one-to-one correspondence with the spectral features of an independently doped sample of trans-$(CH)_x$, a principal conclusion was that in both cases the charge was stored in the form of spinless charged solitons.

Coupling of electronic excitations to nonlinear conformational changes can be an intrinsic feature of conducting polymers. Generalization of this idea and application to the larger family of non degenerate ground-stage, conjugated polymers was done for the first time with poly(thiophene) (26).

The photoinduced absorption experiments on a 0.3 wt.% poly(thiophene) on KBr were carried out by H. Schaffer[27] using an IBM Instruments IR/98 vacuum Fourier-transform interferometer modified to allow an external beam of an Ar^+ laser ($\hbar\omega$ = 2.41 eV) to irradiate the sample simultaneously with the probing infrared beam. The observed spectrum, which is a difference spectrum (spectrum due to laser irradiation minus dark spectrum), consists of four relatively narrow lines at 1020, 1120, 1200, and 1320 cm^{-1} plus a very broad band peaking at ~3600 cm^{-1}. The close correspondence of the four narrow peaks with those reported by Hotta (27) (1330-1310, 1200, 1120-1080, 1030-1020 cm^{-1}), is strong evidence that the electronic structure of the photoinduced species is the same as that of the doped material. Results of similar experiments carried out by Moraes[26] on the same KBr suspension of neutral poly(thiophene) using electron spin resonance as a probe, revealed that contrary to the case of $(CH)_x$, a spin bearing species was generated; strongly implying the formation of radical cations (polarons). In this context, the broad band with a maximum at 0.45 eV (3600 cm^{-1}) can be assigned to the lowest-energy electronic transition of these photogenerated, delocalized radical cations (polarons) (26).

Since we had shown earlier[18] that the species generated upon (dark) doping were delocalized cations (polarons), one can conclude from the above experiments that in poly(thiophene) irradiation generates both cations and cation radicals; in sharp contrast to what is observed in $(CH)_x$.

Finally, another set of preliminary experimetns sowed that Schottky barriers and diodes could be prepared from PT films or pressed pellets. Efficiency of photoenergy conversion by these devices is currently under intense investigation by M. Isogai (28) in our group.

Recent Results on Poly(isothianaphthene)

PITN has recently been prepared in our group by several chemical and electrochemical procedures (29). The reason for its creation is best explained with Scheme I, below.

Scheme I

If both resonance forms 1a and 1b were isoenergetic, then there would be no bond alternation in the hydrocarbon backbone of PT and the polymer would probably exhibit metallic conductivity without the need to dope. In reality, as shown in Scheme I, the two forms are not isoenergetic and the resulting bond alternation in resonance form 1a of PT, gives rise to the observed bandgap of ca 2eV.

On the other hand, benzannulation in the 3,4 positions of PT, would produce the nearly degenerate structures 2a and 2b (Scheme I) because the gain in aromaticity of the benzene ring in 2b is expected to outweigh the loss of thiophene aromaticity in 2a. The result would be amelioration of bond alternation ("Peierls distortion" of the backbone) with a concomitant decrease in bandgap.

Recent results of in situ doping spectroscopy (Figure 2) as well as cyclic voltammetry by N. Colaneri and M. Kobayashi in our laboratory showed that indeed the bandgap of PITN is ca one eV lower than that of PT. An interesting "fallout" of this result is that upon doping, the absorption moves into the infrared region and the material becomes a transparent (depending on sample thickness), yellow conductor; i.e., PITN is a high contrast electrochromic material. The energy gap is now low enough so that Schottky barrier experiments showed Ohmic, rather than diode behavior, as was observed with PT.

Electron microscopy revealed that the morphology of 6% Cl^- doped PITN depends on the substrate on which it is deposited and that it is a relatively "open" structure, although not as open as poly(acetylene). Selected area electron diffraction on the same sample showed the material to be partially crystalline (three diffraction rings could be seen).

Contrary to PT, the fully dedoped PITN is blue-black and is a semiconductor; an observation which is in agreement with the small energy gap of this new polymer.

Conclusions and Outlook

We have presented evidence to prove the structure of electrochemically generally poly(thiophene) from dithiophene both by independent synthesis and spectroscopy. Diodes and photodiodes were fabricated from lightly doped chemically and electrochemically synthesized PT.

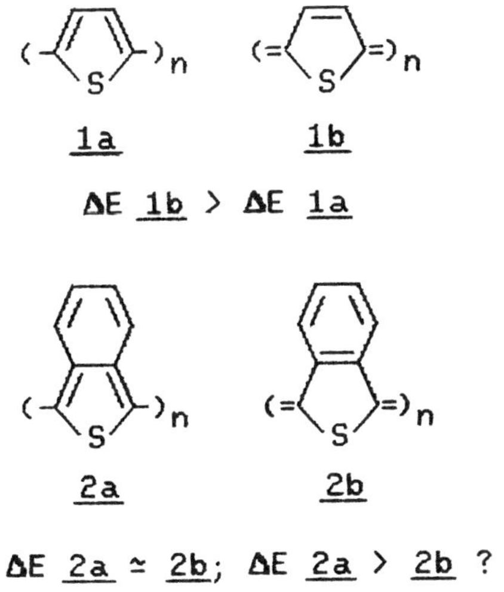

1a 1b

ΔE 1b > ΔE 1a

2a 2b

ΔE 2a $\simeq$ 2b; ΔE 2a > 2b ?

Scheme I

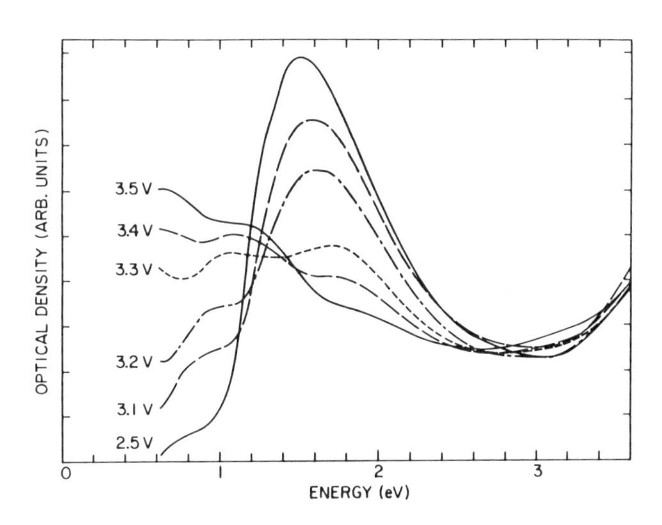

Figure 2. In situ electronic spectroscopy of PITN. Solid line at 2.5V is a 6% $C\ell^-$ doped sample and the 3.5V line corresponds to the same sample, fully doped. Voltages are *vs* Li.

Using physical organic chemical rationalizations, we were able to modify the electronic structure of a poly(heterocycle). The result was poly(isothianaphthene); a polymer which is already a semiconductor in the neutral (undoped) state. In the fully doped state, PITN is a transparent conducting polymer.

Our results with PITN are very encouraging to the development of an organic semimetal (zero gap semiconductor) since a relatively minor modification allowed to reduce the energy gap within a family of poly(heterocycles) by 1 eV (23 Kcal). Further modifications by annulation and substitution (electron donor or acceptor) are expected to further reduce the energy gap of a PT-based polymeric conductor. We are currently actively engaged in the preparation of such neutral organic conductors.

Acknowledgments

We are indebted to Showa Denko K. K. and the Office of Naval Research through grant N00014-83-K-0450 for support of this work.

Literature Cited

1. Lyons, L.E., Gutmann, F. "Organic Semiconductors," John Wiley and Sons, New York, 1967.
2. Bryce, M.R. and Murphy, L.C. Nature, 1984, 309, 119.
3. "Proceedings of the International Conference on the Physics and Chemistry of Synthetic and Organic Metals," J. de Physique Colloque, 1983, C3.
4. Wudl, F. Accounts of Chem. Res. 1984, 17, 227.
5. "Proceedings of the International Conference on the Physics and Chemistry of Conducting Polymers," J. de Physique Colloque, 1983, C3.
6. It is unfortunate that the word "doping" has crept into the language of polymeric organic conductors. This is a misnomer since it does not mean lattice substitution as it applies to current semiconductors science but oxidation or reduction of an electron-rich or electron-deficient chain of atoms. That the atom-chain does not need to consist of π-bonded elements was shown recently by West. West, R.; David, L.D., Djurovich, P.I., Sterley, K.L., Srinivasan, K.S.V., Yu, H.J. Amer. Chem. Soc. 1981, 103, 7352.
7. The word "form" is used here for lack of a better term. The proper word would have been morphology but it has recently assumed a different connotation in polymer science; it usually refers to the microstructure as observed through an electron microscope. The "traditional" concern about fabricability of poly(acetylene) may no longer be a factor if the work of Aldissi and Liepins (Aldissi, M., Liepins, R. Chem. Commun. 1984, 255.
8. Labes, M.M., Love, P., Nichols, L.F. Chem. Rev., 1979, 79, 1.
9. Chung, T.-C., Kaufman, J.H., Heeger, A.J., Wudl, F. Phys. Rev. B, 1984, 30, 702.

10. Diaz, A. Chemica Scripta, 1981, 17, 145.
11. Bargon, J., Mohmand, S., Waltman, R.J. IBM J. of Res. and Dev., 1983, 27, 330.
12. Nazzal, A., Street, G.B. Chem. Commun. 1984, 83.
13. Druy, M.A., Seymour, R.J. Reference 5, p. 395.
14. MacDiarmid, A.G., Heeger, A.J. In "The Physics and Chemistry of Low Dimensional Solids," Alcacer, L., Ed., D. Reidel, Holland, 1981, p. 393.
15. Bredas, J.L., Scott, J.C., Pfluger, P., Krounbi, M.T., Street, G.B. Phys. Rev. B, 1983, 28, 2140.
16. Kobayashi, M., Chen, J., Chung, T.C., Moraes, F., Heeger, A.J., Wudl, F. Synethetic Metals, 1984, 9, 77.
17. Wenkert, E., Leftin, M.H., Michelotti, E.L. Chem. Commun. 1984, 617. These authors exploited this side reaction for the formation of butadienes.
18. Chung, T.-C., Kaufman, J.H., Heeger, A.J., Wudl, F. Phys. Rev. B, in press.
19. Orenstein, J., Baker, G.L., Phys. Rev. Letter, 1982, 49, 1043.
20. Vardeny, Z., Straight, J., Moses, D., Chung, T.-C., Heeger, A.J., Phys. Rev. Lett. 1982, 49, 1657.
21. Shank, V., Yen, R., Fork, R.L., Orenstein, J., Baker, G.L., Phys. Rev. Lett, 1982, 49, 1660.
22. Blanchet, G.B., Fincher, C.R., Chung, T.-C., Heeger, A.J., Phys. Rev. Lett, 1983, 50, 1938.
23. Vardeny, Z., Orenstein, J., Baker, G.L., Phys. Rev. Lett, 1983, 50, 2032.
24. Blanchet, G.B., Fincher, C.R., Heeger, A.J., Phys. Rev. Lett, 1983, 51, 2132.
25. Flood, J.D., Heeger, A.J., Phys. Rev. B, 1983, 28, 2356.
26. Moraes, F., Schaffer, H., Kobayashi, M., Heeger, A.J., Wudl, F., Phys. Rev. B, 1984, 30, in press.
27. Hotta, S., Shimotsuma, W., Taketani, M. unpublished, preprint, 1984.
28. Isogai, M., Kobayashi, M., unpublished results.
29. Wudl, F., Kobayashi, M., Heeger, A.J. , J. Org. Chem., 1984, 49, 3382.

RECEIVED February 6, 1985

17

Photochemical Reactions in Oriented Systems

V. RAMAMURTHY

Department of Organic Chemistry, Indian Institute of Science, Bangalore 560 012, India

Modification of chemical reactions through the use of constrained and/or organized media has attracted a great deal of attention recently. Results from our laboratory in this direction which include a study of photochemical reactions in solid state and in cyclodextrins are presented here. A study of solid state photochemical behavior of coumarins has provided information regarding subtler aspects of topochemical postulates of photodimerization. Results pertaining to geometrical criteria for photodimerization and "chloro" as a crystal engineering group are discussed. As a part of an attempt to correlate chemical reactivity with molecular packing in the solid state, photooxidation of diarylthioketones in the solid state has been investigated. The observed differences in the reactivity of these crystals are rationalized in terms of crystal packing. Though cyclodextrins have been extensively studied, very few photochemical reactions involving molecules complexed to cyclodextrins have been examined. In this connection, the utility of cyclodextrins in bringing about selectivity in photochemical reactions through the study of excited state behavior of olefins and aryl alkyl ketones has been demonstrated in our laboratory.

The photochemistry and photophysics of organic molecules in the crystalline state and in the organized assemblies has attracted considerable attention (1-4). Control of stereo and regiochemistry in photochemical reactions through the use of constrained media such as molecular and liquid crystals, monolayers, micellar assemblies, inclusion complexes and silica gel surfaces has opened new vistas in photochemistry. During the last few years our group has been investigating the environmental perturbations on photochemical reactions (5-19). Such studies have been concerned with solid state, micellar and inclusion complexes. The goal is to achieve selectivity in photochemical reactions using these unusual environments and to understand the features controlling such selectivity. This article which summarizes some of our results is divided into two parts: the first part deals with photochemical dimerization of coumarins and photooxidation of

0097-6156/85/0278-0267$06.00/0

thioketones in the solid state. In the second part, selectivities obtained in the photochemical reactions of olefins and aryl alkyl ketones using cyclodextrin complexes are described.

Solid State Photochemistry

Photodimerization of Coumarins in the Solid State. Studies by Schmidt and his-co-workers on cinnamic acids have demonstrated that photodimerization in the solid state are strictly controlled by the packing arrangement of the molecule in the crystals (20,21). Schmidt has drawn attention to the fact that not only must the double bonds of the reactive monomers of cinnamic acid be within 4.2 Å they must also be aligned parallel for dimerization to occur. Following the initial observation with methyl m-bromo-cinnamate wherein the reactive double bonds are rotated with respect to each other by 28° a few examples have been reported which support the parallelism requirement for photodimerization (22,23). On the other hand, a few cases have also been reported where exact parallelism between reactant double bonds has not been adhered to and yet photodimerization occurs (24-27). It is clear that a reexamination of the subtler aspects of the topochemical postulates is essential. Inspite of growing interest in organic reactions in the crystalline state, the utility of such reactions as a synthetic tool is limited by the difficulty of achieving the desired type of crystal structure in any given case, for the factors that control the crystal packing are not fully understood. Therefore, scope undoubtedly exists for "engineering" organic crystals. In this connection we have embarked on a systematic crystallographic and photochemical study of a large number of substituted coumarins. A study of a large number of coumarins (28 in total) provided an opportunity to derive information regarding various aspects of photodimerization in the solid state. Of these, results pertaining to parallelism criteria for photodimerization and "chloro" as a crystal engineering group are presented here.

Solid state photodimerization of 7-methoxycoumarin is particularly important in connection with the parallelism criteria for photodimerization in the solid state. The crystals of 7-methoxycoumarin are triclinic with a = 6.834(3), b = 20.672(4), c = 12.600(7) Å, α = 108.19(3), β = 95.23(4), γ = 95.22(3), space group $P\bar{1}$ and Z = 4. X-ray crystal structure analysis shows that the potentially reactive double bonds of the monomer molecules within the asymmetric unit are rotated by 65° with respect to each other although the center to center distance between the double bonds is 3.83 Å (Figure 1). However, the dimer yield within twenty-four hours of irradiation of the crystalline 7-methoxycoumarin was 90%. The structure of the dimer was confirmed to be syn head-tail by X-ray studies. We note from the progress of the dimerization with respect to time of irradiation that 7-methoxycoumarin behaves very much like the ones in which the reaction is topochemical. It was observed that there was no perceptible evidence for induction period in the coumarins (including 7-methoxycoumarin) which are believed to be topochemically favorable for dimerization whereas significant induction period was noticed in cases where the photoreaction originated at defects. However as seen in Figure 1 the two double bonds, although within the reactive distance, are not suitably juxtaposed for dimerization. Therefore, presence of a certain degree of inherent orientational flexibility of the mole-

C(5′) O(11) C(4′) C(3) C(3′) C(4) C(5) O(11′)

C(3)····C(4′) = 3.64 Å
C(4)····C(3′) = 4.14 Å
C(3)····C(3′) = 3.67 Å
C(4)····C(4′) = 4.28 Å

center to center distance between double bonds 3.83 Å

Perspective view of the asymmetric unit (crystal coordinates)
(z axis through center of mass perpendicular to the plane of the molecule)

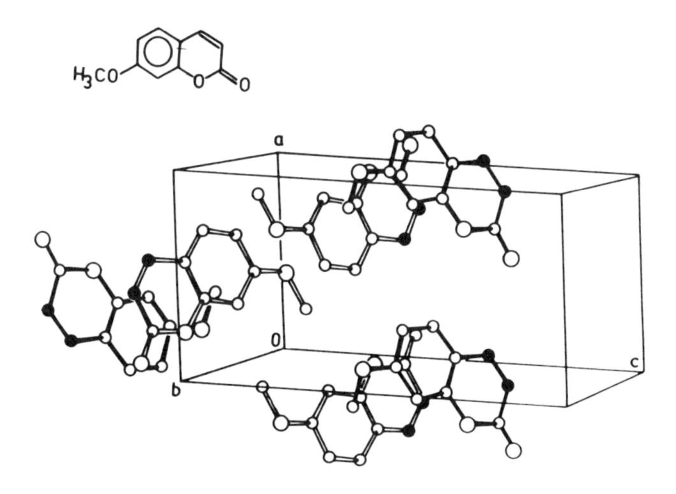

Figure 1. Disposition of the Reactive Double Bonds C(3)-C(4) and C(3′)-C(4′) of 7-Methoxycoumarin in the Asymmetric Unit.

cules in the crystal lattice has to be invoked to explain the topochemical nature of dimerization. Two dimers namely anti head-head and syn head-tail could result if the rotation of the molecules is allowed in the crystal lattice. Formation of anti head-head dimer would require a total rotation of 115° whereas the syn head-tail would require 65°.

It is quite likely that the uv radiation absorbed by the reacting molecules is sufficient to allow the molecules to undergo the required rotation provided the motion is co-operative and extends through the crystal. However, it seems essential to postulate an inherent flexibility within the crystal lattice for these molecules to undergo rotation as this would allow us to understand the large yield of the dimer. Therefore, in order to estimate the inherent orientational flexibility of the molecules in the crystal lattice, lattice energy calculations were carried out using a computer program WMIN developed by Busing (28). Much to our surprise the energy calculations revealed the presence of orientational flexibility in the ground state for both the molecules in the asymmetric unit. Indeed a total rotation of about 20° in the direction to generate syn head-tail dimer in the ground state is possible without much increase in the lattice energy ($\Delta E \sim 9.8$ kcal mole^{-1}) from the minimum energy position as determined by X-ray crystallography. With the increase in attractive forces between the reactive molecules upon excitation, one may expect that the motion of the molecules so as to achieve a maximum π overlap will become possible. We propose that rotation, in addition to that is available in the ground state as indicated by the lattice energy calculations, to generate the syn head-tail dimer is achieved due to the interaction of the excited and ground state molecules. In summary, the mechanism of photochemical dimerization of 7-methoxycoumarin involves a total rotation of 65°, within the crystal lattice. Thus it is demonstrated that topochemical dimerization of non-parallel double bonds is possible once the freedom for motion becomes available due to excited state interaction between molecules.

An aspect of our detailed study on photodimerization of coumarins in the solid state concerns understanding the factors that affect the molecular packing and identifying the groups that may be of value in bringing about the photoreactivity. Substituents such as hydroxyl, methyl, chloro, acetoxy and methoxy were utilized to engineer the crystals of coumarins towards photoreactivity. Of these acetoxy and chloro were found to be useful engineering groups. Results pertaining to chloro are discussed below.

The important points that emerge from the studies on the five chlorocoumarins in the solid state are the following. (a) All chloro substituted coumarins (7-chloro, 6-chloro, 4-chloro, 4-methyl 7-chloro and 4-methyl 6-chloro coumarins) undergo dimerization in the solid state. While four of these give syn head-head dimer as the photoproduct, 4-chlorocoumarin gives a mixture of anti-head-tail and syn head-tail in poor yields (~ 25%). (b) Two of them namely 4-chloro and 4-methyl 6-chloro coumarins exhibit significant induction period for dimerization suggesting that the dimerization in these cases is non-topochemical in nature. (c) Consistent with their photochemical behavior, the packing arrangements, as revealed by X-ray crystallographic analysis, are favorable for dimerization only in the crystals of 6-chloro, 7-chloro and 4-methyl 7-chloro coumarins (Figure 2). Syn head-head dimers are the direct consequence of these

a

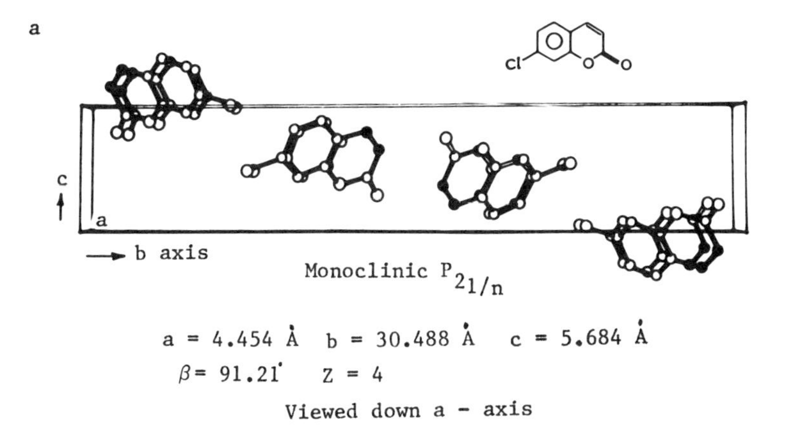

b

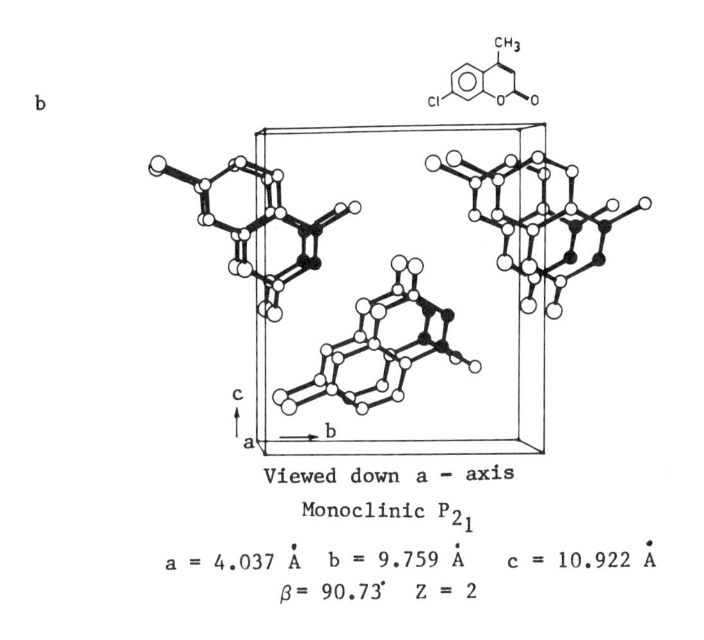

Figure 2. Packing Arrangement of (a) 7-Chlorocoumarin and (b) 4-Methyl 7-chlorocoumarin.

Figure 2 continued

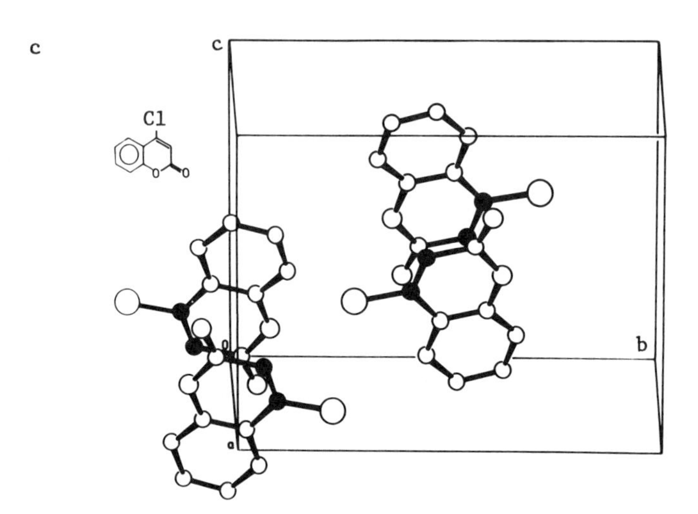

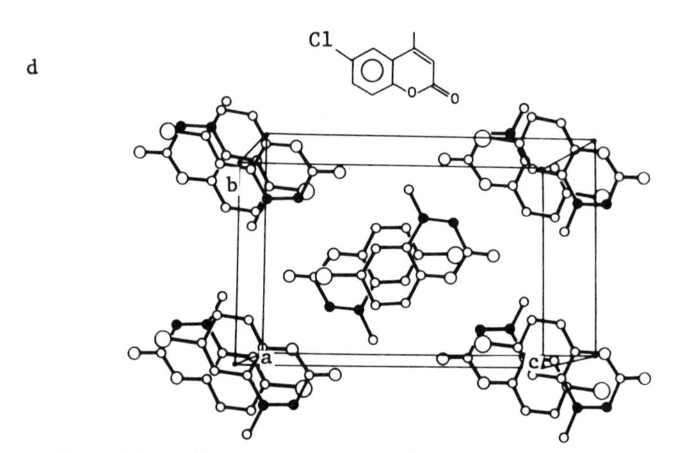

Monoclinic $P2_{1/C}$ a = 7.161 Å, b = 8.903Å, c = 13.775 Å
β= 104.04 Z = 4
Viewed down 'a' axis

Figure 2. Packing Arrangement of (c) 4-Chlorocoumarin and (d) 4-Methyl 6-chlorocoumarin.

packing arrangements. Double bonds in 4-chloro and 4-methyl 6-chloro coumarins, according to X-ray crystallographic analyses, are not suitably oriented for dimerization. It is noteworthy that, whereas coumarin does not undergo dimerization in the solid state, all the five chlorocoumarins undergo photodimerization. It is significant that in the three cases the perpendicular distance between the closest neighbors vary from 3.45 to 4.45 Å while in coumarin crystals it is as large as 5.67 Å.

Although the use of "chloro" as an effective steering device was originally recognized by Schmidt, no systematic study was reported (29,30). Therefore, it was felt that the systematics in the mode of packing in crystal structures containing chloro group attached to aromatic rings is worthy of investigation. The experimental information for our analysis was taken from Cambridge Crystallographic Data Base (Version, December 1981). Out of a total of 132 structures, only 22 did not contain any Cl...Cl interaction within Cl...Cl distance of <4.2 Å. In the 110 structures, there were 341 interactions. The geometrical parameters used in the analysis are shown in Figure 3. When $\chi = 0.0^{o}$, the atoms C_1 and C_2 are in cis configuration, whereas $\chi = 180.0^{o}$ corresponds to trans configuration. Figure 3 shows a plot of Θ_1 vs Θ_2. 60 points lie on the line with $\Theta_1 + \Theta_2 = 180^{o}$ and in these cases the arrangement of the molecule is similar to β-type packing. 66 points are on the line with $\Theta_1 = \Theta_2$ and this condition simulates the packing similar to α-type arrangement. It is noteworthy from Figure 4 which portrays the plot of N (number of interactions) vs d (Cl...Cl distance) that when $\chi = 0^{o}$ most of the Cl...Cl distance lie within a narrow range of 3.8-4.0 Å whereas the range is broad (3.5-4.2 Å) when $\chi = 180^{o}$. The observed smaller width for $\chi = 0^{o}$, may be attributed to the additional intermolecular interactions between close neighbors. One may conclude from the results discussed above that when there is chloro substitution, the chlorine atoms of the neighboring molecules in the crystal lattice tend to come closer to one another within a distance of about 4.2 Å and this propensity of the chlorine atoms to come closer would be of practical value in crystal engineering. The results presented above on chlorocoumarins and the analysis of the packing arrangement of 132 compounds from Cambridge Data Base substantiates the use of chloro group as a steering agent during the solid state photodimerization. Further work to establish the generality of chloro as a "crystal engineering" group is under progress.

Gas Solid Reaction: Photooxidation of Thioketones

Another problem of considerable interest in our laboratory is the correlation of solid state chemical reactivity with moelcular packing. In this connection the photochemical oxidation of diaryl thioketones in the solid state was investigated. Thioketones in general are readily oxidized in solution to the corresponding S-oxides and/or ketones and the rate and product distribution of oxidation are controlled by their inherent electronic and steric features (31,32). A number of examples are known in which substances vulnerable to attack by oxygen in solution or in the melt is indefinitely stable as the crystalline solid. Therefore, it was of interest to investigate the photo-oxidizability of thioketones in the solid state. Thioketones investigated and the results obtained are summarized in Table I. As

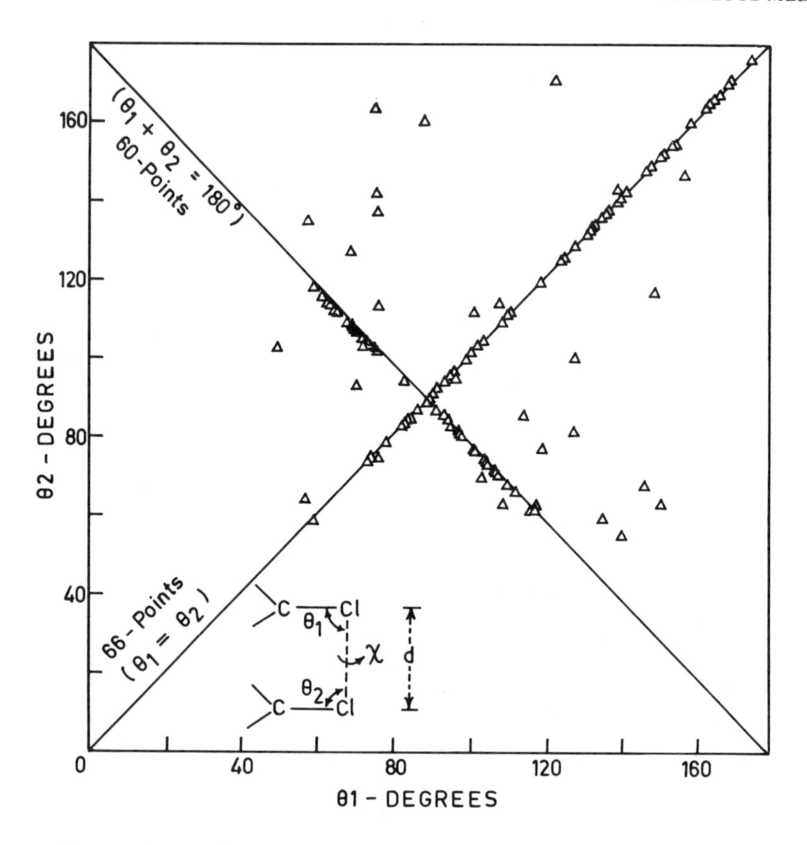

Figure 3. Mode of Packing in Chloro Substituted Aromatic Organic Crystals within 4.2 Å.

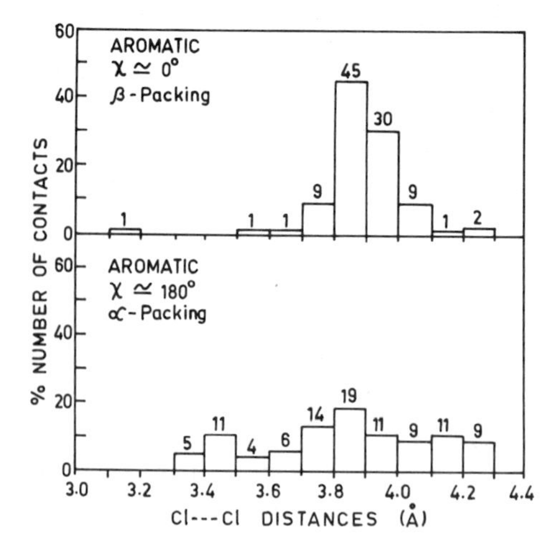

Figure 4. Histograms of Cl...Cl interactions vs. number of contacts.

Table I. Photo-oxidation of Diaryl Thioketones in the Solid State

Crystal studied	Nature of Reaction in Solution	Nature of Reaction in Solid State	Channel	Cross section of the channel Å^2
Thiobenzophenone	yes	yes	b	9
p,p'-Dimethyl thiobenzophenone	yes	yes	-	-
p-Phenyl thiobenzophenone	yes	yes	a	3.7
1-Phenyl naphthyl thione	yes	yes	a	8.3
Michler's thione	yes	no	b	2.3
p,p'-Dimethoxythiobenzophenone	yes	no	-	-
1-Phenyl pyrenyl thione	yes	no	-	-

may be seen from Table I, not all the thioketones are reactive in the solid state and the difference in their behavior between solution and the solid state is noteworthy. For example, 4,4′-dimethoxythiobenzophenone is the most reactive of all requiring 30 mts. of irradiation in solution whereas it is very stable in the crystalline state. This indicated that the electronic properties of the thioketones are not controlling their reactivity in the solid state.

In order to correlate the reactivity pattern of the thioketones with their molecular packing, a systematic crystallographic investigation of a few thioketones, representative examples of reactive and unreactive thioketones, was carried out. Crystal packings along the channel axis for a reactive and an unreactive thioketones are shown in Figure 5. Similar features are preserved in the other thioketone crystals also. It is observed that in reactive thioketones there is a channel along the shortest crystallographic axis with the thiocarbonyl chromophore directed along the channel. Further, thiocarbonyl S...S intermolecular distance between adjacent planes in all these cases are ~ 3.9 Å. On the other hand, in the case of stable thioketones the packing arrangement do not reveal any such channel passing through any of the crystallographic axis. Further, the thiocarbonyl S...S contact distances between the adjacent layers are much larger (~ 4.5 Å) than in reactive thioketones. The measured channel cross sectional area for the thioketones are tabulated in Table I. For the reactive thioketones, the channel cross sectional area is fairly large whereas for the unreactive thioketones either there is no channel or the channel is too small. As typical examples, projection of the crystal packing on a plane perpendicular to the channel axis for Michlers′ thione and 1-phenyl naphyl thioketone are shown in Figure 6.

For the oxidation to be efficient, oxygen, independent of the reactive state, should be able to diffuse into the successive layers of thioketones. As pointed out above, absence of channel in Michler′s thioketone and 4,4′-dimethoxythiobenzophenone might be responsible for their photostability. On the other hand, presence of channel in thiobenzophenone, 4-phenyl-thiobenzophenone and 1-phenyl napthyl thioketone makes them susceptible for oxidation. A simple mechanism that could be visualized for this oxidation involves attack of oxygen at the exposed excited thiocarbonyl chromophore at the crystal surface to form a monolayer of the carbonyl compound. As the carbonyl compound is formed, disorientation of the reacted layer may occur so as to allow the oxygen to diffuse into the next layer where the process is repeated. The presence of thiocarbonyl chromophore at the crystal surface is not the sufficient condition for the oxidation to occur as in all the five cases investigated, presence of thiocarbonyl chromophore at the crystal surface could be identified. Therefore, it is necessary that the thiocarbonyl groups be so arranged that oxidation of one molecule exposes another close neighbor to oxygen molecule. This is evident from the difference in reactivity between the five thioketones whose structural details have been obtained. Thus the observed differences in the reactivity of the crystals can be rationalized on the basis of the crystal packing.

Photochemical Reactions in Cyclodextrins

Cyclodextrins are cyclic oligosaccarides containing six or more D(+)

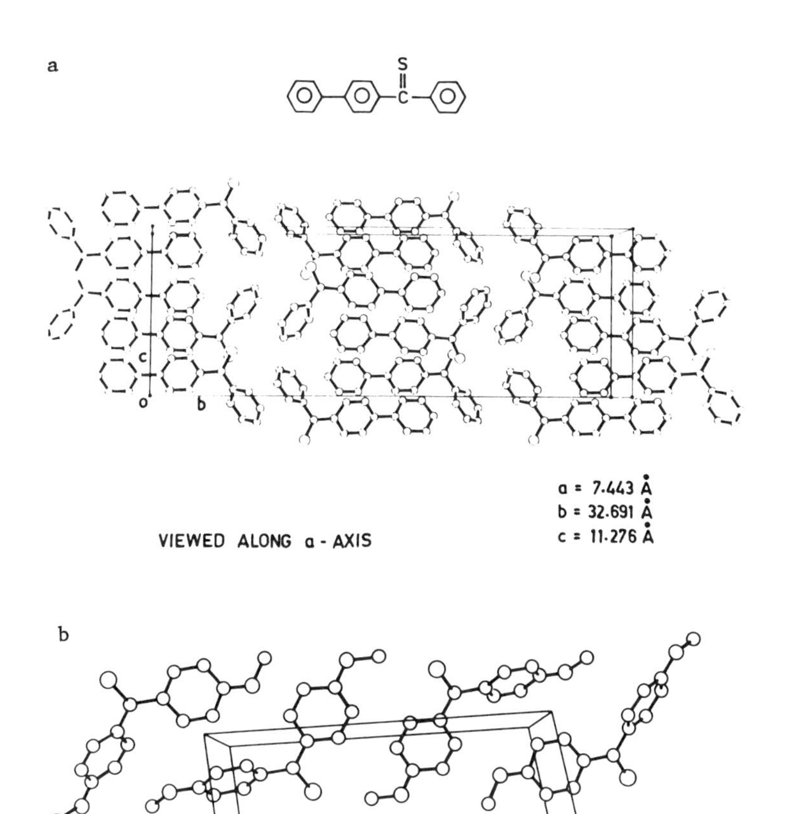

Figure 5. Packing Arrangements of (a) p-Phenylthiobenzophenone and (b) p,p′-Dimethoxythiobenzophenone.

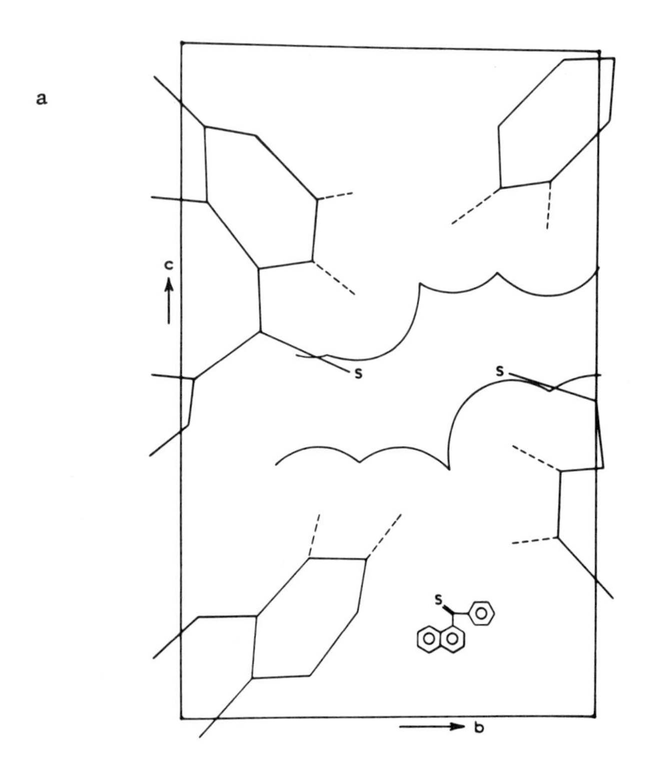

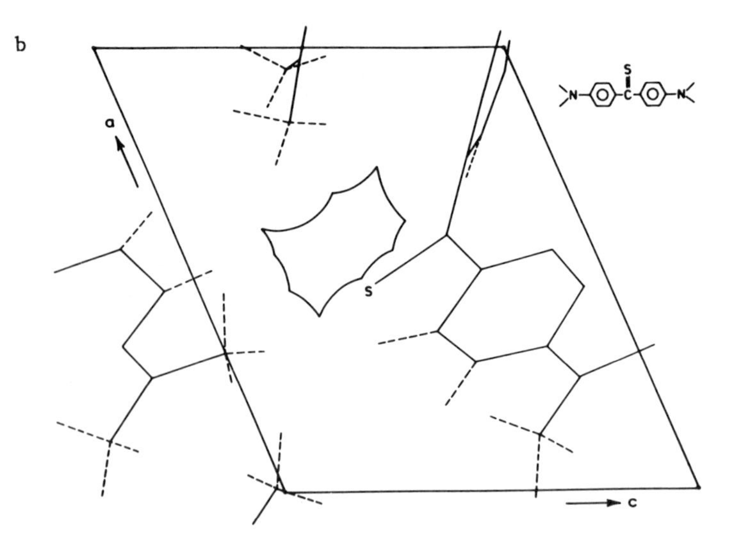

Figure 6. Projection of Crystal Packing on a Plane Perpendicular to the Channel Axis (a) Phenyl naphthyl thione and (b) Michlers thione.

glucopyranose units connected by α-(1,4)-linkages. Cyclodextrins have been very rigid, spatially restricted hydrophobic cavities which have been shown to influence reactions involving large changes in geometry (33-37). Though cyclodextrins have been extensively studied, very few photochemical reactions of molecules complexed to cyclodextrins have been examined (38,39). The specific objective of our program is to demonstrate that cyclodextrins can non-covalently influence photochemical reactions. In this connection, we have been investigating the photochemical behavior of polyenes and aryl alkyl ketones whose solution behavior has been well understood (40-42).

The Type II reactions of aryl alkyl ketones having a γ hydrogen have been extensively investigated (42). The intermediacy of a 1,4-biradical is now well established for this reaction. The α,α-dimethyl substituted phenyl alkyl ketones undergo α-cleavage (Type I reaction) in addition to Type II process. Because of the anticipated sensitivity of these reactions to the environment the photochemistry of phenyl alkyl ketones in cyclodextrins was studied. Results obtained with a few of the aryl alkyl ketones in cyclodextrins (aqueous solution and in solid state) along with those in organic solvents are provided in Table II. The following important points emerge from these studies: (a) cyclodextrins influence the relative yields of Type II and Type I products. Type II reaction is favored over Type I by cyclodextrins. (b) Ratios of products derived from cyclization and cleavage via Type II process are also altered by cyclodextrin cavity in comparison to organic solvents.

It is known that for phenyl alkyl ketones both the Type I and Type II reactions occur only from triplet state. The radical pair formed from α-cleavage undergo disproportionation resulting in Type I products or recombine to regenerate the ground state ketone. The latter process reduces the efficiency of Type I product formation. We suggest that the cyclodextrin cavity provides an environment wherein recombination of the geminate radical pair (from Type I) is favored and this results in lower yield of products from Type I process relative to Type II. Experiments are underway to test this cage effect with other examples.

The quantum yield for Type II reaction and the ratio of products derived from elimination and cyclization are known to be sensitive to the environment. Solvents that are reasonable Lewis bases prevent reversion of triplet generated biradicals to ground state ketone. In the process, they slightly raise the cleavage/cyclization ratio and change the stereochemistry of cyclization. These effects are explicable by a mechanism involving hydrogen bonding of the hydroxybiradical to the solvent. The cavity of the cyclodextrin consists of ether linkages capable of hydrogen bonding to the intermediate 1,4-biradical. The results presented in Table II suggest that this may not be the only factor controlling the E/C ratio. We attribute the variation in the E/C ratio to the steric constraints offered by the cavity of cyclodextrin wherein the 1,4-biradicals are located. While buterphenone and valerophenone are small and fit completely inside the cavity and the resulting 1,4-biradicals do not experience any constraints during closure to cyclobutanol, the biradicals from α,α-dimethyl buterophenone and α,α-dimethyl valerophenone experience considerable steric constraints upon closure. This effect is fairly large in the last compound. We believe that this factor may be responsible for the decreased efficiency of cyclobutanol in α,α-

Table II. Product Distribution Upon Irradiation of Aryl Alkyl Ketones in Organic Solvents and in Cyclodextrins

	BUTEROPHENONE	VALEROPHENONE	α,α'-DIMETHYL-BUTEROPHENONE		α,α'-DIMETHYL-VALEROPHENONE	
MEDIUM	E/C	E/C	E/C	II/I	E/C	II/I
Benzene	6.56	3.85	0.13	1.2	0.11	4.2
t-Butanol	8.55	5.95	0.25	1.8	0.26	10.0
β-CD/H_2O (1:1)	4.10	2.95	0.30	7.3	0.55	16.6
(2:1)	3.80	2.90	-	-	-	-
(3:1)	-	2.90	-	-	-	-
β-CD (Solid)	3.5	3.14	0.37	5.3	0.61	14.0
α-CD/H_2O (1:1)	4.36	3.80	0.51	7.3	0.79	6.25
γ-CD/H_2O (1:1)	-	2.70	-	-	0.21	19.0

dimethyl aryl alkyl ketones. CPK molceular models reveal that α substitution forces the molecule to move closer to the rim of the caivty and thus experience steric constraints during closure to cyclobutanol (Figure 7). Substitution at γ-position should enhance the steric constraints which is found to be the case with α,α'-dimethyl γ-methoxy valerophenone. Efforts are underway to examine other model systems and to understand the features controlling the reactions in cyclodextrins.

Results obtained upon excitation of cyclodextrin complexes of stilbene, cinnamic acid esters and β-ionone are presented in Table III and Figure 8. It is interesting to note that there is significant difference in the behavior of these molecules in solution and when complexed to cyclodextrin. While in organic solvents upon direct excitation the photostationary state of stilbene is heavily favored towards cis-stilbene (~ 85%), excitation of cyclodextrin complexes of either cis or trans-stilbene in aqueous medium results in a photostationary state enriched in trans-stilbene. This behavior is observed to be common with α, β and γ-cyclodextrins. This impressive difference in the photostationary state composition between organic solvents and cyclodextrins is probably due to the influence of the cyclodextrin cavity on the decay ratio of the twisted stilbene. Based on CPK molecular models one can visualise the structure of the cyclodextrin complexes of cis and trans stilbenes as shown in Figure 9. It is inferred from this model that the decay of the twisted stilbene to cis geometry will be restricted by the cavity. This effect arises due to the interaction between the phenyl ring and the rim of the cyclodextrin cavity. Indeed when the phenyl ring is replaced by a smaller group such as cinnamate esters the behavior in solution and in cyclodextrin are identical (Table III).

Impressive difference in the behavior of β-ionone was observed between solution and cyclodextrin complexes (Figure 8). While in organic solvents β-ionone gives rise to products arising from geometric isomerization and 1,5-hydrogen migration, in cyclodextrin (aqueous medium) only 1,5-hydrogen migration occurs. This must be due to the restriction imposed by the cavity on the rotation of the double bond. CPK molecular models of β-ionone-cyclodextrin complex suggest that the rotation of 7-8 double bond is hindered in the cavity independent of how β-ionone is accomodated inside the hydrophobic cavity (Figure 8). Selectivity demonstrated here has wider implication both in terms of synthetic methodology and in understanding the mechanism of isomerization of retinal.

Acknowledgments

The able experimental and intellectual contributions of P. Arjunnan, M.M. Bhadbhade, S. Devanathan, K. Gnanaguru, G.S. Murthy and S. Sharat are greatly appreciated. This presentation would not have been possible without the outstanding contributions by my crystallographic collaborator Prof. K. Venkatesan.

Figure 7. Schematics of Type I and II reactions of Aryl alkyl ketones in Cyclodextrins.

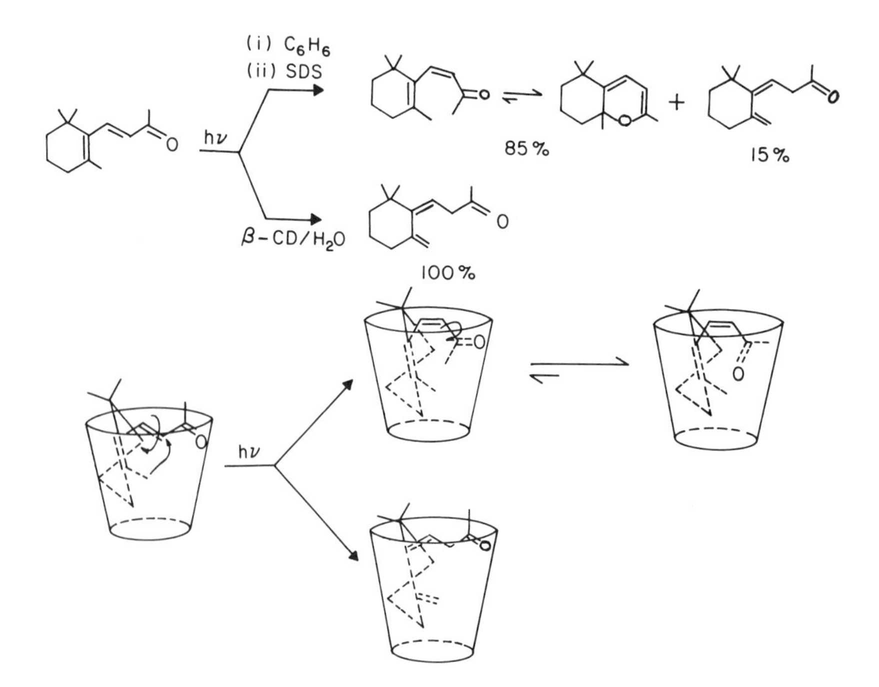

Figure 8. Excitation of β-Ionone in Cyclodextrins.

RO O
O OR
O OR
hν hν
t Twisted c

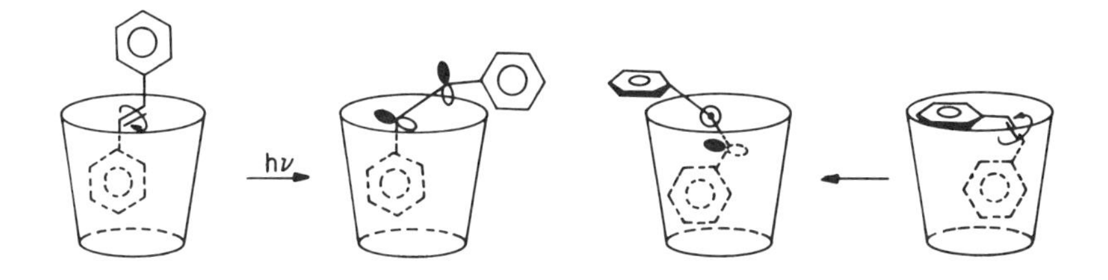

Figure 9. Schematics of Isomerization of Olefins in Cyclodextrins.

Table III Photostationary State upon Irradiation of Stilbene and Cinnamic Acid Esters in Organic Solvents and in Cyclodextrins.

Olefin	Medium	Photostationary State Composition (%)		
		cis	trans	Phe [a]
(a) Stilbene	Benzene	93	7	-
	β-CD/H_2O	5	90	5
	α-CD/H_2O	8	92	-
	γ-CD/H_2O	4	91	-
(b) Methyl Cinnamate	Benzene	45	55	-
	Methanol	45	55	-
	β-CD/H_2O	42	58	-
(c) Ethyl Cinnamate	Benzene	50	50	-
	Methanol	51	49	-
	β-CD/H_2O	50	50	-

a: Phenanthrene

Literature Cited

1. Turro, N.J.; Gratzel, M.; Braun, A.N. Angew. Chem. Intl. Ed. Engl. 1980, 19, 675.
2. Whitten, D.G.; Russel, R.C.; Schmehl, R.H. Tetrahedron 1982, 38, 2455.
3. Thomas, J.M.; Morsi, S.E.; Desvergne, J.P. Adv. Phy. Org. Chem. 1977, 15, 63.
4. Nerbonne, J.M.; Weiss, R.G. J. Am. Chem. Soc. 1979, 101, 402.
5. Muthuramu, K.; Ramamurthy, V. J. Org. Chem. 1982, 47, 3976.
6. Muthuramu, K.; Ramnath, N.; Ramamurthy, V. J. Org. Chem. 1983, 48, 1872.
7. Ramesh, V.; Ramamurthy, V. J. Org. Chem. 1984, 49, 537.
8. Ramnath, N.; Ramamurthy, V. J. Org. Chem. 1984, 49, 2827.
9. Ramesh, V.; Ramamurthy, V. J. Photochem. 1984, 24, 395.
10. Muthuramu, K.; Ramamurthy, V. Ind. J. Chem. 1984, 253, 502.
11. Ramamurthy, V. Proc. Ind. Acad. Sci., 1984, 93, 635.
12. Ramasubbu, N.; Guru Row, T.N.; Venkatesan, K.; Ramamurthy, V.; Rao, C.N.R. J. Chem. Soc. Chem. Commun. 1982, 178.
13. Ramasubbu, N.; Gnanaguru, K.; Venkatesan, K.; Ramamurthy, V. Can. J. Chem. 1982, 60, 2199.
14. Arjunan, P.; Ramamurthy, V.; Venkatesan, K. J. Org. Chem. 1984, 49, 1765.
15. Bhadbhade, M.M.; Murthy, G.S.; Venkatesan, K.; Ramamurthy, V. Chem. Phys. Lett. 1984, 109, 259.
16. Gnanaguru, K.; Murthy, G.S.; Venkatesan, K.; Ramamurthy, V. Chem. Phys. Lett. 1984, 109. 255.
17. Ramasubbu, N.; Gnanaguru, K.; Venkatesan, K.; Ramamurthy, V. J. Photochem. 1984, in the press.
18. Padmanabhan, K.; Venkatesan, K.; Ramamurthy, V. Can. J. Chem., 1984, in the press.
19. Ramasubbu, N.; Gnanaguru, K.; Venkatesan, K.; Ramamurthy, V. J. Org. Chem. 1985, in the press.
20. Cohen, M.D.; Schmidt, G.M.J. J. Chem. Soc. 1964, 1996, 2000, 2014.
21. Schmidt, G.M.J. Pure Appl. Chem. 1971, 27, 647.
22. Leiserowitz, L.; Schmidt, G.M.J. Acta. Cryst. 1965, 18, 1058.
23. Irnagartinger, H.; Ackev, R.D.; Rebaflka, W.; Staab, H.A. Angew. Chem. Int. Ed. Engl. 1974, 13, 674.
24. Kaftory, M. J. Chem. Soc. Perkin II. 1984, 757.
25. Frank, J.K.; Paul, I.C. J. Am. Chem. Soc. 1973, 95, 2324.
26. Hasagawa, M.; Nohara, M.; Saigo, K.; Mori, T.; Nakanishi, H. Tetrahedron Lett. 1984, 25, 561.
27. Theocharis, C.R.; Jones, W.; Thomas, J.M.; Motevalli, M.; Hursthouse, M.B. J. Chem. Soc. Perkin II, 1984, 71.
28. Busing, W.R. (1981) WMIN, a computer program to model molecules and crystals in terms of potential energy functions, Oak Ridge National Laboratory, Tennessee, U.S.A.
29. Green, B.S.; Cohen, M.D. Chem. Britain, 1973, 9, 490.
30. Green, B.S.; Lahav, M.; Rabinovich, D. Acc. Chem. Res. 1979, 12, 191.
31. Jayathirtha Rao, V.; Muthuramu, K.; Ramamurthy, V. J. Org. Chem. 1982, 47, 127.

32. Ramnath, N.; Ramesh, V.; Ramamurthy, V. J. Org. Chem. 1983, 48, 214.
33. Bender, M.L.; Komiyama, M. "Cyclodextrin Chemistry", Springer Verlag, New York, 1978.
34. Griffiths, D.W.; Bender, M.L. Adv. Catal. 1973, 23, 209.
35. Saenger, W. Angew. Chem. Int. Ed. Engl. 1980, 19, 344.
36. Tabushi, I. Acc. Chem. Res. 1982, 15, 66.
37. Breslow, R.; Trainor, G.; Ueno, A. J. Am. Chem. Soc. 1983, 105, 2739.
38. Ohara, M.; Watanabe, K. Angew. Chem. Int. Ed. Engl. 1975, 14, 820.
39. Cox, G.S.; Hauptman, P.J.; Turro, N.J. Photochem. Photobiol. 1984, 39, 597.
40. Liu, R.S.H.; Asato, A. Tetrahedron, 1984, 40, 1931.
41. Saltiel, J.; Charlton, J.L. in "Molecular Rearrangements in the Ground and Excited States"; de Mayo, P. Ed.; Wiley-Interscience, New York, 1980; Chapter 14, pp. 25-89.
42. Wagner, P.J. in "Molecular Rearrangements in the Ground and Excited States"; de Mayo, P. Ed.; Wiley-Interscience, New York, 1980; Chapter 20, pp. 381-444.

RECEIVED January 10, 1985

18

Photoassisted Sonosynthesis of 1,2,3,4-Tetrakis(methylthio)hexafluorobutane

MADELINE S. TOY and ROGER S. STRINGHAM

Science Applications International Corporation, Sunnyvale, CA 94089

The synthesis of isomeric 1,2,3,4-tetrakis(methylthio)hexafluoro-n-butane was finally achieved by photoassisted sonication, after the numerous attempts through the various methods had failed. A successful method described here consists of reacting trans-1,4-bis(methylthio)hexafluoro-2-butene in an excess of methyl disulfide with an added appropriate gas (hexafluoropropane) to increase and control the pressure in the reaction vessel. The heterogeneous mixture was then subjected to the combined photolysis and sonication at 50°C. The ^{19}FNMR and mass spectral data of the product and the by-product 1,2,4-tris(methylthio)-3-H-hexafluoro-n-butane were presented.

This work studies the synthesis of polymethylthio adducts of hexafluorobutadiene. The polymethylthioperfluoroalkanes will then be explored for their oxidation products, perfluoroalkanepolysulfonic acids. The interest in the multifunctional sulfonic acids is based on the oxygen reduction kinetics of trifluoromethane sulfonic acid on platinum being significantly higher than in phosphoric acid (1,2). The improved performance is possible due to the higher oxygen solubility and the lower adsorption of $CF_3SO_3^-$ anions on Pt (3-5). The CF_3SO_3H is also thermally extremely stable to at least 350°C (6), but has the drawback of a low boiling point at 162°C (7). Due to the operation temperature of the present phosphoric acid fuel cell at 150 to 200°C, the possible potential of the higher molecular weight perfluoroalkanepolysulfonic acids becomes apparent. The latter would not only reduce the vapor pressure of the monobasic acid, but also increase its functional groups (>2) for the fuel cell electrolyte applications.

The known synthesis of perfluoroalkanepolysulfonic acids has been limited to the general formula $HO_3S(CF_2)_nSO_3H$ by Ward (8)

0097-6156/85/0278-0287$06.00/0

through the hydrolysis and the oxidation of α,ω-bis(methylthio)-perfluoroalkanes (Eq. 1):

$$1/2\ nCF_2{=}CF_2 + CH_3SSCH_3 \xrightarrow[350^{o}\ C]{600\ psi} CH_3S(CF_2)_nSCH_3 \quad (1)$$

When we reacted hexafluoro-2-butyne and methyl disulfide, the reaction proceeded readily in good yield to form the isomeric 1:1 adducts [i.e., cis- and trans-2,3-bis(methylthio)hexafluoro-2-butene in equal mole ratio], but the formation of the methylthio adducts with hexafluorobutadiene was difficult (9). This agrees with the previous alkyl disulfide and the terminal olefinic bond reaction (Eq. 2, where k_{-1}/k_2 [RSSR] is relatively large) proceeding to a very poor yield of the adduct (10):

$$RS\cdot + CH_2{=}CHR' \underset{k_{-1}}{\overset{k_1}{\rightleftharpoons}} RSCH_2\dot{C}HR' \xrightarrow{k_2[RSSR]} RSCH_2CH(R')SR + RS\cdot \quad (2)$$

$$RS\cdot + HC{\equiv}CR' \underset{k'_{-1}}{\overset{k'_1}{\rightleftharpoons}} RSCH{=}\dot{C}R' \xrightarrow{k'_2[RSSR]} RSCH{=}C(R')SR + RS\cdot \quad (3)$$

In contrast, Heiba and Dessan (10) also observed and explained that the addition of RS• radical to the acetylenic bond (Eq. 3, where k'_{-1}/k'_2[RSSR] is very small) was much less reversible than its addition to the olefin (Eq. 2). Our results of CH_3SSCH_3 reacting with hexafluoro-2-butyne and not with hexafluorobutadiene are in agreement with Heiba and Dessan. One interpretation following the same trend suggests, that the sonic irradiation in Eq. 4 may have functioned in decreasing the reversibility of CH_3SSCH_3 addition to $CF_2{=}CFCF{=}CF_2$ during photoexcitation (resembling Eq. 2, except where $k_{-1}/k_2[CH_3SSCH_3]$ is greatly reduced to favor the adduct formation).

Our several attempted additions of CH_3SSCH_3 to $CF_2{=}CFCF{=}CF_2$ reactions by photolysis alone with prior emulsification, solution photolysis, sonolysis singly, and thermolysis up to $230^{o}C$ for 24 hours were unsuccessful; but the combined photolysis and ultrasound provided a convenient path to synthesize the trans-1,4-bis(methylthio)hexafluoro-2-butene (I, the trans-1, 4-adduct) as the major product in the presence of trace amount of the other 1:1 adducts and minor quantity of oligomers (Eq. 4) (11):

$$CF_2{=}CFCF{=}CF_2 + XS.CH_3SSCH_3 \xrightarrow{UV\ )))} CH_3SCF_2C(F){=}C(F)CF_2SCH_3 +$$

trans-I (major product)

$$CH_3S(C_4F_6)_nSCH_3 +$$

oligomers (minor)

$$CH_3SCF_2C(F){=}C(F)CF_2SCH_3 + CH_3SCF_2CF_2CF{=}CFSCH_3 \quad (4)$$

cis-I (trace) cis- and trans-II (trace)

The four 1:1 adduct products were the two commonly expected cis- and trans-I and the other two were very unusual adducts, which were later identified as cis- and trans-1,4-bis(methylthio)hexafluoro-1-butene (II). Subsequently, we subjected trans-I and CH_3SSCH_3 to very high pressure at 16,000 atm and 200°C for 24 hours. There was certain disappointment in the products, which were cis- and trans-II instead of the desired saturated adduct. A possible explanation for the high pressure experiment was that the trans-I isomerized with complete bond migration from the internal to the terminal olefinic bond (12). This result suggests that the trace amount of cis- and trans-II in Eq. 4 may be formed by a sonication effect, which creates localized very high pressure and temperature. These localized sites are formed, because the very high intensity waves passing through a liquid cause local vaporization of the liquid in a process called acoustic cavitation. This phenomenon involves the rapid formation growth and implosive collapse of gas vacuoles within the liquid, which generates localized hot spots, lasting only a few nanoseconds. During these short periods, some photoexcitation may have initiated certain uncommon activated species. The results were the formation of new products distinct from photochemical and thermal processes.

Boudjouk and co-workers reported the rate enhancements by sonolysis on a number of heterogeneous reactions such as Wurtz type coupling of organic halides (13) and organometallic chlorides (14) in the presence of lithium wire, cycloaddition of activated olefins in the presence of zinc powder (15) and the Reformatsky reaction (16) requiring neither freshly prepared zinc powder (17) nor acid catalysis (18). Significant rate accelerations of the Barbier reaction (19), the synthesis of the thio amides (20), the lithium aluminum hydride reduction of aryl halides (21), and the silicon-silicon double bond formation from dimesityldichlorosilane and lithium wire (22) also point to the considerable synthetic potential of sonolysis. Suslick and co-workers reported the chemical uses of sonication in the homogeneous systems on metal carbonyls to initiate catalysis and to form unusual products, which were not analogous to either photochemical or thermal reactions, such as the ligand association product $Fe_3(CO)_{12}$ from $Fe(CO)_5$ and finely divided iron (23-25). Ultrasound-promoted reactions were also applied to the syntheses and rate enhancements of fluorinated compounds such as perfluoroalkyl iodides in the presence of zinc and carbon dioxide to give perfluoroalkanoic acids (26,27).

Experimental

Materials and Apparatus. Hexafluorobutadiene, hexafluoropropane, and 1,2-dichloro-1,1-difluoroethane were purchased from PCR, methyl disulfide and chloroform from Aldrich, and deuterated chloroform from Stohler. These reagents were checked by infrared spectroscopy and used as received, except for the methyl disulfide, which was redistilled.

A sonicator (Model W-370) was purchased from Heat Systems-Ultrasonic with a cup horn attachment. The horn was the resonant body, which vibrated at 20 kHz (20,000 cycles per second) and served as a second stage of acoustic amplification. The standard tapped titanium disrupter horn was immersed in circulating water at 50°C during sonication.

Standard vacuum manipulations were applied. Pressures were measured with a Heise gauge (0-100 cm Hg absolute with 500 increments) to accuracy of 1 mm Hg. The amount of volatile reactant was determined by P-V-T measurements, assuming ideal gas behavior. A sealed quartz reaction vessel containing the reactants was vertically suspended in the water in the horn cup. A 200-watt high pressure mercury arc lamp was the outside irradiation source, which was focused with a quartz condensing lens. This lens gathered about 40% of the emitted unfiltered light passing through the water level in the horn cup and into the reactants; while the reactants in the 1 mm wall thick quartz sealed tube were under sonication. With the identical conditions but substituting Pyrex for quartz, no reaction products were found.

The UV-absorption characteristics of the reagents described below are at the ground state. The CH_3SSCH_3 absorbs UV-light below 300 mμ and is a strong absorber (e.g., the molar extinction coefficients of 250 l/mole-cm at 240 mμ and of 1500 l/mole-cm at 210 mμ) (28,29). Hexafluorobutadiene absorbs UV below 260mμ with the molar extinction coefficient such as 130 l/mole-cm at 240mμ (30); whereas the UV-absorption of trans-I is below 220 mμ and is a very weak absorber with the molar extinction coefficient of 0.044 l/mole-cm at 200 mμ. CH_3SSCH_3 is apparently the dominant UV-absorber and it may have greatly reduced or totally screened off the available UV-irradiation for trans-I. However, these extinction coefficients are for ground state species. When they are photoexcited within the localized high pressure and temperature sites due to sonication, their values are likely to be different.

The ^{1}HNMR and ^{19}FNMR were recorded on a Nicolet spectrometer operating at 282 MHz and 35°C. The ^{19}F chemical shifts of the products are converted to δ-values upfield from fluorotrichloromethane by using the value of 61.9 ppm for 1,2-dichloro-1,1-difluoroethane. The latter compound CH_2ClCF_2Cl was added as an internal standard. The ^{19}FNMR data of the products show AB patterns indicating restrictive rotation, as the geminal F-F coupling constant was previously reported at J(4A,4B) = 230.0 Hz for $\overset{(1)}{(CF_3)_2}\overset{(2)}{C}\overset{(3)(5)}{F}C\overset{(4)}{F}HCF_2SCH_3$, where chemical shifts of $\overset{(4)}{F}$ and $\overset{(3)}{F}$ were 88.6 and 203.8 ppm from $CFCl_3$ respectively (31,32).

The mass spectra were determined on a LKB 9000 Model GC/MS instrument. The elemental analysis was obtained by a VG-model ZAB double focusing high resolution mass spectrometer. The infrared spectra were measured on a Perkin-Elmer 567 spectrophotometer and with a chromatographic infrared analyzer (CIRA 101). The ultraviolet spectra were determined by a UV-visible spectrophotometer (Hitachi Perkin Elmer Model 139).

Procedure. Trans-I was prepared by an ultrasonic photolysis method and separated by gas-chromatography from the product mixture as previously described (Eq. 4)(11). I (10 mmole) was added to an excess of CH_3SSCH_3 in a quartz tube with a quarter inch (outside diameter) neck, which was attached to the vacuum manifold and evacuated at -196°C using a liquid nitrogen bath. An appropriate amount of hexafluoropropane was then condensed on top of the frozen mixture of I and CH_3SSCH_3 in the quartz tube at -196°C. At ambient temperature the pressure inside the quartz tube was checked to be about 100 psi. The quartz tube was cooled again to -196°C, evacuated and sealed under vacuum. The vacuum-sealed quartz tube was warmed to ambient temperature and suspended vertically with the liquid-liquid interphase in the quartz bulb under the water level in the cup horn of the sonicator. The immiscible colorless liquids easily homogenized under ultrasound and were simultaneously subjected to ultraviolet irradiation for 6 hours. The temperature of the circulating water in the cup horn was maintained at 50°C. At the end of the reaction time, the mercury lamp and the sonicator were turned off and the sealed quartz tube was removed from the horn cup. The homogeneous liquid in the tube was cooled to -196°C and the tube was opened. The liquid content was vacuum distilled to remove the excess methyl disulfide. The conversion of I to the main product 1,2,3,4-tetrakis(methylthio)hexafluoro-n-butane (III) and by-product 1,2,4-tris(methylthio)-3-H-hexafluoro-n-butane (IV) was elucidated by GC-mass spectroscopy to be about 50% yield of III and 40% of IV. The two isomeric ratios of III and IV varied with experimental conditions.

The ^{19}FNMR ($CFCl_3$) of III shows two peaks for each isomer. Isomer A: δ84.0 [coalesced AB system, relative peak area 4,4F,CF_2(1)], 157.2 [broad singlet, relative peak area 2,2F, CF(2)]. Isomer B: δ82.3 [center AB system, J(1A,1B)= 220 Hz, relative peak area 4, 4F, CF_2(1)], 152.7 [broad singlet, relative peak area 2,2F,CF(2)]. The GC-mass spectral data of III show two species, which have different elution times but with identical parent ions at m/e value 350 ($C_8F_6H_{12}S_4^+$) and also with the same mass fragments. These results indicate that there are two isomers in III.

Mass spectroscopic weight of the two isomers of III: Calcd. for $C_8F_6H_{12}S_4$: 349.9726. Found: 349.9726.

The ^{19}FNMR ($CFCl_3$) of IV exhibits four peaks for each isomer. Isomer A: δ 87.3 [center AB system, J(1A,1B)=228 Hz, relative peak area 2,2F,CF_2(1)], 161.3 [septet, J(2,5)=18 Hz, relative peak area 1, 1F, CF(2)], 193.6 [doublet of quartets, J(3,5)=56 Hz, relative peak area 1,1F,CF(3)], 86.0 [center AB system, J(1A,1B)=228 Hz, relative peak area 2,2F,CF_2(4)]. Isomer B: δ84.1 [coalesced AB system, relative peak area 2,2F,CF_2(1)], 165.7 [complex singlet, relative peak area 1,1F,CF(2)], 195.4 [doublet of sextet, J(3,5)=54 Hz, relative peak area 1,1F,CF(3)], 84.4 [center AB system, J(1A,1B)=231 Hz, relative peak area 2,2F,CF_2(4)]. The GC-mass spectral data of IV show two species, which have different elution times but with identical parent ions at 304($C_7F_6H_{10}S_3^+$) and the same mass fragments. These results indicate that there are two isomers in IV.

Mass spectroscopic weight of the two isomers of IV: Calcd. for $C_7F_6H_{10}S_3$: 303.9849. Found: 303.9851.

Results and Discussion

It was originally thought that the sonication approach would promote miscibility at the interface of the heterogeneous system and also would raise the electronic states of a small portion of the reactants above their ground state, which would hopefully enhance the addition reaction. We later found, that only the simultaneous photoexcitation and ultrasonic irradiation plus the addition of an appropriate gas (e.g., hexafluoropropane), which increased the pressure in the reaction vessel, formed the fully saturated products. The added pressure above the liquid phase is another parameter required to form the desired saturated product and should have affected the temperature and pressure at the localized 'hot spots' created by sonication.

Under photoassisted sonosynthesis, the process of initiation may be considered to involve two steps, the first being the homolytic S—S bond-breaking in CH_3SSCH_3 to yield a pair of methylthio free radicals $CH_3S\cdot$

$$CH_3SSCH_3 \underset{}{\overset{UV}{\rightleftharpoons}}\!\!\!)))\; 2\ CH_3S^{\cdot} \qquad (5)$$

and the second the irreversible decomposition of CH_3SSCH_3 by disproportionation at a much lower rate

$$CH_3SSCH_3 \xrightarrow{)))} CH_3SH + CH_2{=}S \qquad (6)$$

where the thioformaldehyde is likely to polymerize or to form an adduct in the presence of $CH_3S\cdot$ radicals,

$$2CH_3S\cdot + nCH_2{=}S \longrightarrow CH_3S(CH_2S)_nSCH_3 \qquad (7)$$

where n=1 or greater. Our unpublished results have shown oligomeric products, which contain no fluorine. The methanethiol (Eq. 6) is decomposed to form hydrogen atom and $CH_3S\cdot$ radical.

$$CH_3SH \xrightarrow[\longleftarrow]{UV\ ))} CH_3S\cdot + H\cdot \quad (8)$$

The product formations or terminations involve the addition of $CH_3S\cdot$ radicals from CH_3SSCH_3 to the olefinic bond of I

$$\underset{I}{CH_3SCF_2CF{=}CFCF_2SCH_3} + CH_3SSCH_3 \xrightarrow{UV\ ))} \underset{III}{CH_3S\overset{(1)}{C}F_2\overset{*(2)}{C}F(SCH_3)\overset{*(2)}{C}F(SCH_3)\overset{(1)}{C}F_2SCH_3} \quad (9)$$

and $CH_3S\cdot$ radical and hydrogen atom from CH_3SH.

$$I + CH_3SH \xrightarrow{UV\ ))} \underset{IV}{CH_3S\overset{(1)}{C}F_2\overset{*(2)}{C}F(SCH_3)\overset{*(3)(5)}{C}FH\overset{(4)}{C}F_2SCH_3} \quad (10)$$

Another of our unpublished results has shown (Eq. 10) the presence of only the product IV by reacting an equal mole ratio of I and CH_3SH in the presence of an excess of CH_3SSCH_3. Although the activated intermediate species caused by the combined photoexcitation and ultrasonic irradiation are still unknown, the proposed reaciton sequence (5-10) is shown to be consistent with the observed products.

The two diastereomers of III with the two similar asymmetric carbon (*) are separable by gas chromatograph and suggested to be the dl-racemic mixture and the meso-form. The two diastereomers of IV, containing two dissimilar asymmetric carbon (*), are also separable by gas chromatograph and suggested to be the two racemates (dl-IV and d'l'-IV), although the actual resolution of the optical isomers was not carried out. The diastereomers under GC-mass spectroscopy show the same parent ions but have different elution times, which indicate the presence of isomers, similar to the previously reported cis- and trans- or geometric isomers (9,33).

Acknowledgments

The authors would like to acknowledge the Electric Power Research Institute for support of this work under Contract RP1676-3, Dr. J. Appleby for helpful discussion, Dr. D. Thomas for mass spectra, Mr. L. Cary for NMR spectra and Professor J. C. Cook for elemental analysis by a double focusing high resolution mass spectrometer.

Literature Cited

1. Adams, A.; Foley, R.; Barger, H. J. Electrochem. Soc. 1977, 124, 1228.
2. Appleby, A.; Baker, B. J. Electrochem. Soc. 1978, 125, 404.
3. Petrie, O.; Vasina, S.; Lukyanycheva, L. Sov. Electrochem. 1982, 17, 1144.
4. Zelenoy, P.; Hobib, M. A.; Bockris, J. O'M. J. Electrochem. Soc. 1984, 131, 2464.
5. Kotz, R.; Clouser, S.; Sarangapani, S.; Yeager, E. J. Electrochem. Soc. 1984, 131, 1097.
6. Dresdner, R. D.; Hoover, T. R. In "Fluorine Chemistry Review"; Tarrant, P., Ed.; Marcel Dekker: New York, 1969; Vol. IV, pp. 19-21.
7. Gramstad, T.; Haszeldine, R. N. J. Chem. Soc. 1954, 4228.
8. Ward, R. B. J. Org. Chem. 1965, 30, 3009.
9. Toy, M. S.; Stringham, R. S. Ind. Eng. Chem. Prod. Res. Dev. 1983, 22, 8.
10. Heiba, E. I.; Dessan, R. M. J. Org. Chem. 1967, 32, 3837.
11. Toy, M. S.; Stringham, R. S. J. Fluorine Chem. 1984, 25, 213.
12. Toy, M. S.; Stringham, R. S. J. Fluorine Chem. 1984, 25, 487.
13. Han, B. H.; Boudjouk, P. Tetrahedron Lett. 1981, 22, 2757.
14. Boudjouk, P.; Han, B. H. Tetrahedron Lett. 1981, 22, 3813.
15. Han, B. H.; Boudjouk, P. J. Org. Chem. 1982, 47, 751.
16. Boudjouk, P.; Han, B. H. J. Org. Chem. 1982, 47, 5030.
17. Rieke, R. D.; Uhm, S. J. Synthesis 1975, 452.
18. Rathke, M. W.; Lindert, A. J. J. Org. Chem. 1970, 35, 3966.
19. Luche, J. L.; Damiano, J. C. J. Am. Chem. Soc. 1980, 102, 7926.
20. Raucher, S.; Klein, P. J. Org. Chem. 1981, 46, 3558.
21. Han, B. H.; Boudjouk, P. Tetrahedron Lett. 1982, 23, 1643.
22. Boudjouk, P.; Han, B. H.; Anderson, K. R. J. Am. Chem. Soc. 1982, 104, 4992.
23. Suslick, K. S.; Schubert, P. F.; Goodale, J. W. J. Am. Chem. Soc. 1981, 103, 7342.
24. Suslick, K. S.; Goodale, J. W.; Schubert, P. F.; Wang, H. H. J. Am. chem. Soc. 1983, 105, 5781.
25. Suslick, K. S.; Gawienowski, J. J.; Schubert, P. F.; Wang, H. H. J. Phys. Chem. 1983, 87, 2299.
26. Kitazume, T.; Ishikawa, N. Chem. Lett. 1981, 1679; 1982, 137, 1453.
27. Ishikawa, N.; Takahashi, M.; Sato, T.; Kitazume, T. J. Fluorine Chem. 1983, 22, 585.
28. Ueno, T.; Takezaki, Y. Bull. Inst. Chem. Research, Kyoto Univ. 1958, 36, 19.
29. Calbert J. G.; Pitts, J. N. "Photochemistry"; John Wiley: New York, 1966; p. 490.
30. Toy, M. S. "Photochemistry of Macromolecules"; Reinisch, R. F., Ed.; Plenum Press: New York, 1970; p. 136.
31. Haszeldine, K. N.; Mir, I.; Tipping, A. E. J. Chem. Soc. Perkin I, 1979, 565.

32. Wray, V. "Annual Reports on NMR Spectroscopy"; Webb, G. A., Ed.; Academic Press: New York, 1983; Vol. 4, p. 23.
33. Toy, M. S.; Stringham, R. S. J. Fluorine Chem. 1976, 7, 375.

RECEIVED January 10, 1985

INDEXES

Author Index

Subject Index

A

D

M

N

O

Q

T

Production by Hilary M. Kanter
Indexing by Karen McCeney
Jacket design by Pamela Lewis

Elements typeset by Hot Type Ltd., Washington, D.C.
Printed and bound by Maple Press Co., York, Pa.